The evolution of melanism

Arctia caja L. f. *brunnescens* Stattermeyer and f. *fumosa* Hörhammer (× 1), (page 275).

                Left   Right

1. f. *typica*.
2–4. f. *brunnescens*, varying degrees of expression.

1. f. *fumosa*.
2–4. f. *brunnescens*, varying degrees of expression.

(Photograph by John Hayward).

The evolution of melanism

THE STUDY OF A RECURRING
NECESSITY

*With special reference to Industrial Melanism
in the Lepidoptera*

BERNARD KETTLEWELL

CLARENDON PRESS · OXFORD
1973

Oxford University Press, Ely House, London W.1

GLASGOW NEW YORK TORONTO MELBOURNE WELLINGTON
CAPE TOWN IBADAN NAIROBI LUSAKA DAR ES SALAAM ADDIS ABABA
DELHI BOMBAY CALCUTTA MADRAS KARACHI LAHORE DACCA
KUALA LUMPUR SINGAPORE HONG KONG TOKYO

© OXFORD UNIVERSITY PRESS 1973

PRINTED IN ENGLAND BY
HAZELL WATSON AND VINEY LTD
AYLESBURY, BUCKS

THIS BOOK IS DEDICATED JOINTLY TO
THE NUFFIELD FOUNDATION
AND PROFESSOR E. B. FORD F.R.S.
WHO TOGETHER GAVE ME AN OPPORTUNITY

Contents

LIST OF PLATES	xii
LIST OF FIGURES	xv
PREFACE AND ACKNOWLEDGEMENTS	xvi

PART I: THE CONCEPT OF MELANISM

1. INDUSTRIAL MELANISM AND ITS PLACE IN NATURE	3
(i) An introduction	3
(ii) Definitions (*polymorphism, heterosis, multifactorial inheritance, super-gene, heterozygous variability*).	7
2. THE ATTRIBUTES OF BLACK COLOURATION	11
(i) General remarks	11
(ii) The physical differences between black and white	12
(iii) The pigments	16
3. EXAMPLES OF MELANISM AND ITS FUNCTIONS IN LIVING ORGANISMS	19
(i) Cryptic Melanism	19
(ii) Aposematic Melanism	30
(iii) Thermal Melanism	33
(iv) Barrier Melanism	34

PART II: MELANISM IN THE LEPIDOPTERA

4. A CLASSIFICATION OF MELANISM	39
IN IMAGINES	
(i) Industrial Melanism	40
(ii) Geographic, Ancient, or Relict Melanism	41
(iii) Recessive Melanism	42
IN LARVAE	
(i) Genetic plus phytoscopic	44
(ii) Genetic only	48
(iii) Aposematic	49
(iv) Crowding effect	49

5. THE PHENOMENON OF INDUSTRIAL MELANISM   51

   (i) Previous work on Industrial Melanism   52

   (ii) The changed environment—air pollution   54

   (iii) The effects of air pollution   56

   (iv) The history of the spread of industrial melanic species in Britain   58

   (v) Behavioural and physiological differences   65

   (vi) The genetics of Industrial Melanism   86

6. THE WORLD DISTRIBUTION OF INDUSTRIAL MELANISM   89

   (i) Europe   89

   (ii) North America   99

PART III: BISTON (SYN. AMPHIDASIS PACHYS) BETULARIA L. (SELIDOSEMIDAE). THE ORIGINAL SELECTION EXPERIMENTS (1952–5). THE FREQUENCY SURVEYS (1952–70)

7. THE CHOICE OF MATERIAL   105

   (i) *Biston betularia* L. and other cryptic species, their advantages and disadvantages   105

   (ii) The genetics of *Biston betularia* L.   106

   (iii) The life-history and behaviour of *Biston betularia* L.   107

8. THE EXPERIMENTS ON THE DIFFERENTIAL SURVIVAL OF *BISTON BETULARIA* L. F. *TYPICA* AND ITS MELANIC FORMS   113

   (i) The preliminary selection experiments in an aviary   114

   (ii) The selection experiments in polluted woodland (Birmingham)   115

   (iii) The selection experiments in an unpolluted woodland (Dorset)   124

   (iv) Release problems   127

9. THE DISTRIBUTION OF *BISTON BETULARIA* L. AND ITS MELANIC FORMS IN GREAT BRITAIN—FREQUENCY SURVEYS BETWEEN 1952 AND 1970 (See Appendix C)   131

   (i) Early spread of the melanic forms   131

   (ii) The three surveys: 1952–6, 1957–64, 1965–70   134

   (iii) Analysis of f. *carbonaria* frequencies   136

   (iv) Analysis of 'f. *insularia*' frequencies, and its distribution, past and present   141

   (v) Frequency changes in the two forms between 1952 and 1970   147

Contents ix

PART IV: NON-INDUSTRIAL MELANISMS
10. GEOGRAPHIC, RELICT, OR ANCIENT MELANISM: A SURVEY AND CLASSIFICATION 155
    (i) Rural or Background Choice Melanism 157
    (ii) Northern Melanism 164
    (iii) Western Coastline Melanism 169
    (iv) Ancient Conifer Melanism (Relict Melanism) 170
    (v) Melanism associated with fire-resistant trees 174
    (vi) Pluvial Melanism 177
    (vii) Thermal Melanism 178

11. *AMATHES GLAREOSA* ESP. AND ITS MELANIC FORM *EDDA* STDGR. IN SHETLAND: AN EXAMPLE OF NORTHERN MELANISM 181
    (i) Choice of material 181
    (ii) Description of morphs 182
    (iii) Life-history 183
    (iv) The genetics of the two forms 184
    (v) Dominance modification 185
    (vi) The distribution of the two forms 187

12. MARK–RELEASE–RECAPTURE EXPERIMENTS ON *AMATHES GLAREOSA* ESP. 195
    (i) Gene flow in the Tingwall Valley district 195
    (ii) Selective predation on *Amathes glareosa* in Shetland 198
        including evidence from mark–release–recapture experiments and bird stomach contents
    (iii) Habit differences 202
        including evidence of differential flight behaviour and of a different time of emergence

PART V: THE ORIGINS OF MELANISM
13. RECESSIVE MELANISM AND RARE MELANIC MUTANTS 211
    A CLASSIFICATION OF RECESSIVE MELANISM
    (i) Melanism in aposematic species 213
    (ii) Melanism in cryptic species 214
    (iii) A first step to polymorphism: *Lycia hirtaria* Clerck f. *nigra* Cockayne 223

14. RECESSIVE MELANIC POLYMORPHISMS 227

(i) *Lasiocampa quercus* L. f. *olivacea*, f. *olivaceo-fasciata*, and f. *lurida*, and its separate polymorphisms in Britain 227

(ii) *L. quercus* ssp. *callunae*: the Caithness polymorphism 231

(iii) *L. quercus* ssp. *callunae*: the Yorkshire polymorphism 244

(iv) *L. quercus* ssp. *quercus*: the Cheshire and Lancashire sand-hill polymorphism 255

(v) A theory for a recessive melanic polymorphism 256

PART VI: MISCELLANEOUS MELANISMS

15. MISCELLANEOUS MELANISMS IN MOTHS AND BUTTERFLIES 263

(i) Aposematic Melanism in the Heterocera 263

(ii) Melanism in the Rhopalocera 264

16. EXCEPTIONAL MELANISM IN THE HETEROCERA 271

(i) Incomplete dominance in cryptic species 271

Polia nebulosa Hufn. 271

(ii) Incomplete dominance in aposematic species 273

Spilosoma lutea Hufn. f. *zatima* 273

Arctia caja L. f. *brunnescens* 275

f. *fumosa* 275

Panaxia dominula L. ssp. *persona* f. *italica* 277

f. *nigradonna*

f. *medionigra* 279

f. *bimacula*

Aglia tau L. 289

17. MELANISM INFLUENCED BY SEX 293

(i) Sex-limited melanism 293

Cycnia mendica Clerck f. *rustica* 293

Hepialus humuli L. race *thulensis* 294

(ii) Monomorphic melanism in a normally sexually dimorphic cryptic species 298

Xylomyges conspicillaris L. 298

18. ENVIRONMENTAL MELANISM 301

PART VII: THE SYNTHESIS

19. A THEORY OF MELANISM ... 313

    The natural history of Industrial Melanism. The evidence of the hybridization experiments ... 318

APPENDIX A. BREEDING TECHNIQUES ... 319

APPENDIX B. A LIST OF SOME BRITISH LEPIDOPTERA WITH THE FREQUENCIES OF THEIR MELANIC FORMS ... 323

APPENDIX C. A TABLE OF PHENOTYPE FREQUENCIES OF *BISTON BETULARIA* AND ITS TWO MELANICS, F. *CARBONARIA* AND F. *INSULARIA*, FROM CENTRES IN BRITAIN, 1952–70 ... 362

LIST OF RECORDERS ... 373

HARDY–WEINBERG TABLES ... 375

BIBLIOGRAPHY ... 378

AUTHOR AND CONTRIBUTORS ... 409

SUBJECT INDEX ... 413

SPECIES INDEX ... 419

List of Plates

(Frontispiece)
Arctia caja L. f. *brunnescens* Stattermeyer and f. *fumosa* Hörhammer

(between pages 56–7)
3.1. Photograph of English bible, 1672.
7.2. Photograph of page 177 of *History of insects* by John Ray, 1710.
4.1. Larva of *Boarmia roboraria* Schiff.
4.2. The 'lichened' form of the larva of *Gonodontis bidentata* L.
4.3. The melanic form of the larva of *Gonodontis bidentata* L.
4.4. The 'intermediate' form of the larva of *Gonodontis bidentata* L.
5.1. Beech trunk, industrial Yorkshire, 1965.
5.2. Pollution obtained from washing 6 oz. leaves from different industrial areas.
5.3. Manchester. Three views 1730–1954.
5.4. Male *Phigalia pilosaria* Schiff. and its black form at rest on lichened tree-trunk.
5.5. Male *Phigalia pilosaria* Schiff. and its black form at rest on heavily polluted Oak trunk.
5.6. Examples of industrial melanic species.
5.7. Examples of industrial melanic species.

(between pages 80–1)
5.8. Examples of industrial melanic species.
5.9. Examples of industrial melanic species.
5.10. *Brachionycha sphinx* Hufn. and its melanic form.
5.11. *Brachionycha sphinx* Hufn. f. *typica* at rest.
5.12. *Brachionycha sphinx* Hufn. melanic form at rest.
5.13. *Ectropis consonaria* Hueb. f. *nigra* at rest on Beech trunk, Stroud, 1957.
5.14. *Biston strataria* Hufn. at rest on lichened trunk.
5.15. Method of testing background choice (barrel experiment).
5.16. Experimental tree-trunk used for testing background choice (1972).

(between pages 120–1)
5.17. *Catocala cerogama* Guenée, f. *typica* and its melanic f. *ruperti* Franc.
5.17a. *Apamea sordens* Hufn. (syn. *basilinea*) melanic form.
5.18. *Biston strataria* Hufn. and its melanic forms.
5.19. *Lymantria monacha* L. and the extreme melanic ab. *atra*: multifactorial inheritance.
6.1. *Achlya flavicornis* L. and its melanic forms.
6.2. *Simyra albovenosa* Goeze f. *typica* and f. *murina* (Finland).
6.3. *Epimecis hortaria* Fab. (syn. *virginaria* Cramer) and its melanic f. *carbonaria* (Pittsburg, U.S.A.).
7.1. The earliest *Biston betularia* L. f. *carbonaria* extracted from collections made in the last century compared with modern heterozygotes.
8.1. *Biston betularia* L. f. *typica* and its melanic f. *carbonaria* on lichened Oak trunk (Dorset).
8.2. *Biston betularia* L. f. *typica* and its melanic f. *carbonaria* on polluted Oak trunk (Birmingham).

List of Plates   xiii

8.3. Redstart with *B. betularia* L. f. *typica* in its beak taken from polluted tree-trunk (Birmingham).
8.4. Spotted Flycatcher about to take *B. betularia* f. *carbonaria* on lichened tree-trunk (Dorset).
8.5. Robin with *B. betularia* f. *carbonaria* taken from lichened tree-trunk (Dorset).
9.1. *Biston betularia* L. and *Biston cognataria*. A comparison of their various melanic forms.
10.1. Examples of Rural (Non-industrial) Melanism (Britain).
10.2. Shetland melanics (on the right) with their corresponding normal British mainland form on the left.

(between pages 176–7)

10.3. *Cleora tulbaghata* Felder and its more extreme melanic form (Cape Province, South Africa).
10.4. Alpine *Setina* spp. (Arctiidae) and their melanic forms from Central Europe.
10.5. *Euphyia bilineata* L. and its melanic f. *atlantica* Kane.
10.6. *Cleora repandata* L. and its melanic forms.
11.1. *Amathes glareosa* Esp. f. *typica* and its melanic f. *edda* Stdg.
11.2. *Amathes glareosa* Esp. f. *typica* at rest in heather (Shetland).
11.3. *Amathes glareosa* Esp. f. *edda* at rest in heather (Shetland).
11.4. Series of *Amathes glareosa* Esp. taken in Orkney (1971).
12.1. Common Gull regurgitating *Amathes glareosa* (Unst, Shetland, 1960).

ERRATUM

Owing to a binder's error the section of Plates 3.1–5.7 has been transposed with Plates 5.17–10.2. The former should appear between pp. 56–7 and the latter pp. 120–1.

13.4. *Lycia hirtaria* Clerck f. *nigra* at rest on London Lime trunk.
14.1. *Lasiocampa quercus* L. ssp. *callunae* f. *typica* and f. *olivacea*.
14.2. *Lasiocampa quercus* L. ssp. *callunae* f. *typica* wild female at rest.
14.3. *Lasiocampa quercus* L. ssp. *callunae* f. *olivacea* female at rest on heather.
14.4. The undersides of male *L. quercus* ssp. *callunae* f. *typica* and f. *olivacea*.
14.5. Two forms of the larva, 'chocolate' and 'normal' (Yorkshire).
15.1. Melanism in *Papilio*.
16.1. *Polia nebulosa* Hufn.
16.2. *Spilosoma lutea* Hufn. f. *typica* and incomplete dominant f. *zatima*.
17.1. *Cycnia mendica* Clerck.
17.3. Three forms of *Xylomyges conspicillaris* L.
14.6. Black-headed Gulls in the act of taking female *Lasiocampa quercus* f. *typica* from heather.
14.7. *Lasiocampa quercus* ssp. *callunae* female f. *typica* and male f. *olivacea* copulating on heather stems.

(between pages 288–9)

16.3. *Panaxia dominula* L. Various expressions of the gene controlling forms *medionigra* and *bimacula*.
17.2. *Hepialus humuli* L.
17.4. *Lasiocampa quercus* ssp. *callunae*. Wild larvae from Caithness (typical and melanic), 1972.

xiv *List of Plates*

(between pages 320–1)

19.1. The breakdown of dominance: third generation from P_1 *B. betularia* f. *carbonaria* crossed to Canadian *B. cognataria* f. *typica*.

19.2. The build-up of dominance: random 'f. *carbonaria* hybrid' (third generation) × *B. betularia* f. *typica* from industrial Birmingham, England (first generation).

19.3. The build-up of dominance: random 'f. *carbonaria* hybrid' (fourth generation) × *B. betularia* f. *typica* of Cornish origin, where f. *carbonaria* does not occur.

19.4. Distribution chart of hybrid phenotypes.

List of Figures

5.1	Frequency Map of *Phigalia pedaria* Fab. and its melanic forms from 88 localities in Britain (Lees 1971)	60
9.1	A frequency map of *Biston betularia* and its two melanics, f. *carbonaria* and f. *insularia* comprising more than 30 000 records from 83 centres in Britain	135
9.2	Diagram showing rate of increase of a melanic (dominant) mutant with a mutation rate of one in a million, assuming a constant advantage of the heterozygote throughout (which in practice will not occur), with a 50 per cent selective advantage over its typical form.	138
9.3	Graph showing frequencies of the three phenotypes of *Biston betularia* in Britain	143
9.4	Graph showing frequencies of *insularia* for the majority of districts in Britain (plotted against *carbonaria* frequencies).	145
10.1	Hatching of *Cleora repandata* and its black form *nigricata*. Black Wood of Rannoch, 1963	172
11.1	A frequency map of *Amathes glareosa* and its melanic form *edda*, comprising 20 000 records from 22 localities in Shetland	188
11.2	Decline in frequency of wild type *Amathes glareosa* and replacement by f. *edda* with distance northwards in Shetland. Phenotype frequencies are represented by X, gene frequencies by O. The dotted line shows the cline in phenotype frequency, the continuous line that in gene frequency	189
11.3	Map of the Tingwall area (the valley proper is shaded) showing the percentage frequencies of f. *edda* in 1961 (except for sites 6 and 18 which were only sampled in 1960)	192
12.1	Histograms of marked recaptures of both forms of *Amathes glareosa* showing the number of days' survival in the wild population (from 19 separate releases), Unst 1960.	201
12.2	Map of Shetland showing the localities mentioned in the text, (and Chap. 11, table 4) with histograms of moths recaptured on each day after release	203
12.3	Model histogram of recaptures on assumption of 70 per cent of moths flying each night and 30 per cent every other night	204
12.4	Graph showing daily percentage of f. *typica* in the wild population in Unst 1960	206
14.1	Distribution map of the morphs of *Lasiocampa quercus* in Caithness 1965–71 (cross-hatching = 1971 samples).	235

Preface and Acknowledgements

DURING many years of work on melanism in the Lepidoptera I have become increasingly aware that I was dealing not merely with a phenomenon restricted to that one order, but with something which contributes to the survival of living creatures in general.

It is time then, that a book, however incomplete, should be devoted to melanism, and I am here attempting to meet this need.

Except for two chapters the work is limited to the Lepidoptera. Because of their great dependence on colouration for survival these insects serve as ideal material for studying the various uses of melanism.

Nevertheless, I thought it necessary even here to refer briefly to melanism in other groups, which I consider in Chapter 3. I decided to do this to show that the same phenomenon exists throughout nature and that advantages may be conferred by dark colouration in a variety of environmental situations. That approach, of considering melanism as a basic recurring need since the inception of life on this planet is, I think, essential in order to appreciate its antiquity. On the one hand, melanism is found in the simplest fungi (sooty moulds); on the other, melanism (maybe to an extreme degree) is a subspecific characteristic of many races of Man.

In approaching this subject, it is important at the beginning to realize that, as an indirect consequence of being black, living things may have had to adjust, by way of changes in the gene-complex, many other characters; in particular, behaviour patterns, sexual selection, temperature-tolerance, and, probably most important of all, the demands of the various lines of chemistry necessary to produce the little-understood complex of pigments that come under the heading of the melanins.

This book then, though primarily intended for ecological geneticists, nevertheless demonstrates practical approaches to field and laboratory problems to all students of natural selection. Many techniques had to be improvised in the field when we met an entirely different set of circumstances to those we had anticipated. Designs of experiments laid down in advance from armchairs by biometricians, without intimate knowledge of the intricacies in the field, frequently failed.

I would like this book, therefore, on the one hand to illustrate the practical difficulties of field work to the theoretical biometricians who have not usually experienced them in wild populations; on the other, I hope that naturalists will be able to harness their own difficulties to the solutions we have had to devise in solving our own problems.

In every instance that I describe, I am anticipating that the reader has some knowledge of biology (though not an intimate one of the Lepidoptera), genetics (though I shall discuss the attributes of polymorphism), and a minimal experience of methods of statistical analysis as well as of present-day scientific aids and approaches. Without knowledge of such techniques, including their applications and limitations in field experiments, it is useless to attempt to analyse the many different uses of melanism in nature.

A considerable amount of information given here is previously unpublished. I have however quoted verbatim certain sections of my original field experiments on the selective predation of melanic and typical morphs in wild populations, both in industrial and non-industrial species. I am indebted to the editors of *Heredity, Lond*, and the *Annual Review of Entomology* for permission to do so. To the few who no doubt will criticize me for that decision, I offer no apologies. I decided to present parts of my work in this way in answer to a number of requests, particularly from biologists in the U.S.A. who had been unable to obtain reprints of the original work. Nor can I see any advantage in rewriting the techniques and results of experiments which I consider were adequately presented at an earlier date.

Much of the data recorded would have been unobtainable without the contributions from two teams who have helped me for many years. Foremost amongst these is the small band of workers who, both by day and by night, have assisted me in field work in Shetland, northern Scotland, and elsewhere—some of them each year for the last ten years. In particular, I would like to thank Dr. R. J. Berry, Dr. J. Cadbury, Dr. D. R. Lees, Dr. P. Harper, Dr. G. Phillips, and various members of Oxford University Scientific Society; similarly Dr. G. Howard, T. Peet, Dr. A. Shapiro, and Dr. and Mrs. C. Perrins, who have helped sporadically when possible. My son, David Kettlewell, has contributed on numerous occasions; apart from shooting a limited number of gulls, Hooded Crows, and Wheatears in Shetland for examination of stomach contents, he has also helped greatly in the mark–release–recapture experiments.

The second and larger group to whom I am profoundly grateful have provided more than 100 000 records of the melanic and typical frequencies in over 50 species of Macrolepidoptera. They consist of approximately one hundred part-time lepidopterists. These are the records of quite exceptional people, for not only is each contributor an expert on British moths, but in addition each holds a responsible position outside entomology, be he a general, an admiral, an air marshal or a bishop; he may be a stockbroker, an ironmonger, or, as is one of the greatest of Lepidoptera breeders, a man who worked in a gas-works; each has contributed. We are proud in Britain that a brotherhood of entomologists exists which is epitomized in the Verrall Association. Through this association, many of us are able to meet once a year, thanks to the perspicuity of G. H. Verrall who in 1826 endowed such a gathering. It is the part-time entomologists such as these who have contributed so much to our knowledge of distribution. Because of the differing types of districts in which each individual lives, be it town or country, collectively they have been able to list local records from the majority of biotopes throughout Britain. To all these recorders I owe so much. I credit each, because of limited space, by initials only†.

I have dedicated this book jointly to the Nuffield Foundation and to Professor E. B. Ford F.R.S., because together they made it possible for me to leave medical practice for research in ecological genetics. This was a difficult decision for me to make at the age of 42, with family responsibilities, even more so because of my happy relationship with a large number of patients. Retrospectively, I have no regrets, and nothing but gratitude to the University of Oxford which throughout has given me optimal facilities.

An inherent love of natural history from an early age led me, as an undergraduate at Cambridge, to take a degree in Zoology at the same time that I worked for medical qualifications. I recommend such dichotomy to all students because it is impossible for any young man to forecast his future interests, still less to foresee the opportunities that may come his way.

I want to thank a number of people individually who have given me so much help: first, my wife, who not only had to re-adapt her own mode of living, but also encouraged me to make such a change. For 35 years, she has assisted me and looked after duplicate stock to that in my laboratory. She has tolerated the incipient spread of papers from my study to a complete 'take-over' of the rest of our home, whilst writing this book. Also she contributed greatly to my earlier field experiments at a time when I had no other assistants. She was in fact the first to observe bird predation of the Peppered Moth through binoculars.

Secondly, I must pay tribute to my old friend the late Dr. E. A. Cockayne with whom I had close contact for over 30 years. I was indeed privileged to work with this leading geneticist of his day, both in the genetics of Man and of the Lepidoptera. The Rothschild–Cockayne–Kettlewell Collection

† At that time under the direction of Mr. L. Farrer-Brown who helped me so much.

of British Lepidoptera was born out of this association, with the consent of the Trustees of the British Museum of Natural History, where the Collection is now stored.

The Collection (now the National Collection of British Lepidoptera) attempts not only to show polymorphisms, local races, rare aberrations, and teratological specimens, but also to give references to their genetic origins when known (Cockayne and Kettlewell, 1953, Riley, 1948, Robinson, 1971).

The Collection came about in the following way. Cockayne and I had previously unified our own series of insects because we both believed that reduplication of identical specimens was a futile effort involving much waste of time. We also could not see sense in the ultimate dispersal and redistribution of rare and mutant forms which inevitably takes place at a sale on one's death. On 10 May 1947, the Trustees of the British Museum signed an agreement not only permitting the amalgamated collections of the donors with that then known as 'The Rothschild British Collection', but also adding the following clause '(4) The Trustees shall cause the Keeper of the Department of Entomology to make available for incorporation in the Rothschild–Cockayne–Kettlewell Collection such specimens at present contained in the collections of British Lepidoptera already the property of the Trustees as may add materially to the value and usefulness of the Rothschild–Cockayne–Kettlewell collection.' This must be one of the few occasions when scientific sense has transcended other considerations. A large amount of scientific data, including much on melanism, will be not only stored here in perpetuity, but also added to, owing to a handsome endowment by Cockayne. The success of the project from conception to fulfilment was largely due to the then Keeper of Entomology, Norman Riley, who throughout has shared our original concept, that of a single unified (national) collection of British Lepidoptera, fully documented with references. Unfortunately, we have run into considerable difficulty since 1947. Leslie Goodson, trained for many years by Cockayne as Curator to the Collection, and well versed in the entomological literature, was to our dismay retired in 1969 on reaching a particular age, specified by the Civil Service. We had hoped for an extension. He had dedicated his life to British Lepidoptera and had in particular a great knowledge of the various melanic forms. To him, not only I, but a majority of lepidopterists, must proffer our gratitude. He and I equally share our disillusion in the present-day bureaucratic outlook which seems not to cater for personal enthusiasms, dedication, and a love of one's work. As an earlier Director once said to me 'why should people be paid for work they enjoy?'

I have dealt with the Rothschild–Cockayne–Kettlewell Collection at length because it is my belief that Lepidoptera in particular demonstrate visually those genetic principles not so easily disclosed in other forms of life, especially the selective advantages and disadvantages of melanism. I think that, given reasonable access, this Collection can in future help generations of lepidopterists, geneticists, and others.

I would like to tender my thanks to others in the British Museum of Natural History for their help and advice, in particular to the late Sir Gavin de Beer (former Director), Dr. J. P. Doncaster (former Keeper of Entomology), and Timothy Tams into whose safe keeping I was firmly put by my mother in 1919, and who has continued to help my co-workers and myself for over fifty years. I also want to thank S. Fletcher for his supervision of the Collection (and D. Carter) and for helping me with his great knowledge of the Selidosemidae, in particular *Boarmia*, from different parts of the world; also the museum staff, who have provided photographs on numerous occasions, and T. G. Howarth and Baron C. de Worms who have sent me records for over 20 years.

There are many to whom I am indebted since my arrival in Oxford in 1957. Foremost amongst these I must again refer to Professor E. B. Ford, who, having enabled me to engage in research at Oxford, has throughout encouraged and discussed with me the various projects I have undertaken.

I was indeed fortunate to arrive at Oxford at a time when there were so many outstanding ecological geneticists there. In particular, I am grateful to Professor P. M. Sheppard, F.R.S., who (with Professor E. B. Ford) was concentrating his efforts on studying population dynamics by field experimentation. He, and my long-standing friend Professor C. A. Clarke, F.R.S., have subsequently shared with me unstintingly the data on their *Biston betularia* investigations centred on the University of Liverpool. Professor A. J. Cain was pioneering taxonomy into the field, and Professor M. Williamson was always asking difficult and provocative questions. Professor B. C. Clarke's snails were frequently challenging previous theories. It was indeed stimulating company. I was greatly helped in various aspects of my work by Professor G. C. Varley, of the Hope Department of Entomology, and by Professor G. E. Blackman, F.R.S., of the Department of Agriculture, who made it possible for me to apply radioactive markers to insect populations in his laboratories.

I look back on my earlier predation experiments as a time of great pleasure and excitement. I shared these with Professor N. Tinbergen, F.R.S., who for two summers joined me in the field and lived with me in my caravan-trailer. Apart from the recorded sequences of selective predation which he took, and his observations on the bird behaviour, I learnt a great deal from him. After a particularly frustrating series of experiments, I remember his dictum, 'That if you get quite different results from those anticipated, the final answer will always turn out to be of much more interest.'

Before I came to Oxford I had felt it my duty to admit that my capacity for statistical analysis was not of the highest order. N. T. J. Bailey, Professor of Biometrics, wrote a book, *Statistical Methods in Biology* (1959), and in it he acknowledges my work in reading the scripts, chapter by chapter. What he does not state however is his reason for this favour in mentioning me. It was in fact that 'if Kettlewell could understand his presentation, anyone else could'. Thus I have depended throughout on help from biometricians and apart from expressing my gratitude to Norman Bailey, I must mention others.

The late Professor J. B. S. Haldane, F.R.S., had always been fascinated by the population dynamics of Industrial Melanism (Haldane, 1924), and from the outset, he encouraged me to discuss problems with him. Such meetings usually took place at sandwich-eating ceremonies in my laboratory. He, along with Professor Maynard Smith, was conscious of the difficulties encountered in the field, and never suggested impossible designs of experimentation.

The late Sir Ronald Fisher, F.R.S., constantly gave me and my co-workers advice on experimental projects, particularly in regard to dominance modification in *Biston betularia* and the *B. betularia* × *B. cognataria* experiments. It was he who forecast that the gene-complex of industrial *B. betularia* would be found to be different from that of insects occurring in districts where melanic forms had not occurred.

I have been most fortunate in my assistants. The indefatigable Dr. C. J. Cadbury is one of the best all-round naturalists I have met, with a vast knowledge of botany, ornithology, and also the Lepidoptera in Britain. Dr. D. R. Lees has a capacity for attacking the basic problems of any project, and also the patience which is so necessary in carrying out routine repetitive analyses.

That all-important part of my work, the breeding of Lepidoptera, has been for many years in the capable hands of Miss G. Brookes, who came to me with little knowledge and no experience of the subject. Yet her perspicacity has enabled her to devise successful methods of large-scale breeding, not only of routine stock such as *Biston betularia, Lycia hirtaria, Phigalia pedaria,* and *Panaxia dominula,* but also of *Brachionycha sphinx,*

Utetheisa pulchella, and many others. She designed light-period/temperature techniques, enabling us to reduce the normal two year life-history of *Lasiocampa quercus* ssp. *callunae* to a univoltine one. She has been in charge of the temperature experiments on *Arctia caja*, and has for years carried out the arduous duty of moving the pupae into different temperatures three times a day. Without the differing contributions of all these people, my efforts to understand melanism in the Lepidoptera would have been futile.

I must thank a miscellaneous body of people, without whose help the rendering of our work would have been impossible: for paintings, Miss Christine Court; for photographs, J. S. Hayward, R. Tanner, the late F. Tilby, and M. Lyster, of the Department of Zoology, Oxford; for secretarial work, Mrs. K. Davies, Mrs. B. Garms, Miss J. Milsome, and Mrs. S. Fox, who contributed to the earlier scripts, and in particular Mrs. R. Wickett, who, as a biologist, has not only checked the script, but also accomplished the irksome undertaking of compiling the large bibliography. I am able to continue the present work under optimum conditions in the new Department of Zoology, due to the facilities offered by Professor J. W. S. Pringle, F.R.S., together with the sensitivity to our requirements of Miss D. H. Statham, the Administrator of the Department of Zoology, and Mr. K. Ford.

There are others who have helped and encouraged me in my work. In particular Sir Julian Huxley, F.R.S., on whose knowledge I have drawn so much in regard to melanic polymorphisms in birds. More important was it to me earlier, that he sponsored my request for extra help in the various projects I was to undertake. He has joined me in 'field work' in the London squares and again at my home near Oxford. His particular brand of original scientific thought, his humour, and his unique ability to recount his personal experiences with his earlier friends and associates stretching as they do over a period of 80 years has been a joy to me. Dr. Miriam Rothschild who has contributed so much to recent advances in our knowledge of toxicology in aposematic insects has been a constant inspiration. She has helped me not only by sampling, but also by discussion, and I have always valued her ruthless criticism greatly.

Without the help I have been given by my friends in other countries I could not have attempted to give résumés of melanism outside Britain. In the U.S.A. I have been privileged to know Professor Th. Dobzhansky, not only working with him there, in the South-western Research Station, Arizona, but also in this country. I have followed many of his suggestions on particular problems. Dr. Lincoln Brower and his wife Dr. Jane van Z.

Brower of Amherst College have twice spent the greater part of a year with our genetic unit at Oxford. I am grateful to them for advice on techniques for testing selective bird predation in captivity. I also learnt much from Dr. Charles Remington of Yale University on a similar visit. I must also thank the following who have helped me in one way or another on my visits to U.S.A. Dr. Sid Hessell, who introduced me to the New Jersey Pine Barrens, Dr. Alexander Klots. Dr. D. F. Owen, who is the first to have attempted a survey of North American melanism, and Professor J. Moore.

I have been introduced to Canadian Lepidoptera by Dr. Eugene Monroe, Dr. C. A. Sheppard of Montreal, and Professor Brian Beirne of the University of Belleville.

In Europe I am grateful for colloboration with the following: Dr. B. J. Lempke, of the Netherlands, Bishop S. Hoffmeyer of Denmark, Professor Suomalainen of Finland, Dr. Douwes at the University of Lund, Sweden, and also Professor V. Orel, Head of the Gregor Mendel Department of Genetics, Brno, who invited me to visit Czechoslovakia in 1965, on the occasion of the Mendel Centenary. This enabled me to see, albeit briefly, the incredible number of industrial melanics that occur in that country. My thanks are also due to Dr. Guy Howard who spent two seasons sampling Lepidoptera around Stockholm and on the Baltic Archipelago (latitude 60°N), and also in Lapland, and who gave me his records for inclusion in this book.

I would also like to express my gratitude to a large number of people who though not mentioned by name have contributed so much to the successful study of melanism.

Part I. The concept of melanism

1. Industrial Melanism and its place in nature

(i) An introduction

Amongst all living things it has fallen to the Lepidoptera to provide evidence of the most striking evolutionary change in nature ever to be witnessed by Man. In the phenomenon of Industrial Melanism we have been able to see, within the period of our life-time, the replacement of light colours and patterns in many different species of moths by dark or black ones.

For this reason Industrial Melanism has, in recent years, been recognized as being of the greatest interest to the biologist in general. Through field experiments, it has provided information on selective advantages and disadvantages in the wild; in the laboratory it has demonstrated to geneticists the various mechanisms which can contribute to the inheritance of an all-important character. The chemistry of the pigmentation (the melanin complex) is still a challenge to the biochemist.

Because of the unique opportunities it offers to the various lines of research into principles of evolution, and because we have reached partial answers to many of the problems, I have decided that, after delaying publication for many years, the time has now come when a book, covering the main features of melanism, would be of interest to a wide range of scientists and naturalists. Apart from the more usual lines of biological inquiry, Industrial Melanism has led me into less frequented paths: a study of air pollution and its distribution in many countries, the survival of lichens, the behaviour of birds and, in particular, that of gulls, and, indirectly, the distribution of lung carcinoma and bronchitis in Britain.

At each step new lines of experimentation have had to be initiated as each species demanded a different approach. For example, mark–release–recapture experiments originally designed for analysing the structure of *Polyommatus icarus*, the Common Blue, in the Isles of Scilly by Dowdeswell, Fisher, and Ford (1940) have been redesigned to establish selective advantages and disadvantages of two morphs in a continuous population, the one in an industrial area, the other in pollution-free countryside. Artificial colonies of a known gene-frequency of two or more forms have been established in certain districts where a particular species had recently become eliminated—in theory because mutation to a melanic form had failed to occur in time. A technique for marking larvae in the wild by feeding them on radioactive isotopes has been devised in order to discover the mortality-rate during this stage. More important is it that I have established a liaison with over 100 part-time lepidopterists in Britain. They have furnished me yearly with figures (not merely percentages) of the various forms of many polymorphic species of Lepidoptera which occur in their own districts, from northern Scotland to southern England, and from the eastern counties to western Ireland (see Appendix B).

In my field work, because maximum recapture figures were essential, I found it necessary to devise automatic trapping methods, particularly for night-flying moths. These have been of two kinds: mercury-vapour light traps and sex-assembling devices. In using such methods, it has of course been necessary to ascertain in each instance that there are no behavioural differences between the forms that were being studied. On one occasion, such a behavioural difference was indeed established between two populations; it rendered months of field-work futile in regard to demonstrating a differential survival due to selective predation. On this occasion, the answer to whether or not differential predation occurred was given only after an examination of the stomach contents of the predators themselves had been carried out.

New techniques are constantly having to be designed and two out of three of these have to be discarded for one reason or another. I find the challenges of such work stimulating and exciting; I am certain that we are as yet only on the threshold of designing experiments on population analysis. As a naturalist who has spent most of his time observing living things in the field, I believe it is essential to undertake such observations for 24 hours in a day whilst investigating the intimate life of any one species. One has to live on the spot alongside it both day and by night. I can only look with amazement on the efforts of those who attempt to formulate in Xs and Ys the total forces acting in a complex situation involving, for example, density-dependent factors. The computer which eventually does this will have to be fed by data from scientists of many different kinds. Industrial Melanism most certainly demands this same treatment also, for no amount of field work by one man can cover all the components and their variables in his life-time.

During such studies, I have had to spend long periods of time in the central London squares (such as Thurlow, Stanhope Gardens, etc.) which appear to me, because of their isolation, to offer the facilities of outdoor laboratories. By contrast, I have lived for months in Perthshire, central Scotland, and the Shetland Isles. Only by observing the same species under two different sets of conditions can a comparison be made of such basic characters as colour-pattern and behaviour. Only by choosing such contrasting environments is it possible to come to an understanding of how the same species can become adapted to live in entirely different niches.

In the past, the Lepidoptera have tended to be discounted as scientific tools. The reason for this is not hard to find. Industrial Melanism lies on the borderland of the lepidopterist on the one hand, who is usually an amateur or part-time collector, and often an extremely good naturalist in the field, and the scientist on the other, who frequently is not so. The former may not have had any scientific training whatsoever; the latter does not appreciate the highly complex and individual requirements of each species in this order. In Britain alone 780 species of Macrolepidoptera (apart from approximately 1600 species of Microlepidoptera) demand 780 different techniques in breeding in each of their four stages of metamorphosis. No wonder this rewarding field has been largely neglected! Its history records a series of lost opportunities.

It is essential also to consider the part played by Lepidoptera in nature. Here they provide a constant source of food throughout the World for innumerable birds, small mammals, reptiles, and for other Insecta. Many Lepidoptera depend for their survival on concealment alone, and crypsis is their one line of defence. Their wing patterning, unlike that in smaller insects such as *Drosophila*, may be their most important character. Their gene-complex is, in part, written into the architecture of their wings. Night-flying moths in particular must, in order to survive, pass the day motionless and invisible on particular backgrounds. The fact that Industrial Melanism is found, with rare exceptions, only among cryptic insects is therefore a strong indication that the main reason for its success is that dark colouration gives them better camouflage under the conditions imposed. The colour-patterns of cryptic insects are constantly having to vary in order to conform with environmental changes, as they have always had to in the past; for example, under conditions of high rainfall, cloud, and lack of light, selection must favour dark individuals; conversely drought and desert must favour light ones.

I shall provide evidence that there is correlation between melanism and industrial areas in two ways. First, the number of species maintaining black forms is greater around such centres; secondly, within each of these species the frequency of such forms increases with the degree of pollution fall-out.

The phenomenon of Industrial Melanism has occurred as an *indirect* effect of a recent environmental change: the pollution of the air from the combustion of coal, coke, and oil. Probably no other previous cataclysm in the history of the World has produced such immediate and widespread effects upon natural history. In the past, the eruptions of volcanoes must have altered environments but usually these were of limited extent. By contrast, air pollution by smoke has a multitude of diverse origins which have continually poured out effluvia into the atmosphere, both by day and by night, for more than 200 years. Evidence will be provided that smoke particles can travel great distances on weather fronts but that their maximum effects on living things take place around the centres of origin. Since smoke follows the prevailing wind, the degree of damage inflicted on vegetation is inversely proportional to the square of the distance from the source. It is, in fact, essential to have some knowledge of the distribution of pollution fall-out and its effects before attempting to postulate theories for the absence or presence of melanic individuals. This is discussed in Chapter 5.

It is important to note that, in attaining a cryptic advantage by becoming dark, such individuals cannot avoid the other basic consequences of their colour. These may, under certain conditions, such as dry heat, nullify the over-all selective advantage of cryptic melanism. The physical aspects of blackness are discussed in Chapter 2.

The phenomenon of Industrial Melanism has frequently in the past been considered an exceptional event. I hope to show that this is entirely incorrect and that it reflects but one of a number of situations which, in the Lepidoptera, demand black colouration for survival. What is so striking is the *speed* with which black forms have replaced the previous light-coloured populations. Moreover, in Britain alone, there are over one hundred species of moths in widely diverse genera which manifest Industrial Melanism. Similar situations are now being recorded around all industrial areas in the Northern Hemisphere; wherever, in fact, they have been looked for. Industrial Melanism occurs also, but rarely, amongst the Arachnida and the Coleoptera.

One of the major points I want to stress in this book is that melanism is found commonly amongst Lepidoptera in entirely different conditions to those due to industrialization today, for example in the Subarctic regions during the summer months, on high mountain-tops, and in rain forests, and I believe it always has been in the past. Geographic or Ancient Melanism is discussed in Chapters 4(**ii**), 10, 11, and 12.

Furthermore, I believe that melanism should be considered on an even wider basis—one involving most living creatures. For this reason I have included in Chapter 3 examples of melanism occurring in nature in orders other than the Lepidoptera. The hypothesis I wish to present is that dark and light forms of many organisms have possessed advantages or disadvantages under varying environments since the inception of life on this planet. Such a situation could have led to the earliest visible polymorphisms; the history of melanism is written deep into the past. Only by approaching Industrial Melanism by this route can one hope to come to an understanding of its place in nature and the genetic mechanisms which control it.

(ii) Definitions

It is essential in this first chapter to explain these mechanisms briefly and to define clearly my conception of the various categories into which melanic variability can be classified.

Polymorphism. Polymorphism has been defined by Ford (1940a) as 'the occurrence together in the same habitat of two or more discontinuous forms of a species in such proportions that the rarest of them cannot be maintained merely by recurrent mutation'. This situation is certainly the most frequent one found in industrial melanic species in Britain. The word 'discontinuous' implies disruptive selection which has favoured either one form or another and the elimination of intermediates (Mather, 1955). Here then, there is a strong tendency for the heterozygotes to become indistinguishable from one or other of the homozygotes through the evolution of dominance; yet it is the heterozygote which may be largely responsible for maintaining such a polymorphism.

It is as well to reflect here on the factors which contribute to a polymorphism. When a mutation occurs the overall effects can be either advantageous, disadvantageous, or neutral.

Considering neutral survival first, Fisher (1930a) has shown that it is unlikely for a gene of neutral value to be maintained in a population because of the violent fluctuations in selection pressures which each species regularly experiences in its changing environment. Secondly, Fisher (1930b) states that, except in very small populations, the number of individuals possessing such a gene cannot, if it is derived from a single mutation, differ materially from the number of generations since that mutation occurred. This can, at best, be an exceedingly slow process. Genes conferring neutral advantages must therefore be excluded from our list of those having contributed to evolution.

Neutral survival must be clearly differentiated from the situation in which two advantageous genes compete in a population. Here high selection pressures for two or more successful alleles, each having a number of different advantages and disadvantages, and each capable of rapid adjustment to genetical or environmental changes, bring about an equilibrium which leads to a state of balanced polymorphism. These pressures, until 30 years ago, were usually considered to be small (in the order of 0·1–1·0 per cent). More recently, selective advantages and disadvantages for a particular character have been shown to be greater (30–60 per cent) and our work on Industrial Melanism has confirmed, indeed contributed, to this.

The third situation is one in which a mutation is disadvantageous to a greater or lesser extent. Here its future depends on the degree of detriment it bestows; if severe it will be eliminated rapidly, but if some of the characters which such a gene controls give advantages in the heterozygous state but disadvantages as a homozygote, it will continue to survive in the population as a rare recessive mutant. Nevertheless, such a state must constitute a polymorphism.

In industrial melanic species the black forms have dominance and in *Biston betularia* there is evidence that f. *carbonaria* (the blackest one) continues to increase rapidly in the population to a point where the maximum number of heterozygotes can be maintained in view of the contending advantages and disadvantages involved (see Chapter 9). A successful melanic mutation will, therefore, follow a sequence. At first its spread through the population will be slow as it replaces its 'normal' allele, then more rapid, and finally slow again because of the disadvantageous effects of the new homozygous mutant (Chapter 9, Fig. 9.2). The frequency of the three genotypes in a balanced polymorphism must depend on the overall success or failure of each. Ford has referred to the period of such spread and adjustment as *Transient Polymorphism*.

Heterosis. The concept of heterosis or heterozygous advantage is implicit in polymorphisms; the heterozygote is favoured because it will tend to accrue the advantageous characters of both alleles, which will be balanced by disadvantages in the mutant homozygote. There are two reasons for this. First, lethal or semi-lethal genes can accumulate when positioned on a chromosome close to a new successful major gene. Here, as heterozygotes, they are given protection from selective elimination. Yet the homozygotes of the new mutant form must suffer from the deleterious effects of these disadvantageous genes when a double dose of any one may be lethal.

A second reason for the advantage of the heterozygotes is that every mutation so far investigated has been shown to be responsible for controlling more than a single character. But some may be advantageous while others are detrimental. Natural selection, by rearrangement of the gene-complex, will favour the former, which will become dominant, whilst the latter will tend to be recessive. The *complete* replacement of a 'normal' allele by a highly successful new mutant must indeed be a slow process.

Multifactorial inheritance—polygenic. Frequently, however, more complicated situations are found and these can be classified into two groups. A second and different melanic form may arise at another locus, or it may be allelic (an alternative form) and situated at the same point on the chromosome. In the first instance then, it is possible for nine phenotypes to appear in wild or laboratory crosses of double heterozygotes, eight of which will be melanic. The genotypes will be AABB (the double homozygous melanic), AAbb, AaBB, AABb, AaBb, aaBb, Aabb, aaBB, and aabb (f. *typica*, non-melanic) in the proportions 1:1:2:2:4:2:2:1:1.

In the second group, when the two forms are allelic, only six genotypes can occur, five of which produce melanic phenotypes, mm (f. *typica*), M^1m, M^2m, M^1M^2, M^1M^1, and M^2M^2.

It is the second of these groups which is likely to have been the main contributor to melanic polymorphisms because it permits a highly sophisticated mechanism to evolve, comparable to that of the super-gene.

The super-gene. If a number of genes affecting melanism have occurred due to mutation in the past they are likely to have been situated at random on different chromosomes. Yet one in association and combined with another will have provided certain advantages, and such an association of advantageous genes will therefore be favoured. Sectional interchanges will therefore, in time, have moved genes controlling melanism on to the same chromosome. Not only will natural selection have favoured these genes, but also their positioning in close proximity. Such close linkage can be brought about by reducing chiasma formation in the immediate vicinity or by the production of an inversion. By these means the chances of crossing-over within the relevant section at meiosis are diminished so that the favourable units tend to be kept together. The nearer together two co-adapted genes may be the smaller is the selection pressure needed to hold them together. Disadvantageous recombinants are produced, of course, at a low level, but they can be eliminated in the same way as a disadvantageous mutant would be. This permits a switch mechanism to control a number of alternative allelic forms from a common centre, which is protected against cross-over. The creation of such a specialized unit must reflect a selective pressure in the same direction usually over eons of time.

I believe that in the past such units have been built into the gene-complex of many species which have melanic forms today, and I shall discuss this later.

Heterozygous variability. It occasionally happens that a species may exhibit a range of forms from black to light with every imaginable gradation. This, of course, can be the result of multifactorial control at a number of separate loci. I have just pointed out that two such loci alone can give rise to nine phenotypes. The visible effect in sampling may well be a colour-cline from the darkest to the lightest. In practice, individuals in the nine classes are indistinct.

A similar picture can be brought about when a single gene controls melanism, but with incomplete dominance. Here the effects of disruptive selection have taken no effect. This can reflect one of two somewhat different situations. In the first place it could be that a new melanic mutant is in the process of becoming fully dominant, where natural selection favours dark or light individuals, but disruptive selection has not yet become finalized. Alternatively, there are ecologies which can favour graded variability. Such differences in heterozygote expression are decided by the effects of the whole gene-complex.

Apart from rare instances where the environment is responsible for pigment formation (see Chapter 18), the control of melanism is genetic and falls into one of these three categories (see Chapter 4). However, on initial field-sampling it is frequently impossible to decide into which category a particular insect falls. Only laboratory breeding from it can decide. It is important to recognize at the commencement that darkening, both of pattern and wing colouration, can be brought about and controlled by several entirely different evolutionary processes. It is these I shall be discussing in the following chapters.

2. The attributes of black colouration

(i) General remarks

BLACK forms of living things are found throughout most phyla. These are held in their population by natural selection, and by their frequency they reflect the degree of advantage or disadvantage of the melanic form at a particular time; they recur by mutation. Organisms maintaining alleles controlling black and light forms are likely to have more diverse (but not necessarily more intense) pressures brought to bear on these characters than, for instance, those which exhibit red and yellow. For example, the latter cannot be recognized by colour-blind predators, though their tone may be. Considering the matter retrospectively, there must therefore have been some order of precedence in the origin and the importance of these characters. The alternative of a black or a light form is likely to have become of consequence to survival at an earlier date than red or yellow, during the period before colour vision was established (though subsequently this appears to have evolved quite separately on at least three occasions). This may be reflected today in the widespread occurrence of alleles controlling different degrees of black colouration throughout the Animal Kingdom. Because of the fundamental and recurring need for melanism, it is likely that gene-complexes have been built up in the past which ensure that, on mutation today, nearly full expression of melanic forms is ensured throughout the range of living creatures.

Blackness can, in fact, offer advantages in a number of situations, because of several quite different physical attributes it may confer: for example, crypsis, the capacity for absorption of heat, or a barrier against ultra-violet light.

Dark colouration offers cryptic advantages to the Lepidoptera in many entirely different environments: in Europe alone we see this in pine forests, on peat moors, on the western coastlines, and in northern latitudes where there is no darkness during the summer months. Elsewhere it is common in rain forests such as South Island, New Zealand, and on the summits of high mountains.

In mammals, zoologists have found a correlation between humidity of the environment and the frequency of melanic forms in many different species. Nocturnal crypsis due to melanism is found throughout the World in animals, both predators and prey. Anyone who owns a black spaniel dog and exercises him at night in woodland countryside will appreciate this! On most nights he is completely invisible at a distance greater than five yards.

In bird species, a melanic 'phase' is constantly referred to in the literature but I shall discuss only a few of these by way of example in the following chapter.

A rather interesting demonstration of a selective advantage for dark birds was recently given me by a most astute observer of nature (personal communication) from Glasgow. She wrote to me to record what she referred to as 'Industrial Melanism in pigeons'. Some years ago the pigeons in Glasgow had contained a high proportion of white or white-marked birds. These rather rapidly disappeared and after a short time few remained. The population was now composed of dark-coloured birds. An investigation revealed that one of the reasons for this rapid change was surprising. Pigeons had become so numerous that the authorities decided that steps had to be taken to reduce their numbers. The pigeon catchers worked by day but so great was the public outcry against this that it was decided the pigeons must be destroyed by night only, as they roosted on buildings. The dark individuals were most difficult to see in semi-darkness, but any which had white markings were snared without difficulty. Man was the predator selecting against white pigeons (though breeding success also is associated with certain dark forms, thereby favouring their increase (Murton, in press).

Melanism is found in both the Amphibia and the Reptilia which have, in common with the Arthropoda, and as distinct from the homoiothermic mammals and birds, a system of temperature regulation largely controlled by the external environment. Heat absorption replaces the method of heat production by metabolism in the homoiotherms; it is less efficient because the metabolic rate itself (and hence activity) is dependent on the extrinsic environmental conditions.

The smaller and darker a creature is, the more rapidly is it affected by external heat sources in daylight. Very small arthropods are likely to perish from dehydration under these conditions; hence they have developed special behavioural patterns to avoid this. In small living things, the internal temperature can be recorded by thermo-coupling but the physical properties of black objects are analysed more easily by using inanimate material in such experiments.

(ii) The physical differences between black and white

If one takes two coins of the same vintage and of the same weight, and paints one black and the other white and places them both on a block of ice in daylight, a very clear demonstration of one of the basic physical differences between blackness and whiteness can be shown. After a few hours, the black coin will be seen to have sunk into the ice block whilst the white one is still resting on the surface: black absorbs heat while white reflects it. This physical property affects, of course, living things similarly, but the bulk (or volume), shape, and surface area of each creature must also be taken into account. Conversely, in darkness, black objects may lose heat more rapidly than white. This simple observation demonstrates one of the main environmental influences which are imposed on living creatures, particularly small ones.

The experiments of K. Schropp, using thermo-coupling, showed that when black and white surfaces were exposed to direct sunlight in still air, whilst insulated on cork (of 5 cm thickness), the temperature of each was very different: black surfaces attained 45°–55°C, white ones 15°–20°C, and reflecting surfaces, such as aluminium, only 15°C. By night all the surfaces had a temperature of 2°–4°C lower than the surrounding air, but he does not refer to a difference between the black and the white ones.

It is a fact that the temperature of poikilothermic animals may frequently be higher in direct sunlight than that of homoiothermic ones, particularly if dark-coloured. The temperature of any body in equilibrium with its surroundings, animate or inanimate, is that at which its net heat exchange is zero. Parry (1951) has shown that radiation from the sun increases the temperature of any dark object, unless it is a very small one (such as *Drosophila* sp.), whilst convection by air currents and conduction by contact are the main contributors to heat loss. He also showed that any insect's metabolic heat production is insignificant and that heat loss due to evaporation is negligible (though death by dehydration is a most important consideration in the economy of every insect).

Parry used small, blackened, brass spheres and discs as models. The spheres were of three diameters ($\frac{1}{3}$ in., $\frac{3}{8}$ in., or $\frac{1}{2}$ in.) and insulated against heat loss by conduction. He placed these at fixed distances under $\frac{1}{3}$ in. from the ground surface in direct sunlight, which was, however, at times, partially screened by cumulus cloud. Because of the protection of surrounding vegetation, convective heat loss due to wind was at a minimum so close to the ground. By this means he obtained the following data:

(1) The temperature of the spheres rose, on a clear day in June, to 40°–45°C.
(2) There was a significant difference in the equilibrium temperature (heat debit and credit account) between discs of various sizes under the same conditions.

14 The attributes of black colouration

(3) Very small changes in height above soil surface-level affected the equilibrium temperature.

(4) The temperature of the discs at a distance within one inch of the ground surface was more influenced by their background than by the air temperature surrounding them. Comparable temperatures must be experienced by many insects living under these same conditions; in particular by Locusts (and other Orthoptera), Coleoptera, and certain Lepidoptera.

Parry then proceeded to investigate the situation at a height of about two feet above ground level, and showed that here the change in temperature with height was very small, 'less than 1°C between 18 and 30 inches' (but a black disc almost touching the ground was about 15°C hotter). By contrast, at this height the temperature of his models was much more affected by the air temperature and the convection currents surrounding them.

His conclusions are so important to an understanding of the survival of small creatures when colour is taken into account, that I quote verbatim:

> The temperature of arthropods in sunlight is determined by the balance between radiative and convective heat exchanges, compared with which the heat associated with metabolism and evaporation is normally insignificant. The radiation load is made up of short-wave radiation received from the sun directly, diffusely from the sky, and by reflexion from the ground, and an exchange of long-wave radiation between the animal and its surroundings, including the atmosphere. The heating due to solar radiation is affected by the colour of the animal, and there may be variations of up to 50 per cent in the total radiation load due to this factor alone. The total mean radiation load per unit area will also depend on the shape of the body, and for bodies (such as most arthropods) with markedly unequal surfaces, it may be affected by up to about 50 per cent by changes in orientation. The net gain or loss of long-wave radiation is made up of a loss of heat to the atmosphere due to its having an equivalent temperature of approximately 0°C, and a small heat exchange with the ground due to the difference of temperature between it and the animal. Screening the sky by vegetation will reduce the net loss of heat to the atmosphere but will at the same time reduce diffuse solar radiation, so that the resultant effect of changes in the radiation conditions are unlikely to affect the heat balance significantly.
>
> Convective heat loss depends upon the shape of the animal, the wind-speed and the animal's orientation to it, and for a given shape, orientation, and wind it also varies inversely as a fractional power of the linear dimension. This means that in any given conditions the smaller an animal is the lower will be the temperature at which convective heat loss balances the radiation load.
>
> It will be evident that the factors determining the temperature of an animal in sunlight, even though they may be reduced to radiation and convection, are very complex and difficult to measure, so that even in the simple case of an animal suspended in free air above the ground, a calculation of its temperature will only be very approximate. On the other hand, the insignificance of the 'biological'

factors of metabolism and evaporation implies that the temperature of a living animal in a particular situation is likely to be very similar to that of a model of the same approximate size, shape, colour, and orientation. Such a model, constructed so that its temperature can easily be measured, may be looked upon as a type of thermometer giving information of considerable biological significance.

When the sun is strong, an arthropod, unless it is very small, has at its command temperatures which equal or exceed those of a warm-blooded animal. It remains to be considered to what extent such temperatures are necessary to some animals, and to what extent they are prohibited by the accompanying high rate of evaporation and so represent an important constraint upon the evolution of the habits of terrestrial arthropods. It has already been remarked that, despite the extensive literature on the water relations, particularly of insects, no information seems to be available concerning the rate of evaporation in natural habitats where the animal may be warmed considerably above air temperature, where local humidity, no less than temperature, may be very different from what is reported by the meteorologist, and where sources of water may be available to replenish what is being lost.

Parry specifically designed his experiments to ascertain the temperature of small animals under natural conditions, and they were conducted outside the laboratory in those same complex environmental conditions that the majority of small terrestrial animals have to live in. Digby (1955) has pointed out that Parry's conclusions are not applicable to the smallest insects, such as *Drosophila*, for here the bulk is so small that they take on the temperature of their immediate background.

Larger creatures, particularly warm-blooded mammals, must always be melanic for reasons other than that of heat absorption which, because of their bulk, must be relatively less important than in smaller ones. They may have black colouration for crypsis in dark or shade, or the pigment melanin may act as a barrier to ultra-violet rays penetrating the deeper layers of the skin. Only in small living things of a certain size is it possible for temperature control to be attained through melanism. It would appear then from the work of Parry and Digby that neither the smallest nor the largest creatures are able to make use of melanism as a means of heat production; only those insects larger than the Diptera, such as Orthoptera, Lepidoptera, and some Hymenoptera on the one hand, and those vertebrates as small as some lizards and snakes on the other, can do this because of the physical barriers imposed by size. I am personally not convinced, however, that some of the smallest creatures, and the largest ones, have not overcome their respective limitations. Amongst the smallest animals this might be accomplished by the use of hair-scales which diminish air flow, along with specialized positioning in specialized microclimates and amongst the largest by increasing their surface areas as in some fish. Though it replaces the method of heat production by metabolism in the homoiotherms, direct heat absorption from the sun is less efficient because the metabolic rate itself (and hence activity) is dependent on extrinsic conditions. It must be added, however, that temperature increase is regularly produced within the body of winged insects by rapid wing-vibration prior to and during flight (Adams and Heath, 1964)*.

Pigment and its distribution in the Insecta is therefore of over-riding importance to their activity, quite apart from other advantages and disadvantages to which it contributes. It also demands the use of exact and complicated behaviour patterns in climatic crises, such as particular orientations with respect to the sun in excess heat or, conversely, in a deficiency of heat (Tinbergen *et al.*, 1942), an appreciation of shade in resting positions, and recognition of the degree of humidity. A detailed analysis of microclimates and the ability of living things to recognize them is discussed by R. Geiger (1958) in his book *The climate near the ground*.

In spite of adaptations to immediate requirements, a period of hibernation or aestivation has normally to be undertaken at some period of the life cycle by the majority of poikilothermic species which live in temperate climates. This reflects their complete dependence on the external environment.

(iii) The pigments

Dark colouration can be produced by several distinct biochemical paths. First, in the Insecta, by the tanning of a cuticle which, in hardening, forms an exo-skeleton to which muscle bundles can be attached. This is comparable to the tanning of leather. During the process the colour becomes darker through various shades of amber and brown to black. The extremely hard, plastic-like substance it forms is scelerotin. The chemical paths leading to this involve derivation from phenylalanine, $C_6H_5.CH_2.CH(NH_2).COOH$, by way of hydroxyphenylalanine, $HO.C_6H_4.CH_2.CH(NH_2).COOH$, and tyrosin. Secondly, but more pertinent to this book, there are other tyrosin substances which polymerize and produce the little understood complex of pigments, the melanins. These are widespread in the animal kingdom and are found not only in insects as eumelanin but also in the black feathers of birds and the black hair of mammals (eumelanin and phaeomelanin). Both pigments are of high molecular weight and are derived by oxidation of phenols.

Melanin in the Insecta is exceptional in that it is laid down *within* the

---

* Recently, my wife (H.M.K.) observed that male *Lasiocampa quercus* caught in flight, and held between cold fingers, 'felt hot'. This was confirmed immediately by others.

cuticle itself; other pigments, when present, are seen *beneath* the cuticle. Melanin, frequently in the form of granules which vary in size, appears to be insoluble in the normal solvents. In the wings of Lepidoptera, the colour unit is a scale and here, once again, the melanin is found within the external chitin. In most melanic Lepidoptera, each scale is coloured thus. In the patterning of others, black scales may be interspersed with yellow ones due to the presence of xanthopterin, thereby producing an olive-green colouration. This process is used, for instance, in procuring the underside hindwing crypsis in the butterfly *Pontia daplidice* (Wigglesworth, 1928), which passes wet days resting on low-lying green foliage. In 1945 I caught a specimen of this species in Britain in which the xanthopterin was missing; the pattern appeared as a grey-black etching (=ab. *infragrisea* Goodson). Onslow (1916) worked on *Pieris brassicae*, the Large White Butterfly, which has contrasting colours of white with black patterning. He demonstrated that the undeveloped wings in the pupa of the species, when treated with tyrosine, became completely black. In another series of tests when the embryonic wings were subjected to the enzyme tyrosinase only, the normal pattern developed. This must imply that tyrosinase is normally present throughout the wing scales but that the specific black pattern, genetically controlled, is brought about by the flooding of these areas by tyrosine. A most enlightening résumé on pigment is given by Robinson (1971).

The ommochrome pigments, usually yellow, red, or brown, as for example in the Nymphalid butterflies, are at times responsible for dark colouration. They were originally, in fact, mistaken for the melanins but, unlike them, are present as granules within the cell and not in the cuticle. They are responsible for the dark colouration of many Lepidopteran larvae, grass-hoppers, and other insects.

From this brief summary, it is evident that dark colouration can come about by several different chemical processes. In most, the deposition of pigment is controlled genetically and is inherited. More rarely it is under the influence of the external environment and is affected by the degree or length of daylight, background colouration, or temperature. It can be reversible and this is brought about by the migration of granules containing ommochrome pigments to or from the surface of an individual; however, this does not occur in the imagines of the Lepidoptera.

3. Examples of Melanism and its functions in living organisms

ONE of the objects of this book is to emphasize the universal distribution of dark-coloured animals throughout nature. I have done this deliberately in order to underline the antiquity of melanism due to the physical differences between black and white. In some species, black-coloured individuals form 100 per cent of their population; in others, as in humans, there are recognizable black, white, or yellow races, each varying in a number of other characters, perhaps sufficient to qualify as subspecies. Frequently polymorphisms exist, particularly in birds, insects, and mammals, in which a proportion of a species is black. With such phases it is certain that the frequency of the morphs will vary both over periods of time and also geographically due to the different environmental conditions in which they live.

There are four main functions of melanism: (1) crypsis (found throughout the animal kingdom); (2) warning colouration (a moving black object is conspicuous to all animals including those without colour vision); (3) heat absorption (in smaller creatures only); (4) a barrier and a defence against ultra-violet light (as seen in humans and slugs).

In order to demonstrate this universal distribution, I shall refer to a few examples only of each in certain phyla.

(i) Cryptic Melanism

Melanic crypsis in Insecta other than Lepidoptera

A capacity of insects in various orders to take on dark colouration is widespread in nature and is commonly under the direct influence of the environment. Such stimuli as overcrowding, dark background, or increased humidity can each lead directly to melanism *not* necessarily controlled genetically.

Some of the best examples are found in the Locusta, which probably existed in the Carboniferous period, two hundred and fifty million years ago. Hoppers of *L. migratoria*, when at a high density, are all black; the resulting adults are also black but with yellow markings. In the *solitary* phase, both hoppers and adults are green, and match the foliage on which they rest. Individuals in the *swarm* phase, apart from dark colouration with conspicuous markings which might be considered aposematic, vary in other characters: they are restless, intensely gregarious, and they develop more rapidly than those of the *solitary* phase which are cryptic and sluggish. The latter condition is probably brought about by the excessive activity of the corpus allatum with consequent increase of juvenile hormone (Wigglesworth, 1964). Melanization in the Locusta seems to be comparable to that found in the larvae of some species of Lepidoptera, when occurring in swarm conditions, such as *Plusia gamma* L. and *Diataraxia oleracea* L. (Long, 1953) (see Chapter 4), but in the Lepidoptera it appears that there is a genetical component also. The work of Plotnikov (1924) showed that the dark phase in the Acrididae is independent of daylight as it occurred in darkness when a mechanical activator was used. Recurring stimulation leading to increased activity seems to be the cause. It is hard to understand the selective principles involved in this. Though the *solitary* phase is undoubtedly cryptic and a gregarious individual would normally be considered aposematic, locusts in swarm are regularly a main source of food for many migrating birds, (such as White Storks, *Ciconia ciconia*).

The Locusta are also capable of becoming black under an entirely different stimulus. If adult light-coloured grasshoppers are put on to a burnt area of ground, they rapidly become melanic and highly cryptic. This is probably due to a mechanism under the control of the nervous system which is triggered off by a visual appreciation of the background and a consequent migration of pigment granules. As a result of the same mechanism, dark-coloured grass-hoppers are frequently found on peat moors and also on urban commons (such as Wimbledon), where this phenomenon has somewhat misleadingly been referred to as 'Industrial Melanism'.

Isely (1938) demonstrated that crypsis in different species of Acrididae was important to their survival,, by introducing their dark and green phases onto different backgrounds in the presence of Bantu Chickens, domestic Turkeys, English Sparrows, and Mocking Birds. Each of these four insect predators devoured more of the conspicuous hoppers than of the concealed ones.

In other orders (for example Coleoptera or Hymenoptera), though melanism is widespread, I have little knowledge of this contributing to crypsis. There are, however, several examples of melanism in the Arachnidae; for example *Arctosa perita* Latr., which may be found on old colliery tipping sites at Pooley Fields, Warwickshire, England (Arnold and Crocker, 1967). Here the melanic individuals vary in the degree of melanism, and intermediates occur. In a somewhat similar position is the spider *Salticus scenicus* Clk. recorded by Mackie (1964) from old walls near gas-works at Stockport in Lancashire. Other species showing Cryptic Melanism are *Ostearius melanopygius* O.P.-C. and *Drapetiscea socialis* Sund. which again is variable. There seems to be general agreement amongst arachnologists that dark forms in each of these species have only arisen in the last few years and also that they are only found in the most heavily polluted industrial areas. Their mode of inheritance is unknown.

There are, then, a number of different methods of attaining melanic crypsis throughout the Insecta (and Arachnida), controlled either genetically, environmentally, or, as is the more usual, by a contribution from both.

More important is it to realize that in the Lepidoptera melanic crypsis has been exploited genetically since their genesis. Their wings have been equally important to them in providing cryptic protection as in providing flight.

Melanism in Birds—Environmental and Cryptic (non-aposematic)

Melanism is common amongst birds but it must be admitted in the first paragraph that the reasons for it may be obscure; indeed even in clear-cut polymorphisms the selective advantages conferred are frequently not understood. Also, 'Environmental Melanism' occurs; thus a survey of Wood Pigeons, *Columba palumbus*, by Miriam Rothschild showed that a high proportion of melanic individuals suffered from avian tuberculosis (McDiarmid, 1948). Landsborough Thomson (1964) points out that this 'involves the adrenal glands, which cause an increase in pigmentation notably reminiscent of the pigmentation of Addison's disease in human beings.'

Abnormal diet will turn the normal colouration of some birds completely black. Deficiency in vitamin D causes various breeds of poultry to deposit abnormal quantities of black pigment in the feathers. Even Buff Orpington chickens, which normally have no black pigmentation, showed 'intense black' under this treatment (Decker and McGinnis, 1947). They conclude by saying that 'vitamin D exercises some influence over the degree of oxidation of melanin compounds in the developing feathers'. The best known wild example is the Bullfinch, *Pyrrhula pyrrhula*, when fed on hemp seed which have a high oil content (Sage, 1962). The Goldfinch, *Carduelis carduelis* (Stevenson, 1866), and the Hawfinch, *Coccothraustes coccothraustes* (Newman, 1855), appear to behave similarly. When the hemp seed is removed from the diet the melanic birds will gradually return to their normal plumage. Such examples as these can contribute nothing to survival and they must be considered as the product of abnormal chemical and physiological processes (Staples, 1948).

Melanic phases in birds are found in a number of widely divergent genera. In the majority of instances, the selective advantages and disadvantages are not yet fully known. In many species melanism is inherited as a clear-cut one gene difference and the morphs may vary in their frequencies geographically from 100 per cent of the one form to 100 per cent of the other. Such an example is the Australian Goshawk, *Accipiter novae-hollandiae*, which is dimorphic in Australia, about 80 per cent being of the melanic form in southern Queensland; in Tasmania the whole population is of the white form; in New Guinea the whole population is of the dark form. Yet in spite of this there is no evidence of subspeciation (Huxley, 1955).

On the other hand, the Great Blue Heron, *Andea herodias*, shows an increasing white-morph frequency in North America from Yucaton eastwards to Florida Keys, where 100 per cent are of this form and where it has been separated as a distinct species, *A. occidentalis*; in fact, it is probably a subspecies (Mayr, 1955).

It is indeed remarkable that in these instances there appear to have been no experimental approaches in the field whatsoever with the object of assessing the various advantages and disadvantages of the two morphs at the extremes of their range. Nor have differences in behavioural characters been noted between them. In spite of this both must exist.

Other examples of melanism, the selective advantages of which we have no knowledge, include the Little Egret, *Egretta grazetta*, and a number of polymorphisms in predatory species, for example: the Buzzard, *Buteo buteo*; Montagu's Harrier, *Circus pygargus*; Eleanora's Falcon, *Falco eleanorae*, in the Mediterranean; the Gabar Goshawk, *Micronisus garbar*, in East Africa; the Rough-Legged Buzzard, *Buteo lagopus*; and the Short-tailed Hawk, *Buteo brachyurus*, in North America (Sage, 1962). In none of these instances is it likely that heterosis alone is the cause of the polymorphism and that the colour phase can be discounted.

A little more is known about the melanic polymorphisms in skuas and petrels. In *Starcorarius longicaudus*, the Long-tailed Skua, the dark phase is very rare and possibly does not occur in adult plumage, nor does it show a cline (Southern, 1944). In two other related species, *S. parasiticus* and *S. pomarinus*, the proportion of light and dark morphs varies greatly from area to area and even from colony to colony. 'In Shetland less than one fifth of the Arctic Skuas, *S. parasiticus*, are light, but elsewhere the proportion is higher, up to about 100 per cent in Spitzbergen and arctic Canada' (Landsborough Thomson, 1964). Southern (1943) states that there is a cline of increased melanic frequency to the north, and that dark birds preponderate in oceanic areas.

O'Donald and Davis (1959b) showed that pale, intermediate, and dark individuals occurred in Shetland and that from an analysis of the breeding data it appeared that the dark form was dominant to light and that the intermediates were incomplete dominants. The dimorphism is therefore controlled by two alleles at a single locus. The cline frequencies appear at present to be stationary, in fact the cline itself might be due to degrees of dominance.

O'Donald (1959) also suggested that the morphism was in part controlled by assortative mating and that non-random mating occurred in Caithness and Shetland; thus there would be a tendency for similar morphs to pair together, and such bonds would continue over a number of years. He concluded by saying that such homogamy might contribute to a more rapid evolution of dominance.

More recently Berry and Davis (1970) have presented data which could account for the cline in melanism. They observe that both melanic and intermediate males accept females more easily, and therefore earlier, than the pale males, which are aggressive for a longer period. This behaviour difference ensures the perpetuation of the melanic dominant unless counterbalanced by other factors.

In regard to petrels, Mackworth-Praed (personal communication) informs me that many species in the Southern Hemisphere show melanic polymorphisms. Landsborough Thomson (1964) states 'the species nesting in high latitudes tend to be white all over . . .; those breeding in warmer climates tend to be dark-brown or black above . . .; some groups of closely related species or races show much geographic variation in conformity with this trend'.

Fisher (1952) records that in the Fulmar Petrel, *Fulmarus glacialis*, the melanic form's frequency is low in the southern Atlantic breeding range but high in the north. Yet in the Pacific subspecies the relation is reversed. We have as yet little knowledge as to what environmental factors favour or disfavour the morphs. Yet there is one factor which must contribute to the perpetuation, namely imprinting. For each chick must be biased in its eventual choice of mate by the colouration of its own parents as in the Arctic Skuas.

That a black mutant can have over-riding selective advantages to a species is seen in the West Indian Bananaquit, *Coereba flaveola*. For here an uncommon melanic became widespread on the Islands of St. Vincent and Granada (Lowe, 1912) in a comparatively short time; we know not why.

More easy is it to speculate on the cryptic advantages of melanism in other species. For example, in the Desert Lark, *Ammomanes deserti*, blackish forms occur on black lava terrain in various parts of its range (Sage, 1962). Similarly, the Partridge, *Perdix perdix*, has in north-west Germany a melanic form adapted to its particular habitat, black peaty soils (Sage, 1962). A similar form was exhibited to the British Ornithological Club in 1915 by W. Rothschild who recorded that such individuals had been shot near Tring, Hertfordshire for 10 successive years. The cryptic advantages of melanism in such species as Snipe, *Gallinago gallinago*, Pheasant, *Phasianus colchicus* (where we know the melanic mutant is dominant), and Red Grouse, *Lagopus lagopus scoticus*, during incubation and when at rest under local conditions, can clearly be anticipated, though unfortunately such simple concepts have not been tested by experimentation in nature.

The rate of spread of the colour forms of the House Sparrow, *Passer domesticus*, has been successfully studied by Johnston and Selander (1964); this species has increased in many parts of the World, where it has built up large populations. It has colonized North America, New Zealand, South Africa, Hawaii, and elsewhere. More important is it that we know the dates of their first introductions and we can today gauge the changes which have taken place. In regard to colour (this is only one of many variants), they found that in some places a pale form with absence of dark markings had developed (as in the sandy wastes of south California and Texas). In other areas the House Sparrow had become particularly dark (as around Vancouver on the North Pacific coast where the species arrived in 1900, and again in Mexico City where the species was unknown prior to as recent a date as 1933). In such contrasting ecologies 100 per cent of each population have now become respectively light on the one hand, or melanistic on the other. Evidence from Mexico City suggests that the species is well on the way to subspeciation in as short a period as 32 years.

It has been pointed out to me by Ford (personal communication) that, only 40 years ago, it was a widely held view that the evolution of geographical races of birds could only take place in a period of 5000 years or more (Moreau, 1930). This is only one of the many examples of our underestimation of selective pressures before the mathematical theories of Haldane (1924) and Fisher (1930*a*) were shown to be correct in practice.

I must also mention the observations of Hardy (1937) and Rollin who provided evidence of a darkening of the plumage of House Sparrows, in the Liverpool area, and that washing their feathers did not change their appearance. They referred to this as Industrial Melanism in birds. There can be little comparison here with the situation in moths nor is there any question of morphism; furthermore, it is hard to see what the selective advantage of the darker individuals would be in a district comparatively free from predators, except for domestic cats.

We see then that melanic forms, frequently single gene morphs, are widespread throughout bird species. We can also accept the fact that many immature sea birds, for example gulls (*Larus* spp.), have an increase in dark pigmentation (G. Phillips, 1962). We find clines which can only be maintained by intensive selective pressures; yet we know not why. Only in a few instances can we speculate that the advantages of melanism are in providing crypsis. Though such examples of crypsis may 'appear obvious' in such contrasting geology as black lava rock and white sand, I know of no experimental proof of this.

Melanic crypsis in mammals

A large number of mammalian species have black forms and many of these reflect a single gene difference. Also, we have here, as we do not in birds, some evidence of their selective advantages. Cryptic Melanism can be divided into two clear-cut types, namely situations in which there are advantages to the predator and those that benefit the prey. Thus nocturnal predators, or those which hunt under dense forest canopy, are frequently black. The common Leopard, *Felis pardus*, is a good example of this. In the open savannah ecology of southern Africa the black mutant is rare. I believe I came across the type skin of this in Salisbury Museum in 1951. In south-east Asia, however, this form may constitute 100 per cent of a population. R. Chipperfield (personal communication) who has bred this species extensively in captivity states that the black form is dominant to the light, spotted form. There are many other examples amongst nocturnal carnivores where melanism occurs or where melanics may form 100 per cent of a population. The chemical processes in such dimorphisms have been discussed by Fox and Vevers (1960), Foster (1966), and Fitzpatrick *et al.* (1958).

The sensitivity of melanin deposition in certain animals to external conditions is seen in the Siamese Cat. *Felis cattus*. One of the most obvious characteristics of this species is that the extremities—feet, tip of the tail and ears—are dark, whilst the rest of the animal is fawn. Also it has been noted that when an area of the body had been shaved the new hair which grew was dark, similar in tone to that of the extremities. Melanization takes place in fact only in conditions of lowered temperature. As the difference is in order of only 1°–3°C, this underlines how sensitive is the pigment deposition in this species.

Hershkovitz (1968) has attempted to show that in South American marmosets and tamarins (primates of the family Callithricidae) the process of melanin saturation, reduction (bleaching), and eventual elimination of the pigment is inevitable and irreversible. He states that such 'metachromic pressures' are in conflict with the natural selection forces regardless of the environment. It would appear to me that he has taken no account of the various genetic mechanisms which ensure the maintenance of melanic forms of different intensity in each population. More likely is it that each colony of each species reflects the requirements of its own special environment.

Amongst prey, Cryptic Melanism is found in a large range of mammals. For example, the Deermouse (the best documented) and also rabbits, bush opossums, hamsters, and squirrels; these I shall refer to briefly.

Peromyscus maniculatus (the Deermouse) is widespread in North America. Individuals vary in colour from pale straw to near black, according to the particular locality. Thus in southern New Mexico, as early as 1930, Dice worked the Tularosa basin, where dark Malpais lava runs into the silver sand, thereby producing contrasting backgrounds. Here he recorded a local pale race of *Peromyscus* occurring on the white sands. Subsequently, Blair (1941, 1943) found that both dark and light forms of *P. maniculatus* were confined exclusively to their correct backgrounds. Furthermore, of the score of other endemic small mammals occurring there, five were characterized by having dark fur which matched the laval rock. Four of these species were also found in the white sand desert near by (but screened from it by a partial barrier of creosote bushes) and here, as in *Peromyscus*, each of the four were represented by pale forms only. Blair concludes, 'Selection on the Malpais lava appears to be sufficiently rigorous to overcome the nullifying effects of a considerable interchange of individuals between this and neighbouring populations.'

Working farther afield Dice (1936–1942) showed that, throughout the range of the animals, the colour of the fur followed the colour of the soil. Thus in the Columbia Basin, the Snake River is a complete barrier yet the separate populations of *Peromyscus* have the same colouration on either side; in arid sage-bush terrain of sand the pale forms always preponderated. By contrast the darkest Deermouse occurred in the most humid elevations of the Blue Mountains, where the surface soils were dark. He finds similar situations in Nebraska, where the pale forms were confined to small isolated areas of sand. The darkest forms occur along the coasts of Oregon and California.

Having shown convincingly an association between soil and fur colouration, Dice (1947) proceeded to undertake a series of controlled laboratory experiments in which the forms of *P. maniculatus* were released on to contrasting backgrounds, and he was able to show conclusively that elimination took place selectively. Using owls as predators under conditions of low light intensity, a significant excess of melanic Deermice were caught on a white silver-sand background, and conversely of the pale form on a dark one. These important experiments stress the selective intensities of background crypsis in prey species. Though there are many similar frequency clines in other small mammals, equally convincing proof of crypsis being an essential factor to survival is not forthcoming. I therefore refer only briefly to four other examples of melanic polymorphisms, though the underlying selective pressures are not fully understood.

The Rabbit, *Oryctolagus cuniculus*, is, due to introduction by Man, widespread throughout the world. The jet-black mutant is common in some localities. Usually this melanic form is controlled by a single pair of alleles and has dominance. In Britain it is particularly frequent on the northern isles and headlands. In the past, this has been attributed to the introduction of variable domestic stock by lighthouse-keepers for food. This may be true in part, in that variability has been available there on which natural selection could work. Small populations have not had to await advantageous mutations. Once feral, the disadvantageous forms must have been rapidly eliminated; yet the black forms constitute a proportion of the population in many places. I have met them commonly on heath-moors in north and west Scotland (Glenelg), and on the Shetland Mainland where they certainly appeared more cryptic than the wild type.

A well documented account of the frequencies of the black morph following the introduction of the species is given by Barber (1954). The Rabbit was introduced into Tasmania around 1803. Of 10 000 counted in 1952–3 in the eastern part of the island, melanics constituted 20 to 35 per cent of the total population, but only in regions of high rainfall and at heights over 1500 ft. The melanic frequency-cline falls rapidly towards the coast where it is only 1 to 2 per cent. Some of the clines are 70 miles in length, possibly 'steepest in the middle of the range'. In central Tasmania also, in savannah country, the black form is rare (less than 2 per cent). In open Eucalyptus forests around Great Lake, it rises to about 10 per cent. Frequencies of over 20 per cent are always associated with cleared rain-forest or Eucalyptus rain-forest with rainfall of more than 50 inches. There is no correlation between the frequency of the black and the density of the population. Barber's survey demonstrated that during 150 years intensive selection must have been taking place, probably by visual predators, which abound. It would be interesting to know whether different habits (such as diurnal or nocturnal activity) have arisen in the contrasting light and dark populations.

In a somewhat similar way Guiler (1953) has recorded that in the black morph of the Tasmanian Bush Opossum, *Trichosurus vulpecula*, there is a pronounced ecological correlation, the black phase again being associated with rain-forest terrain (Pearson, 1937).

Hamsters, *Cricatus cricatus*, so popular as pets, are difficult to obtain in nature because their activity is limited to crepuscular and nocturnal surfacing. In southern U.S.S.R. melanic hamsters may form up to 75 per cent of a population in the humid forest-steppe regions. 'In dry forest-steppe or mountain-forest, the frequency of black falls to zero' (Gershenson, 1945). Gershenson showed that this form was controlled by a pair of alleles and had dominance. He has suggested that it may be homologous with the black E^D of Punnett in the Rabbit and the Black Rat, *Rattus rattus*. As for other prey species, a distribution map of southern U.S.S.R. shows rapid cline-differences in the morphs, probably due to intensive selective predation.

The last two examples of melanic polymorphisms in prey species are squirrels, *Sciurus vulgaris* L., the European Brown Squirrel, and *S. carolinensis* Gmelin which is native to North America. On the mainland of Europe *S. vulgaris* has a melanic mutant which is widespread in certain districts of France, Germany, and Switzerland. In the Fümen region of Denmark, Spärck claims there is a homozygote population of some 72 000 dark squirrels (Shorten, 1954). In Britain, however, *S. vulgaris* is a subspecies and no dark forms have been recorded. I think that somewhat different pressures have been responsible for the morph distribution of *S. carolinensis* in North America. Schorger (1949) suggests that melanism developed in this species on the borders of glaciations during the Pleistocene, when the climate was moist and cold. According to Kennicott (1857), an earlier disappearance of the black morph was a result of hunting by Man because of the popularity of this form for fur coats. He states, in fact, that at the time of the earliest settlements melanics were commoner than the grey form, and that in 1856 there were places in Illinois where this form constituted 100 per cent of the population. These were selectively eliminated by Man and today, the fashion for black squirrel coats having passed, *S. carolensis* is free to take up its morph-frequency undisturbed by selective hunting. Yet it appears to me that its present distribution in the larger cities of Canada is surprising, for here I saw a high proportion of melanics in some and none in others. In Britain, a few melanics occur in Bedfordshire and Hertfordshire, possibly following a deliberate introduction of a dozen at Woburn. The black form is again dominant to grey, 'though intermediates have been reported' (Shorten, 1954), but these are probably immature individuals.

I have discussed melanic polymorphisms in five species of mammals which, certainly in the past, if not today, have been prey species. In each there is, with rare exceptions, clear-cut segregation. In each there are frequency-clines both steep and gradual, and the melanic form in each is inherited as a dominant. That such Cryptic Melanism should occur in both birds and mammals can but point to a common denominator, and the antiquity of such polymorphisms.

Cryptic Melanism in other orders

Apart from Cryptic Melanism in birds and mammals, there is an abundance of evidence of similar situations in lizards, snakes, and fish. Thus in *Lacerta vivipara*, the Common Lizard, melanism is not uncommon, though it is alleged to be 'more frequent in the female than in the male' (Smith, M., 1964). Nor is the function of melanism here necessarily limited to crypsis, as he goes on to state, 'when basking, the body is flattened and tilted and the limbs extended, to catch as much heat as possible'. The pale-coloured Sand Lizard, *L. agilis agilis*, has melanic forms, where sand dune abuts on moorland, though only on the mainland of Europe and not in Britain.

In snakes a good example of melanism is the Adder, *Vipera berus berus*, though there is some confusion as to the origin, inheritance, age, or sex. Probably each contributes part but at least one melanic form is inherited, because black females have been known to give birth to black offspring. Moreover there are certain localities in Britain, such as Puck Pits in the New Forest and Port Eynon in the Gower peninsula of South Wales, where 'Black Adders' can regularly be found. This melanic form could equally well have thermal advantages, as the Adder hibernates gregariously, and their degree of activity in the spring depends on the temperature provided by the environment. A large number of other species have black forms, some as polymorphisms, others as subspecies, and, in the mambas (*Dendroaspis*) for example, as species. In both *D. angusticeps* (the Green Mamba) and *D. polylepis* (the Black Mamba) the young are green, which may have led to the confusion between these two distinct species. *D. angusticeps* depends on concealment in bushes and is relatively arboreal. *D. polylepis* is dark brown to black in colour and, being one of the most aggressive snakes in the world, might be considered aposematic.

I do not intend to discuss Cryptic Melanism in fish in any detail because so little is known, indeed I know of only a few melanic polymorphisms, though they do exist (McPhail, 1969). Certain species, such as the Mosquito Fish, obtain perfect crypsis by adapting over a period of time to a light or dark background. Sumner (1934) in a series of experiments using predators obtained the following data. With the Galapagos Penguin the selective index against the more conspicuous fish was 0·351 ($\chi^2 = 6\cdot661$), with the Night Heron the selective index in favour of protectively coloured fish was 0·229 ($\chi^2 = 18\cdot978$), and with the Sunfish it was 0·355 against the more conspicuous form ($\chi^2 = 30\cdot894$). Dice (1947) states the following in reference to this work: 'In these experiments of Sumner the selection index against the animals that were conspicuous when viewed against their background ranged from 0·23 to 0·36. This is a very considerable amount of selection against the conspicuous fish as compared with the concealingly coloured ones. In other experiments the deviation of the selection index from zero is statistically highly significant.'

From the short résumé of Cryptic Melanism it is apparent that melanic polymorphisms and subspecies exist today throughout the animal kingdom due to the selective advantages of camouflage to both predator and prey: further, that, certainly in prey, dark forms can spread and can replace an earlier form with great rapidity; but finally that, unfortunately, in only a few instances have we experimental proof of the advantages and disadvantages, still less of behavioural and other differences which must accompany such genetic changes.

(ii) Aposematic Melanism

Both black and white can at times be signals of distastefulness in many species in different orders. I have pointed out elsewhere that such melanism may have existed before the evolution of yellow–red vision by predators (page 11) and may therefore be of great antiquity. In insects, birds, and mammals, special behavioural patterns have evolved, in particular gregarious living, though this is not an essential to many species. A further point is that species with black warning colouration frequently occur in the same genera as others with bright red or yellow patterning. This is particularly well seen in the Coleoptera. In the genus *Necrophorus*, different species of burying beetles are black, as for example *N. humator*

Goeze, or black and red, as is *N. interruptus* Stephens, and each on attack gives off a nauseating fluid. At the same time they may stridulate and the sound spectrum emitted has been shown to have similar components to those in the buzzing of certain bumble bees. *Bombus* spp. (Rothschild 1961, 1966). Solitary black beetles are notoriously conspicuous and most of them have glands which secrete or even project offensive fluid. Of the many species which do this, *Blaps mucronata*, the Cellar Beetle (*Tenebrionidae*), is a good example in Britain. The black-coloured Bombardion Beetle, *Brachinus ballistarius*, is more specialized and forces a spray of repellent secretion from the end of the abdomen, not at random but directionally. The beetles can discharge as many as 20 shots consecutively (Wigglesworth, 1964). The highly adaptive chemistry producing these explosions is brought about by mixing hydroquinones with hydrogen peroxide, in the presence of appropriate enzymes, causing the sudden liberation of oxygen from the vesicle of the gland. That a beetle can have evolved so complicated a chemical defence must suggest long continued evolution in that direction and the efficiency and antiquity of aposematic melanism as a deterrent signal. Similarly the North American Darkling Beetle, *Eleodes* sp., is jet black, and when attacked takes on a genupectoral position; whilst thus conspicuously up-ended, it sprays a defensive secretion from its abdominal glands. So successful is this warning defence that *Eleodes* is mimicked by a similar looking beetle, *Megasida* sp., also jet black, which takes on the same offensive stance but, in fact, has no defensive secretion (Wickler, 1968). That a Batesian mimic should have evolved is a clear-cut proof of the efficacy of the black *Eleodes* sp.

There can be no surprise, therefore, that some brightly coloured aposematic species maintain melanic polymorphisms with frequency variations from one locality to another. In particular the Ladybird, *Adalia bipunctata* L., has been successfully studied by Creed (1966, 1971). This species is normally red, with two black spots on the elytra. When attacked it emits globules of a distinctively smelling fluid which may be a quinoline derivative and which is reminiscent of the scent of other aposematic creatures, for example the Tiger Moth, *Arctia caja* (see page 214) (Rothschild, 1961). *Adalia bipunctata* has three melanic mutants, all dominant to wild type, and these may contribute up to 95 per cent of a population at a particular time. Though usually found singly in the breeding months of the summer, *A. bipunctata* hibernates gregariously in crevices throughout the winter. Timoféef-Ressovsky (1940) has stated that around Berlin the frequency of the melanic forms was lower in the spring than in the autumn. This likely differential mortality rate has been confirmed by some but not by others, so that it might well simply depend on the type of winter—warm and wet or cold and dry.

Creed has taken samples from over 200 areas in Britain and about 40 in France. He has demonstrated the following points. There is a correlation between the frequency of the melanic form and some types of industrial areas. It is found at its highest in districts with a heavy smoke (for example Lancashire) as opposed to a high SO_2 content. Lusis (1961) states that more melanic × melanic pairings take place than any other mating. The interpretation of these data is difficult, but any advantage due to the crypsis of melanic forms is indeed unlikely; more reasonable is it to suggest that they reflect a balance of aposematic and physiological advantages and disadvantages.

Aposematic Melanism is common in other insects: in Hymenoptera in the tropics, where they are commonly mimicked by Lepidoptera; or in the Diptera; also various Trichoptera are black and highly gregarious. Every fisherman has seen Swallows and martins taking 'sedges' and other flies on the surface of the water, but always avoiding the clouds of these black flies. I have seen the same on Scottish lochs, where Pied Wagtails, *Motacilla alba yarrellii* Goud, have left the black Trichoptera untouched.

In birds warning melanism is widespread and again is frequently associated with flocking. Thus many Corvidae are black, black and white, or brightly coloured as is the Jay, *Garrulus glandarius rufitergum* Hart. All of these when mature are inedible and consequently by the recognizability of their plumage avoid predation. By contrast, the young are frequently good eating during their early life; hence the delicacy of 'rook pie', made from fledgings of *Corvus* which are shot on 1 May. Here the juveniles, being black, depend on mimicking their parents for comparative safety, though I have seen this defence exposed. I witnessed an old Carrion Crow, *Corvus corone corone* L., drop into a flock of Jackdaws (*Corvus monedula spermologus* Vieill.), seize and kill a young one, and proceed to eat it in spite of attacks from other members of the flock. Other birds which are largely black such as swifts (*Apus* spp.) and swallows (*Hirundinidae* spp.) are distasteful and free from attack.

Cott (1946) pioneered the theory of aposematic colouration in birds and provided clear proof of his theories during a five-year study in Africa. His interest arose through his observation that predators, in particular the Hornet, *Vespa orientalis*, and the cat, showed preference for the flesh of cryptic and not black and brightly coloured birds which he was skinning. He lists over 350 species and shows convincingly that amongst desert species their colouring is bimodal: cryptic or conspicuous black. Buxton (1923) had pondered on this previously and in fact could not convince himself that melanic species, living side by side with their highly cryptic confrères for example the chats, (*Oenanthe* spp.), had adaptive significance. Only when Cott tested the flesh of these birds by presenting it to hornets, cats, and Man did the answer become clear. Black Chats and other species were inedible to man and beast alike. Cott concludes by saying: 'In short they are aposematic animals, and they flaunt their feathers as though a living advertisement of the Levitical law of old: "And these are they which ye shall have in abomination among the fowls: they shall not be eaten, they are in abomination" ' (Leviticus). Yet he omits the specific examples which follow: 'The eagle and the ossifrage and the osprey. And the vulture, and the kite, after his kind: every raven after his kind: and the owl, and the night-hawk and the cuckow . . . and the little owl, and the cormorant . . . and the heron after her kind . . . and the bat' (Plate 3.1). This interesting list of distasteful species, including at least two demonstrating Aposematic Melanism, was compiled *c*. 500 B.C., at a time when protein ('flesh' and 'carcases') was in great demand.

I shall mention but a single example of distasteful melanism in mammals. Many shrews (*Sorex* spp.) live in the same habitats as mice (*Mus* and *Peromyscus* spp.) But the shrew is blackish and gives off a stench, and attracts attention by constantly squeaking. They are occasionally killed by cats and other predators, but with the exception of owls (*Strigidae* spp.) they are not eaten. For this reason they are commonly found dead because few animals can face the rank musty odour. By contrast, voles and field mice are cryptic, do not attract attention, and their corpses are not found lying about because they are edible. Clearly shrews must suffer a death-rate from uninitiated predators, but it is unlikely that an experienced fox would mistake a dark-coloured shrew for a vole. In Borneo, five species of Tree Shrews (*Tupaiidae*) are highly unpalatable. They are even mimicked by at least five species of edible squirrels and can only be recognized by skull differences (Shelford, 1916).

Warning melanism is found commonly in snakes and probably in fish also, though I do not intend to discuss this. I have, however, in this brief résumé of Aposematic Melanism in animals, given sufficient examples to demonstrate that a conspicuous black colouration can be an efficient warning signal of distastefulness, both at a morph and at a species level.

(iii) Thermal Melanism

The heat absorption of all small animals exposed to light, both vertebrates and invertebrates, varies according to their colour (see Chapter 2). Because black absorbs radiation more rapidly than white, this must accrue advantages to some and disadvantages to others.

Of the latter there is some evidence, though not significant, that samples of *Biston betularia* collected at the height of a heat-wave are deficient in the blackest form, f. *carbonaria*. I have also recorded that there is a critical temperature at which some highly cryptic *Catocala* spp. with dark forewings, which normally rest on the boughs of high canopy oak forest, take flight simultaneously and seek shelter at lower levels at the base of the trunks—indeed a precarious exercise. In spite of having many contacts around the Equator, I have obtained no evidence whatsoever that Industrial Melanism occurs there, and J. B. S. Haldane, who specifically inquired on my behalf, could find none in India.

The advantages of black colouration leading to increased activity are in fact confined to certain types of creatures (Rose, 1967); for example small diurnal poikilotherms in places of high rainfall and fitful sunshine, or in cold climates at high latitude or altitude; or hibernating animals when it is an advantage to become active early in the year (but rarely is this so).

It is likely that many Diptera, Coleoptera, and other insects make use of Thermal Melanism, but of this there is little evidence and practically no experimental proof, and the same statement can be made about Amphibia and Reptilia and possible candidates from other orders. Thus the eggs of the Frog, *Rana temporaria*, which are laid in March and April, are in part black, and this affects the speed at which the tadpoles emerge from the spawn.

In spite of little evidence we can confidently state that Thermal Melanism must have contributed to survival for about 500 million years since the Pre-Cambrian period.

(iv) Barrier Melanism

When melanin is deposited in the superficial layers of the skin it provides a barrier against the penetration of ultra-violet light which can be harmful to the underlying tissues, though such a screen is normally provided by scale, fur, or feather. Amongst 'naked' animals are species as far apart as slugs (Limenidae) and Man, in which melanism permits extension into open countryside on the one hand, and the ability to live in the tropics on the other.

Eight species of British slugs (in the genera *Arion*, *Milax*, *Agriomilax*, and *Limax*) maintain melanic polymorphisms (Williamson, 1959). *Arion ater* L. is one of the commonest and most variable, with the jet black form most frequent on open downland and the paler forms increasing in frequency inside shady woodland (as in Wytham Reserve, Oxford). On the continent the red form is frequent in the south. No doubt many factors contribute to the maintenance of these polymorphisms, for example 'warning', because slugs are not readily accepted as an article of diet by birds. My Mallard, *Anas platyrhyncha platyrhyncha* L., which flew around my home for many years, refused *A. ater* after the first initial encounter, but other animals, for example the Hedgehog, *Erinaceus europaeus* L., will eat them freely. Because of the different morph frequencies in open and wooded countryside, it is likely that the main contribution of the melanism is a pigment barrier against ultra-violet, though I know of no experimental work on this.

More is known of the advantages and disadvantages of pigment deposition in the skin of Man, but as this book is not the place to discuss melanism in Man in detail (for this see Th. Dobzhansky, 1962, and its extensive bibliography), I will limit my observations to a few generalizations and comments.

Allowing for population migrations, there is a correlation between the degree of melanization and the distance from the Equator (with the exception of a few races which live in Arctic snow). Some of the darkest races are found in West Africa, Melanesia (New Guinea), and Australia, around latitude 0°, and amongst them such diseases as epithelioma (carcinoma of the skin) and rodent ulcer are little known. By contrast, the incidence of cancer of the skin amongst Europeans who have emigrated to countries with maximum sunshine, for example Australia, is increased compared with that in northern Europe. Corroboration of this is seen in that among the Nordics, these diseases can be said to be occupational, for they are found more frequently amongst farmers, gardeners, and others who have to expose their skin surfaces in the course of their work. It might therefore be deemed remarkable that this mechanism of defence against the sun's rays should become a centre of conflict under the heading of 'colour prejudice' or 'colour discrimination'. For 'colour' is only one (but the most obvious) of a number of other characters by which the races of Man differ.

The various degrees of pigmentation are controlled by a number of genes, and each of these must also be responsible for other characters such as resistance to diseases, physiognomy, viability, the emotions, sexual prowess, or body odour, to name but a few. This has led to the formation of ethnic groups with different cultures. The particular degree and type of melanization in Man has therefore become a signal for association with such characters, maybe pleasing or unpleasing.

The white European, in his arrogance, may refuse to accept the suggestion that the average Chinese or Lap has a higher I.Q. than he has; similarly he finds it hard to admit a greater spiritual fervour in the Mohammedan, or the sexual capacity of the Negro. When we add to this the rule that in nature, in situations where one genetic pool differs from another in a number of genes, leading to different complexes, barriers against cross-mating will be established, it is implicit that compulsory legislation forbidding discrimination against black or white cannot work. Only by voluntary mixing, and at the level of sex and mating can *Homo sapiens* become homogeneous as postulated by Bernard Shaw (1903) in *Man and Superman*. Such miscegenation has happened in Brazil, for example, which prior to the sixteenth century, was inhabited by the copper-coloured Central American Indians only, but was then invaded by the Portuguese, who brought with them over three million West African Negroes to work the sugar crops.

After several years sojourn in the Union of South Africa (whose country and people I love so much, but whose rigid, antagonistic and completely unscientific segregation into 'blankes' or 'nie-blankes' I deplore), it came as a great surprise and happy contrast to visit Brazil in 1958, for there, colour barriers are largely broken down, and free relationships exist between people of all colours—as between blondes and brunettes in Europe. Melanisms and their associated gene-complexes in *Homo sapiens* have certainly not led to speciation, but have stopped far short of this, as I think, somewhat surprisingly (with the exception of aposematism), they have in most other species, be they birds, mammals, or insects. I find it strange that such a major change as melanism, be it polygenic or polymorphic, has not attracted different gene-complexes and that speciation has not taken place in the past more frequently (as it has for instance in white polar animals).

This chapter is an incursion into a book primarily intended to discuss melanism in the Lepidoptera. Some of my friends have attempted to persuade me to omit it on the grounds that it was irrelevant. I disagree with them because I am convinced that only by discussing melanism in its widest context, and in other phyla, can a comprehensive understanding of its various uses, its ubiquity, and above all its previous history, be acquired.

Part II. Melanism in the Lepidoptera

4. A classification of melanism

IN IMAGINES
- Industrial Melanism — 40
- Geographical, Ancient, or Relict Melanism — 41
- Recessive Melanism — 42
- Miscellaneous melanisms (see Part VI) — 263

IN LARVAE
- Genetic plus phytoscopic — 44
- Genetic only — 48
- Aposematic — 49
- Crowding effect — 49

IN IMAGINES

IF a biologist who has little knowledge of the Lepidoptera examines any extensive collection of Palaearctic Heterocera, he is immediately impressed by the large number of species which exhibit melanism. This statement must refer to the Heterocera only; by contrast, the Rhopalocera but rarely have black forms.

In the moths, not only are such forms found in species which depend for survival on passing the day motionless on specialized backgrounds, such as dead wood, tree trunks, or dead leaves, but also in other cryptic species which camouflage themselves by non-specialized disruptive patterns. More surprising is it that melanism occurs in the majority of aposematic genera, such as the Arctiidae, Zygaenidae, and *Abraxas*.

Nevertheless, a classification of melanism usually falls clearly under one of three headings, though with certain rare exceptions (Chapters 15–18).

In order to substantiate this, it is essential to have knowledge on two issues: first the habits of each species, and in particular how and where it passes the daylight hours; secondly the method of inheritance of the melanic forms. In regard to the first, it is a fact that in a very large number of instances we are completely ignorant of the intimate behaviour of species in nature, nor do we know the exact resting positions they take up by day. These again may vary according to whether the weather is hot or cold, dry or wet. Modern and highly efficient methods of collecting specimens, such as the ultra-violet (mercury vapour) trap, is producing a race of younger naturalists who are largely ignorant of the sensitive requirements demanded by most species. On the other hand, the genetics of many of the melanic forms are known.

There are a few species in which darkening is brought about by the external conditions imposed. Such Environmental Melanism is discussed fully in Chapter 18.

Apart from these exceptions, melanism in the Lepidoptera can be classified broadly into *Industrial Melanism*, *Geographic, Relict or Ancient Melanism*, and *Recessive Melanism*.

(i) Industrial Melanism

Definition. This must include dark forms which are distributed in and around industrial areas and far beyond in the direction of the prevailing smoke-drift. In some widely distributed species this frequently leads to clines. It is found in cryptic night-flying Heterocera only. Inheritance of melanic forms is usually dominant, less frequently multifactorial or polygenic; more rarely the character has no dominance. In Britain it is recessive in a few instances only; these are discussed in Chapters 13 and 14.

From the middle of the nineteenth century onwards black forms of many species of moths in widely different genera have been recorded in and around industrial areas throughout the Northern Hemisphere, but strangely not in the Southern; nor have they been reported from industrial regions near the Equator.

The history of their present distribution, their early spread, and the recent environmental changes are given in Chapters 5, 6, 7, 8 and 9.

Industrial Melanism in the Lepidoptera is confined to cryptic species and is in the majority of instances controlled by a single gene, which is dominant in its effect on pigment production. This dominance is usually complete and the heterozygotes cannot be distinguished visually from the melanic homozygotes. Occasionally, as in one form of the Noctuid moth, *Polia nebulosa* Hufn., the heterozygotes are intermediate and distinct from both homozygotes. In other species, as in certain North American Selidosemidae, there is a gradation of forms from light to dark; dominance is incomplete and the heterozygote highly variable. It is likely that similar situations existed during the early history of many industrial melanics in Europe. I hope to provide evidence that in a comparatively short period, full dominance can and has been achieved by natural selection in conditions where an industrial environment has continued to move in the same direction, but only in those cryptic species with previous experience of melanism.

A second mechanism for ensuring the presence of the most successful phenotype at a particular moment in time is a very different one, the substitution of one melanic form by another, frequently, I think, allelic. This is discussed in Chapter 5 and in Chapter 19. It is a short cut method of adaptation and it is due to the effects of the past on the genetic code in the incorporation of successful alleles.

In view of the likelihood of a connection between present-day Industrial Melanism and the past, it seemed essential to look for melanism in Lepidoptera in non-industrial regions of the world, in conditions which may be similar today to those experienced in previous periods, and to see whether such melanics had any features in common with those occurring so freely today in and around centres of urbanization. If substantiated, it would suggest a direct association between Industrial Melanism and the past. It could also explain the almost universal dominance of the industrial melanics found today as being due to earlier successes and to the moulding of the gene-complex by natural selection during previous periods in the history of each species (see Synthesis, Chapter 19).

(ii) Geographic or Ancient Melanism

Definition. Melanic polymorphisms occurring today in several quite different types of habitat, which are similar to those which have existed at earlier times. Inheritance of melanic forms is usually dominant. In certain instances this Ancient Melanism has present-day links with Industrial Melanism where the two ecologies overlap.

The fact that melanic forms of cryptic species are found in areas far removed from the effects of air pollution has served to confuse the issue of Industrial Melanism. Nevertheless, there is here a clear-cut group occurring in widely separated species of Lepidoptera which either maintain melanic polymorphisms or have a gradation of forms from dark to light; occasionally 100 per cent of a species may be black in a particular locality. I have referred to this group as Geographic or Ancient Melanism, and each adjective emphasizes a particular aspect of its origin. 'Geographic' implies that the members of this group are confined to certain regions in the world; this, in the majority of instances, is true. 'Ancient' underlines their antiquity.

We can, in fact, define these special regions, past and present, and classify them as (1) Rural (and Background Choice Melanism), (2) Northern (latitude 60°) Melanism, (3) Western Coastline when the prevailing winds are from the west, (4) Ancient coniferous forests, in particular Caledonian pine in Britain and Pine Barrens in North America, (5) Melanism associated with fire-resistant trees, (6) Mountain forests with high rainfall (i.e. Pluvial), and (7) Thermal Melanism.

In each environment the black forms occur for entirely different reasons; reasons necessary for the survival of the species today, which are the same as those which have constantly occurred in the past. Most of these ecologies have remained but little changed over periods of hundreds or maybe thousands of years. Natural selection has therefore had time to perfect a gene-complex which habitually has to be geared to full expression at mutation. The distribution of Geographic Melanism is discussed in Chapter 10 and the experimental approaches that we have made in Chapters 11 and 12.

(iii) Recessive Melanism

Definition. Melanics occurring at a low frequency, minimal around mutation rate. They are found in many genera of aposematic Lepidoptera, but less frequently in cryptic species when they have spread in Britain only in a few instances, under special circumstances. The degree of pigmentation is usually extreme. The homozygous recessive melanics are, in the majority for instances, sub-viable or semi-lethal.

It is true to say that the recessive inheritance of melanism suggests that in the past it has been disadvantageous and that the gene-complex has been adapted to obstruct the heterozygous expression. Retrospectively, it could imply that a major effect of such a mutation had previously not contributed to the survival of the species. As Recessive Melanism is largely found in aposematic species, this is not surprising, when we consider the total and complicated requirements for survival that these insects have evolved. It is futile to attract attention by warning colouration (with the exception of a minority—the Batesian mimics) unless a repellent chemical system has developed at the same time. Though black may be the simplest warning colouration (as in some birds), it is likely to be an inferior one for those predators which have colour vision. Also, to be effective, special behavioural patterns must have been developed, such as gregarious flight. The forfeiting of highly specialized colour mechanisms is, I think, but rarely likely to occur in aposematic Lepidoptera; nor in those species which exhibit warning colouration on the hindwings—red, yellow, or blue—is a substitution by black generally advantageous, except in specialized ecologies. I know of only two instances in which this may have happened in aposematic species, one being the species *Arctia caja* f. *fumosa*, which, in regard to the hindwing, has a variable heterozygote red to black, and the other being the species *Spilosoma lubricipeda* f. *zatima* where melanism affects all four wings (here the black mutant is dominant). Both of these are discussed in Chapter 16.

Recessive Melanism is occasionally found in cryptic species, in particular amongst those which match dead, brown leaves such as bracken. In one instance, *Lasiocampa quercus* ssp. *callunae*, the recessive melanic form, f. *olivacea*, is spreading over moorlands whose ecology has recently changed in central industrial England. This results either from the blackening of heather by air pollution, or from an increase in predators such as gulls, or from a combination of both. In this same species, a similar recessive melanic phenotype contributes up to 70 per cent of the population in northern Scotland, which is far from the effects of industrialization. This exciting situation is discussed in Chapter 14.

Classifications of this kind are frequently convenient for workers but only occasionally provide additional data on their own merits. This clearly defined classification of melanism does, however, indicate certain limited generalizations.

For example, we can state that, unless the pigments present in the melanics of the Recessive Melanism group are chemically different from those of the other two groups, it cannot be the effects of saturation by melanin which is responsible for the differential viability evident when one compares, on the one hand, the semi-lethality of the rare, black forms of aposematic species and, on the other, the hardiness of the melanics in the other two groups. The degree of blackness bears no relationship to the degree of physiological fitness. More likely is it that a gene controlling recessive melanism and with little exposure to natural selection in the past (or other genes closely linked to it) affects other viable characters in a disadvantageous way. Recessive Melanism is discussed in Chapters 13 and 14.

Summary of classification

Recessive Melanism stands as a distinct class on its own, in which such melanism has in the past usually offered little or no protection. On the other hand, there are many features in common between the other two groups, Industrial and Geographic Melanism. In both, the species involved are cryptically coloured, the melanic forms are inherited, usually as dominants, but rarely as recessives, and the viability of the phenotypes is not grossly affected one way or another. This could suggest some kind of relationship between the two and my efforts to substantiate this theory—and theory only it must be at present—and to provide data bearing upon it, are summarized in Part VII, the Synthesis.

IN LARVAE

The main purpose of this book is to discuss melanism in the Lepidoptera in the imaginal stage. Nevertheless, dark colouration may be equally important to larvae.

A classification of melanism in larvae must be treated in an entirely different way to that in the imagines. For here we were able to arrive at fairly satisfactory divisions in two of the three classes, by reference to their particular (and usually contrasting) ecologies (that is, Industrial and Geographic Melanism).

Larvae, being static, have had to adapt to the local opportunities offered by plant, bush, or tree. Their survival, in all but a few internally-feeding species, depends on successful colouration; larvae are a main source of food for a large number of predators, in particular birds which search primarily by vision.

We must remember also, that many more of each species must succumb as larvae than as imagines.

The degree of larval mortality

In order to indicate how great this mortality rate is, I undertook extensive experiments in marking larvae with the radioactive isotope ^{35}S. Because of ecdysis it is impossible to label them other than by isotopes. These experiments showed that between 85 and 95 per cent of last instar larvae or pupae were eliminated by one means or another (Kettlewell, 1952a, Kettlewell and Cook, 1960). Selection pressures must indeed be high in the larval stage though visual predation, in my opinion, usually contributes less to this than such other causes as, for example, polyhedrosis, 'hardiness' to temperature fluctuations, humidity variations, and mycological diseases.

(i) **Genetic plus phytoscopic**

In the majority of known instances, the dark forms of cryptic larvae arise, not by inheritance only, but by a finely adjusted switch-mechanism which is triggered off at a critical point in the larval life history. The colour form is decided by the background on which the individual rests at a particular time in its life history. When once this is fixed, it can but rarely be changed if a different background is enforced on it subsequently. The advantages of such a system over one of fixed, inherited colour-patterns becomes immediately apparent if one considers the requirements of polyphagous cryptic insects.

It has been shown by de Ruiter (1953) and others (Tinbergen, L. 1960) that cryptic Geometrid larvae depend for their survival on copying perfectly the twigs on which they rest (Plate 4.1). Now such species as *Gonodontis bidentata* and *Biston (Amphidasis) betularia* feed regularly on a wide range of trees, and these have stems and twigs of entirely different colours from one another. Thus, for example, those of *Salix* are green and those of *Betulus* (Birch) purplish-black. The larvae of these two Lepidoptera feed on both these trees and a host of others, and, in order to survive, larvae must attain a perfect crypsis on each. Still more remarkable is it that these two species, and many others in widely different genera, are capable of producing a perfect 'lichen-form' when given lichened twigs to rest on. The Noctuid moth *Allophyes oxyacanthae* L., a Hawthorn and *Prunus* feeder, and the Bombicid, *Gastropacha quercifolia* L., which feeds on *Rhamnus* (Buckthorn) or Apple, are examples. In each, the larva is variable; it can be light, dark, or it can produce the complicated lichen pattern; its camouflage is highly efficient. Wigglesworth (1964) points out that it is improbable that such larvae possess 'copy mechanisms' enabling them to match such complicated backgrounds on which they have to rest. More likely is it that these various larval forms represent a definite inherited polymorphism, which has developed in the past, but in which the external environment provides a switch mechanism by way of eye recognition of the particular background. This stimulates only the appropriate genes and not the inappropriate ones. Presumably this could be controlled through the nervous system and the discharge of particular hormones. It is, in my opinion, of interest that a lichen form should be held in the stock of so many species in this way, for it suggests that in the past lichened twigs must from time to time have been very much commoner than they are today, probably during pluvial periods.

Let us consider the advantages for larvae of such a mechanism as distinct from the more frequent one, in nature, of direct genetical control by inheritance. This would lead inevitably to a situation in which large numbers of larvae would find themselves conspicuous on the wrong tree-species. Alternatively, many Lepidoptera would have had to become monophagous and would therefore have been unable to exploit the niches they have today. Such melanic phases may perhaps reflect the results of natural selection during an earlier period when conifers and birch were the predominant trees and when their black twigs had to be copied. This would be prior to approximately 10 000 B.P. Certainly most of the larvae of *G. bidentata* I have collected from pine trees have been of the dark form, as have those of *B. betularia* from birch.

In the large-scale breeding of this last species, it is customary for us to sleeve the larvae out of doors on to various trees. We usually do this at about the end of the second instar, at a time just before they tend to take up resting positions on the main stems. In a series of experiments, I put half of each brood on to *Salix* and half on to *Betulus*. They were not crowded, with about 100 individuals to a six-foot sleeve. In all broods the majority of larva at full growth were green on *Salix* and black-brown on birch. One to five per cent were out of phase. In nature, I have little doubt that these individuals would be rapidly eliminated; they probably represent the absence of an allele responsible for producing the correct colour phase.

More recently we have commenced work (unpublished) on the lichen form of *G. bidentata*. This is a truly remarkable larva (Plate 4.2). This individual was found because it commenced to move on the lichened twig of a *Prunus*. The locality was Strathdon, North Scotland, where all the trees, including *Pinus*, are covered with lichens. The female which hatched in 1969 was paired for me by P. Harper with a male from Central Liverpool, where one hundred per cent of the larvae are black. Here they are common and feed on the privet hedges along the street sides, resting on the blackened twigs.

On hatching, the larvae were separated into two groups. Half were placed in a container which was kept in daylight and which was furnished with black twigs. The other half contained lichened twigs (actually the same ones on which the original female larvae had been found). They were fed on leaves of *Salix*. On assuming full growth they were scored; all twenty of those on black twigs were black and highly cryptic (Plate 4.3), while, of the 28 survivors which had been reared on lichened twigs from the day on which they had hatched, five were of the distinctive lichen form (similar to the original parent). Of the remaining twenty-three, six were black (as in the control cylinder). The remainder were of a different form in which the last five segments had a green-grey patterning laterally (Plate 4.4). Certain individuals, however, graded imperceptibly into the black form. The lichened form was, however, quite distinct. The apparent segregation in a 1:2:1 ratio must be accidental and cannot evoke a simple genetic explanation.

Poulton (1903) carried out many years of research on larval colouration but I am convinced that he was throughout thinking in terms of a direct environmental effect only. He ignored the possibility of genes expressing their characters differently under varying stimuli at a critical period.

That genetic inheritance plus phytoscopy must indeed play a part is proven, at least to my satisfaction, by the record of a single larva by Oertel (1910). *Deilephila elpenor* has a green and a dark-brown form of its larva. Federley (1916) at first thought green was recessive to brown, but later work did not confirm this. Oertel's individual was symmetrically divided down the middle into half brown and half green. We must assume that both sides of the larva had been subjected to the same environmental experience. We can admit therefore that this halved mosaic resulted from differing genes in the respective halves, each no doubt reacting to an independent switch-mechanism at a critical period. (See footnote on p. 50.)

Further evidence of a partial genetic control of larval colouration is given by the Geometrid moth *Rhodometra sacraria* L., the Vestal. This species has many reservoirs of permanent breeding but all are south of latitude 40°. It is primarily a desert insect, living in Africa in wadis and around oases. The larva feeds on *Rumex* and *Polygonum* and is usually green. During periods of drought, however, both plants and larvae become brown. This little moth fairly regularly arrives in northern Europe following southern winds, it being borne, no doubt, on convection currents. I have, on several occasions, bred large numbers from such migrants to Britain. When the larvae are kept in the dark on green food, the proportion of differently coloured larvae per brood closely follows a genetic ratio, so that either 50 per cent or 75 per cent, for instance, are of the dark form, with but a few intermediates. This is in contrast to the larvae of *B. betularia* which, when kept in the dark, show no clear segregation.

The importance of dark colouration at a particular moment in the life history is clearly seen in a large number of lepidopterous larvae which have to wander considerable distances when fully grown in order to find the special requirements needed for the pupal period. The large, maggot-like larva of *Cossus cossus* L., which feeds in the depths of decaying tree-trunks in comparative safety for two, or possibly three, years, has to experience a hazardous 24 hours whilst in search of a pupation site. This is frequently situated in the nest-mounds of the ant *Lasius flavus*, which may be at a distance of 50 yards or more from the tree on which the larva fed (Fraser, 1947; Kettlewell, 1965f). During the short period of its peregrinations, its pale areas take on a dark brown colouration, which makes it much less conspicuous when travelling overground. This colouration results from the deposition of large amounts of ommochrome in the epidermis. Many other larvae in widely diverse genera, which have to wander distances prior to selecting special pupation sites, behave in the same way; for example, the green larva of *Cerura vinula*, the Puss Moth, whose colour is a mixture of three reduced ommatins which are under the control of the prothoracic gland. When these larvae are ligatured in the middle, the darkening only takes place in the anterior portion, the hind-part remaining green (Fox and Vevers, 1960).

The darkening of many other green larvae during the pre-pupation period is probably controlled similarly—in particular that of the green Sphingid larvae. It is unlikely that darkening is fortuitous, and more likely that it is the result of intensive selective pressures which take place during only one of the 365 days of its total life (in the case of the Goat Moth, one out of 730).

(ii) Genetic only

There is only one instance known to me in which the same gene confers melanism on the imago and is also clearly responsible for a black larva. This is in f. *fumosa* Hörhammer of *Arctia caja* L. (Chapter 16). In the imaginal form the heterozygotes vary from individuals with the normal red colour of the hindwings replaced with brown (f. *brunnescens* Stattermayer) to occasional specimens nearly as extreme as the homozygous melanics; the homozygotes have black forewings and hindwings with the usual patterning showing through, etched in darker black (see Frontispiece). Typical *A. caja* larvae (the common 'Woolly Bear') are blackish with a conspicuous band of rufous hairs laterally. Larvae of f. *fumosa* are completely black and can be segregated. I am unable to distinguish the heterozygous larvae from those of f. *typica* (Chapter 16).

A further example of melanic inheritance in larvae is in *Lasiocampa quercus* L. and its ssp. *callunae* Pal., but here we have evidence that a gene controlling one such black form is linked to that responsible for f. *olivacea*, which is a greenish-black form of the imago. This is discussed more fully in Chapter 14.

More recently Lempke (1959), Bretherton (1970), and others have shown that those broods of *Dasychira pudibunda* L. which contain the melanic forms *concolor* and *obscura* also contain a proportion of black larvae. Unfortunately, the evidence of the method of inheritance of both is fragmentary and conflicting, but Bretherton states (personal communication) that a brood in which 100 per cent of the larvae were black produced 20–25 per cent of f. *typica* in the imagines. The same gene cannot therefore be responsible for controlling both. It is more likely that separate genes, linked on the same chromosome, contribute to their respective melanisms. This is similar to the situation in Yorkshire in *L. quercus* ssp. *callunae*. I have examined Bretherton's series of moths and I think it is probable that f. *obscura* is the result of incomplete dominance. The black forms of this moth have occurred at a low frequency in south-east England as far north as the Thames Valley (but not north of this) since 1934 and they seem to be increasing in frequency. Thus Bretherton reports that in Surrey, between the years 1946 and 1963, 1 per cent ($n = 570$) were melanic, and between 1964 and 1969, 5 per cent ($n = 126$).

It is not surprising that genes controlling larval colouration only rarely affect the imaginal state. The requirements of a comparatively static, cylindrical creature, geared for eating green leaves, have little in common with those of the eventual imago. All the evidence goes to show that evolution has proceeded separately in each.

(iii) Aposematic

I know of no instance in which a hairless, distasteful caterpillar relies on all-black colouration as a warning signal. More usually they exhibit conspicuous red or yellow colouration. There are species, however, whose larvae are black, but these live gregariously and usually have spines or hairs: for example, there are many in the genus *Vanessa* in the Rhopalocera, and in the *Arctiidae* in the Heterocera. But these closely packed assemblages of larvae in the *Vanessa* certainly produce another effect. Mosebach-Pukowski (1937) recorded that crowded colonies of *V. io* and *V. urticae* lead to a local rise in temperature to as much as $1 \cdot 5°$ to $2°C$ above that of the surrounding herbage. This he showed led to a more rapid metamorphosis than under solitary conditions.

(iv) Crowding effect

Long (1953) demonstrated that larvae of *Plusia gamma* L. are darker when in swarm than those living in isolation. But the experiments he undertook also tell us something else more important. Long found that amongst characters other than darkening that were consequent on crowding were a more rapid metamorphosis, decreased weight at full growth, an increase in fat content (but a decrease in water), and a smaller number of ecdyses.

He also observed that there were usually a small number of individuals which, when in a crowd, did not lay down dark pigmentation but maintained the same, normal, white-green colouration as the controls which he kept singly. He mated *P. gamma* imagines both from the lightest and from the darkest and produced different results in the larval offspring in the two lines. Of the five categories into which he classified larval colouration, from light to dark, the F_1 offspring of *light* × *light* fell into groups 1 to 3 only; there were no darker ones. However, *dark* × *dark* produced the darker forms 3 to 5. Yet both lines had identical white-green larvae when kept solitarily. This proves that a genetic component must also contribute to the melanism and, no doubt, the other characters which occur in cultures of high density.

It is unlikely that this will happen unless the phenomenon as a whole produces advantages; one can clearly see that, for those species which, early in the year, build up large populations prior to migrating, it offers many rewards—in particular a more rapid metamorphosis and a larger fat body (which is all-important to a migratory species). Faure (1943*a* and *b*) has recorded that larvae observed in the marching columns of the two migratory species *Laphygma exigua* Hübn., and *L. exempta* Walk., the 'army worms' of South Africa, are darker and more active than when observed in periods of non-swarm. This is also the case in *Spodoptera abyssinia* Gwen. in regard to pigmentation, but Faure could not detect any behaviour difference in this species, nor have we data on the fat content. Apart from *P. gamma*, Long used nine other species with green larvae, eight of which were non-migratory (the ninth, the migratory *Pieris brassicae* L., has a complicated larval pattern which is difficult to analyse). Using artificial crowding in the laboratory, he obtained pigmented larvae in such species as *Orthosia cruda* Schiff., *O. gothica* L., and *Diataraxia oleracea* L., but not in some of the others. Darkening, sometimes to an extreme degree, is, in fact, common when larvae, normally green, are confined in a small space. Their colour can change in a few days in wild-collected last-instar larvae. This I have found in *Mamestra persicariae* L. and *Cucullia asteris* Schiff.

That increased activity is itself responsible for this is shown by Long (1953) who used a mechanical activator on the solitary larva of *P. gamma* and obtained a similar result to that of crowding. Yet many solitary green larvae of other species immediately prior to their prepupal wanderings take on the same dark colouration; here colour and activity appear to synchronize. The most likely explanation is that a hormone is responsible for both and that it can be released by either genetic or environmental stimuli or, more likely, a combination of both.

NOTE (see p. 47). Curio (1965*a*, *b*) showed that in Galapagos there were three forms of the Sphingid larva *Erynnyis ello* and that these rested by day on the leaves, stems, and trunks, respectively, of the Mancanillo tree. He suggests that there is a true genetic polymorphism and he demonstrated that heavy selective predation took place by finches and warblers. Similarly, Sevastopulo (1967, 1968) has pointed out that dark forms of the larva of *Acherontia atropos* L. feed at night only and pass the day resting on the ground. Green ones feed by day.

5. The Phenomenon of Industrial Melanism

Introduction

MELANISM in the Lepidoptera must refer to any instance in which there is an increased deposition of dark pigmentation in the wings or other structures of the insect. Usually melanism in the imago is controlled genetically; more rarely it can, in certain species, be due to the direct effect of the environment during the larval or pupal period (Chapter 17). This latter situation plays no part whatsoever in Industrial Melanism, which is brought about by the indirect effects of industrialization through natural selection. Attempts have been made in the past to subdivide melanic Lepidoptera into those that were 'true melanic', 'melanistic', or 'melanochroic' (Harrison 1927*a*) according to whether they were all-black, blackish, or dark whilst maintaining their normal pattern. Examples of all these are found amongst the industrial melanics—moths that are changing their light-coloured patterns to dark ones because of the effects of industrialization.

Industrial melanics have manifested themselves only in the last 120 years and have frequently, but not always, been recorded in the first instance in or near industrial areas. They are characterized by rapid spread, by increased vigour (Ford, 1940*b* and Onslow, 1920*a*, 1920*b*), and, in the majority of species, by having no intermediate forms. They occur only amongst those Lepidoptera that depend on crypsis for their survival and that spend the hours of daylight sitting motionless on backgrounds such as lichened trunks, rocks, or dead wood. Their inheritance (except in a few instances) is one of simple Mendelian dominants. Industrial Melanism has rightly been referred to as a transient polymorphism by Ford (1955*a*). In this it contrasts with Ancient (Geographic) Melanism (Chapter 10), for then the polymorphisms are balanced. I hope to show that the two types, though usually separated spatially, have common denominators.

It is indeed unfortunate that Industrial Melanism was regarded previously as a unique phenomenon governed by laws of its own and dissociated from other kinds of melanism occurring in nature (Kettlewell, 1961*a*).

(i) Previous work on Industrial Melanism

Towards the end of the last century, the effect of Industrial Melanism made itself felt and was mentioned in many scientific journals in Britain. It must be remembered that at that time Mendel's laws of particulate inheritance (1866), though published, had been overlooked and were in fact virtually unknown until 1900. It is not surprising then that in attempting to explain the origin of Industrial Melanism the emphasis was placed at that time on environmental causes. Towards the end of the last century, Chapman (1888), Cooke (1877), and others vied with each other in suggesting different explanations, each with a Lamarckian bias. Tutt (1890), on the other hand, preferred an explanation on the basis of natural selection. The issue was further complicated by a failure to appreciate the differing requirements of day-flying butterflies and night-flying moths. Hence the temperature experiments on the former by Merrifield (1888, 1889), Weismann (1882), and others, in which melanic butterflies appeared, were quoted as proof of the direct effects of temperature and therefore applicable to Industrial Melanism in general. More usually the humidity factor was stressed.

In 1900, the Evolution Committee of the Royal Society commenced an investigation into the problem of Industrial Melanism. It is unfortunate that, in subsequent publications, actual figures were given only on two occasions, and that percentages or pious comments on the matter were the general rule (Doncaster, 1906a and b). The same unfortunate shortcomings are found in the investigations of Mera (1925) and of Adkin (1925–6). In fact, the frequency totals of samples, which would have helped so much at a later date, were never given. Nevertheless, Industrial Melanism was destined to come to the fore of biological science in the next few years in a way that was to attract considerable attention.

In 1926, Professor Heslop Harrison (1926a and b, 1927a, b, c and d, and 1928a, b and c) reported that he had fed stock of *Selenia bilunaria* Esper and *Ectropis bistortata* Goeze on foliage impregnated with lead nitrate and manganous sulphate and that he had produced a number of melanic offspring (Plate 13.2). He attributed this to the direct effects of the salts on the 'soma', which in turn changed the 'germplasm'. This was pure Buffonism (incorrectly called 'Lamarckism') and came at a time when the biological world was as yet noncommittal on the possibility of the inheritance of acquired characters; hence, the importance of Harrison's experiments. As evidence of an increased (and induced) mutation rate, Harrison stated that his melanics had occurred in a non-Mendelian ratio; in fact, there were far too few. He also reported that they were less viable than f. *typica*. Ford (1937) pointed out that the melanics were inherited as Mendelian recessives and that, in view of their admitted subviability, the deficiency was to be expected. Fisher (1933) also observed that if one had to accept Harrison's work as evidence of increased mutation, it would demand a rate of 8 per cent, whereas the highest known rate in natural populations was one in 50 000 (Haldane, 1935).

Hughes (1932) repeated Harrison's experiments and bred 3265 individuals under conditions identical to those of Harrison; Hughes concluded that 'no melanic individuals have appeared either in the treated or the controlled broods'. Thomsen and Lemche (1933) also repeated Harrison's experiments on a large scale (1185 controls: 735 treated) and stated, 'The question which we originally posed, whether the methods used by Harrison could be explained as mutations with any degree of certainty, must therefore be answered in the negative' (translation). Others have repeated his work but on no occasion with positive results. Harrison's interpretation cannot therefore be accepted. Furthermore, as both his melanic forms were inherited as recessive characters they bore no relationship to the industrial melanic situation in which this type of inheritance is extremely rare. Later, Harrison (1956) referred to a specific 'melanogen' as being the cause of Industrial Melanism, but he was unable to substantiate this. Induced melanism must therefore be discounted as being responsible either for Industrial Melanism itself, or for its spread.

In Germany at about the same time, Professor Hasebroëk (1925) was conducting experiments on a rather similar line. He used the gaseous fraction, however, and not the solid portion of an air pollution sample. Thus, the pupae of eight Rhopalocera and four Heterocera were subjected to the gases hydrogen sulphide, ammonia, and pyridine in varying proportions. He claimed that melanism was produced in treated stock (and he produced photographs of these) but not among the controls. In no instance does an insect resembling a melanic appear, nor could any of his material (with the exception of two species) have any connection with Industrial Melanism. He failed to recognize the natural history of the phenomenon and hence primarily used butterflies as the subjects of his experiments. Furthermore, there is no evidence given as to the subsequent inheritance of his so-called melanic forms. He concluded that melanism was caused by the inhalation of various gases during respiration. While admitting the possibility that the conditions imposed by his experiments could have directly affected the pigment chemistry of individual pupae, Hasebroëk's conclusions on the origin of melanism can have no foundation in fact.

Therefore, the claims of Harrison and Hasebroëk—that the effects of industrial environmental conditions were directly responsible for Industrial Melanism—cannot be accepted.

From 1930 onward, more and more melanic forms of Lepidoptera were recorded, but little was written on the theory and origin of the industrial melanics. Ford (1937) put forward suggestions for the cause of Industrial Melanism, which for the first time conformed with known laws that were applicable to the evolution of other living creatures. These suggestions depended on two main evolutionary mechanisms: first, a constantly recurring mutation rate; and, secondly, natural selection, which decided the destiny of the new mutants. Ford argued that black individuals did not find themselves at a cryptic disadvantage on modern industrial backgrounds, but that industrial melanics in general have a superior viability. They were hardier, hence their advantage. This was a great step forward; no longer did new laws have to be invoked to account for the phenomenon of Industrial Melanism. Ford (1940) and Onslow (1919a and b, 1920a and b, 1921c) were the first to show that industrial melanics had, apart from the colour change, a different physiology leading to an increased viability. Ford worked on *Cleora repandata* L. and Onslow on *Ectropis consonaria* Hübner (1919b), *Boarmia punctinalis* Scopoli (1920a), and *Cleora ribeata* Clerck (1920b). All were true industrial melanic species. This was indeed a great advance, but it left many aspects unexplained and unproven. What was the cryptic advantage or disadvantage of the industrial melanics from the point of view of predators? Why the rapid spread, and why was the melanic character nearly always dominant?

(ii) The changed environment—air pollution

Pollution of the air, man's first major sacrilege of nature, commenced over 200 years ago. The burning of coal, providing energy for industry, started in several centres in Europe in the second half of the eighteenth century: the counties of Lancashire and Yorkshire in Britain, the Valley of the Rhine, the Netherlands and eastwards to Hamburg in northern Europe, and, to the south, the area of Czechoslovakia and southern Poland and elsewhere. At a somewhat later date it took place in North America. Air pollution has proceeded insidiously since that day and with it the adaptations of living things to the new environment it has created.

It is a strange thought that the *Calmytes* trees which lived in these places 250 million years ago and which have conserved the energy derived from the sun at that time, should today exert their influence on so many living organisms; this is due to the combustion of their fossilized remains and the consequent ecological changes due to pollution fall-out.

It is as well then that we consider air pollution itself, first in its origin, secondly in its ultimate destination, and thirdly in its direct effects. The Warren Springs Laboratory (1963–7), which is responsible for measuring the amount of air pollution in Britain, classify it under two main headings —a gaseous portion and a solid fall-out; the two behave quite differently.

The gases SO_2 and SO_3 diffuse into the air or are dissolved by rain, producing dilute sulphuric acid, H_2SO_4, which, as such, can be carried considerable distances. It is this which is largely responsible for plant damage (Bleasdale, 1952, Fritz Went, 1955) but, with the exception of the lichens, only when in high concentrations. The severest damage therefore occurs within or immediately around industrial centres.

The other portion, the solid fall-out, is arbitrarily divided into 'heavy' and 'smoke'. The former is measured in tons per square mile per month in and around the point of origin, and many industrial centres receive up to 50 tons per square mile per month. Heavy fall-out, therefore, is, under most circumstances, limited to the industrial centres themselves.

The second portion, the 'smoke', is measured in milligrams per 100 cubic metres of air, which is sucked through a filter. It is the insidious effects of this second portion—drift smoke—that have largely been overlooked, first in regard to the biological significance and secondly (and even recently this has been discounted) in regard to its long-distance effects. There is evidence, in fact, that 'smoke' travels great distances and that after day and night fall-out for a hundred years or more its effects are profound. The idea of widespread dissemination of small particles is endorsed by an accidental experiment—The Windscale reactor incident.

Here, on 10 October 1957, radioactive particles escaped into the atmosphere near ground level (Murgatroyd, 1957). This is probably the first unplanned marker experiment in which radioactive elements were released into the air in a way other than by atomic explosion. The ultimate destination of the particles then bears some relationship to the distribution of the lighter portion of air pollution and the way in which it is carried. Subsequent analyses showed that the radioactive count was increased (though not to danger limits) as far away as Scandinavia, France, northern Italy, and West Germany. This then is an indication of the distances that small particles can travel. It is likely, with the prevailing south-westerly wind in Britain, that the majority of our smoke falls into the North Sea. It is not surprising then that we have found evidence of pollution effects throughout the eastern counties of Britain, though they are far removed from the centres of origin of pollution.

(iii) Effects of air pollution

The smoke particles get caught up on the leaves and twigs of trees and bushes, which act as filters, and they are also precipitated on the ground. The amount of deposition is affected by many variables, including the degree of pollution fall-out, rainfall, and the number of aphids present on the leaves. It follows that evergreens will become dirtier than deciduous trees because they receive both winter and summer precipitation.

The canopy of all trees is washed by rain, and the pollution is carried down the boughs and finally the trunk. The first indications are seen on the underside of boughs, which show dark 'rain runs'. All vegetative lichens are rapidly killed by this, at low dilutions. The trunks become bare and finally black. I have dug, to a depth of two inches, sample cores of the earth around the foot of trees in industrial Yorkshire (Plate 5.1). Up to a distance of two feet from the trunk certain samples consisted of nearly pure pollution. This accounts for the fact that no vegetation will grow around the foot of trees in industrial centres. Pollution drift also cuts down the amount of light, and this has a considerable effect on the natural history in and around all manufacturing districts. The contribution of a tree canopy to the denuding of trunks in industrial areas can be seen in many ways. I recently came across the near horizontal trunk of a Willow (*Salix* spp.), in a wood fifteen miles from central London, which had been blown down several years previously. No vegetative lichens were to be found on the standing trees near by, yet, as its canopy had not drained down it, the trunk of this fallen one was covered on its upper surface with several species of lichens. Furthermore, lichens continue to grow today on rocks, roofs, and walls near industrial areas, but not on tree trunks there.

In the last few years, we have attempted to analyse the direct effects of air pollution. By collecting leaf samples (six ounces) from the same tree at monthly intervals and washing them, we have found that in deciduous trees there is a relation between the amount of pollution and the lateness in the season, thus the leaves are dirtier in September than in April (Plate 5.2). Other considerations must also be taken into account. We have found that when a tree becomes infested with aphids, apart from the sooty moulds which grow in their sticky secretion, pollution itself is held more readily. We have also been studying the direct effects of pollution on lichens. By using a transparent tape-measure to encircle the trunks of randomly chosen trees (only the same tree species can be compared) we

PLATE 5.18 (p. 76). *Biston strataria* Hufn.
1. f. *typica* (× 1).
2. f. *robiniaria* Frings (× 1).
3. f. *robiniaria* Frings (× 1).
4. f. *melanaria* Koch (× 1).

PLATE 5.17 (p. 72). *Catocala cerogama* Guenée f. *typica* and its melanic f. *ruperti* Franc. (× ¾).

PLATE 5.17a (p. 65). *Apamea sordens* Hufn. (syn. *basilinea*) melanic form. Dr. S. L. Sutton, Scholes, E. Leeds, 1968 (photograph by A. Holliday) (× 1½).

PLATE 5.19 (p. 86). *Lymantria monacha* L. (× 1) and the extreme melanic ab. *atra*: multifactorial inheritance.

PLATE 6.1 (p. 92). *Achlya flavicornis* L.
1. f. *typica* ($\times 1\frac{1}{2}$).
2. race scotica Tutt ($\times 1\frac{1}{2}$) (? syn. race finmarchica Schörgen).
3. f. *pseudoalbigensis* Franz. ($\times 1\frac{1}{2}$) (Netherlands).

PLATE 6.2 (p. 94). *Simyra albovenosa* Goeze f. *typica* and f. *murina* ($\times 1$) (Finland).

PLATE 6.3 (p. 101). *Epimecis hortaria* Fab. (syn. *virginaria Cramer*) ($\times 1$) and its melanic f. *carbonaria* (Pittsburg, U.S.A.).

PLATE 7.1 (p. 106). The earliest *Biston betularia* L. f. *carbonaria* ($\times \frac{3}{4}$) extracted from collections made in the last century compared with modern heterozygotes (from a painting by Miss Christine Court, Oxford).

PLATE 8.1 (p. 113). *Biston betularia* L. f. *typica* and its melanic f. *carbonaria* (×1) on lichened Oak trunk (Dorset).

PLATE 8.2 (p. 113). *Biston betularia* L. f. *typica* and its melanic f. *carbonaria* (×1) on polluted Oak trunk (Birmingham).

PLATE 8.3 (p. 122). Redstart, *Phœnicurus phœnicurus.*, with *B. betularia* L. f. *typica* in its beak taken from polluted tree-trunk (Birmingham). This species took 43 f. *typica* to 15 f. *carbonaria* whilst under observation by Tinbergen (Photograph by N. Tinbergen).

PLATE 8.4 (p. 126). Spotted Flycatcher, *Muscicapa striata* L., about to take *B. betularia* f. *carbonaria* on lichened tree-trunk (Dorset). This species was seen by Tinbergen and myself to take 81 f. *carbonaria* to 9 f. *typica* (Photograph by N. Tinbergen).

PLATE 8.5 (p. 126). Robin, *Erithacus rubecula* L., with *B. betularia* f. *carbonaria* taken from lichened tree-trunk (Dorset). This species took 12 f. *carbonaria* to 2 f. *typica* whilst being watched by Tinbergen (Photograph by N. Tinbergen).

PLATE 9.1 (p. 142). *Biston betularia* L. and *Biston cognataria* (× 1). A comparison of their various melanic forms.

Left
1. f. *insularia* (=ins.⁵).
2. f. *insularia* (=ins.⁴).
3. f. *insularia* (=ins.³).
4. f. *insularia* (=ins.²).
5. f. *insularia* (=ins.¹).

Right
1. *Biston betularia* L. f. *carbonaria*.
2. *Biston betularia* L. f. *typica*.
3. *Biston* (syn. *Amphidasis*) *cognataria* f. *swettaria*.
4. *Biston* (syn. *Amphidasis*) *cognataria* f. *typica*.

PLATE 10.1 (p. 160). Examples of rural (non-industrial) melanism (Britain) (× 1).

Left
1. *Xylophasia monoglypha* Hufn. f. *typica*.
2. *Xylophasia monoglypha* Hufn. f. *obscura* Th.-Mieg.
3. *Xylophasia crenata* (syn. *rurea* Fab.) f. *typica*.
4. *Xylophasia crenata* (syn. *rurea* Fab.) melanic form? f. *alopecurus* Esp.
5. *Meristis trigrammica* Hufn. f. *typica*.
6. *Meristis trigrammica* Hufn. f. *obscura* Tutt.

Right
1. *Allophyes oxyacanthae* L. f. *typica*.
2. *Allophyes oxyacanthae* L. 'f. *capucina*' (an extreme example).
3. *Nonagria dissoluta* Treit. f. *arundineta* Hatch.
4. *Nonagria dissoluta* Treit. f. *dissoluta* Treit (=the type).
5. *Nonagria geminipuncta* Hatchett f. *typica*.
6. *Nonagria geminipuncta* Hatchett f. *fusca-unipuncta* Tutt.

PLATE 10.2 (p. 166). Shetland melanics (on the right) with their corresponding normal British mainland forms on the left ($\times 1\frac{1}{4}$).

1. *Diarsia festiva* Schiff., f. *typica*, English.
2. *Diarsia festiva* Schiff. f. *thulei* Stdg., N. Shetland.
3. *Hadena conspersa* Schiff. f. *typica*, English.
4. *Hadena conspersa* Schiff. f. *hethlandica* Stdg., Shetland.
5. *Lygris populata* L. f. *typica*.
6. *Lygris populata* f. *masanaria* Freyer, Shetland.
7. *Amathes xanthographa* Schiff. f. *typica*, English.
8. *Amathes xanthographa* f. *obsoleta-nigra*, Shetland.
9. *Entephria caesiata* Schiff. f. *typica*.
10. *Entephria caesiata* f. *glaciata* Germar.
11. *Xanthorhoe fluctuata* L. f. *typica*, English.
12. *Xanthorhoe fluctuata* f. *thules* Prout, Shetland.

have been able to undertake comparative lichen-coverage counts. A correlation between the total number of lichens present (and the number of species) and the distance from an industrial centre was first recorded by Jones (1952).

More recent work on the presence or absence of lichens in the Midlands and on correlation with the frequencies of the melanic forms of *Phigalia pedaria* has been recorded by Lees (1971). Brodo (1960) has studied the effects of air pollution on lichens on Long Island, New York, and similarly Fenton (1964) for Belfast, Northern Ireland; Skye (1968) has written a monograph on the subject. These investigations confirmed my belief that air pollution is, and has been responsible for profound ecological changes.

Recently Hawksworth and Rose (1970) have carried the study a stage further and have calculated the degree of pollution against the number of lichen species and the coverage of the trunks; the importance of their methods is reflected by a leading article in the *British Medical Journal* (1970). In a somewhat similar manner, I had, in 1952, to attempt to assess the degree of pollution fall-out retrospectively for the previous 50 years. The circumstances were these. The Professor of Medicine at a well-known Yorkshire University had been carrying out research into the effects of air pollution on elderly patients and its association with chronic bronchitis. This had demanded not only extensive investigations on the inhabitants of Chapeltown, which is five miles from Sheffield, the most heavily polluted town I have met, but also similar investigations on a 'control population' living in a supposedly pollution-free area. For this he had chosen the village of Aysgarth, in the Yorkshire Wolds, which had little industry within forty-five to fifty miles of it. Pollution gauges had been installed in Chapeltown throughout the period of his research and these had recorded the extent of the contamination. Unfortunately no such gauges had been present in Aysgarth. It had been assumed that this pleasant area, situated on the eastern side of the Pennines, was pollution free. Yet results of his survey had raised doubts about this. Could the present local ecology reflect the pollution history in the Aysgarth area retrospectively?

The immediate impression on arrival was that the countryside was uncontaminated and the air clean. Also, many of the inhabitants who lived there were unaware of any fall-out. Yet on examining the tree trunks we found that vegetative lichens were absent from all the Oaks (*Quercus* spp.) and Beeches (*Fagus*) though present on Ash (*Fraxinus*) and Apple (*Pyrus*). We also noted that the undersides of the Oak and Beech boughs were blackened and the blackening extended on to the trunks beneath as 'rain-runs'.

The Lepidoptera confirmed that there must have been smoke fallout over a long period of time. In two weeks we collected local samples and nine species of these (Selidosemidae and Caradrinidae) had industrial melanic forms, five at a high frequency. *Biston betularia* f. *carbonaria*, which in my opinion always reflects the degree of pollution with the greatest sensitivity, formed 85 per cent of the population in a sample of exactly one hundred. Aysgarth is situated approximately fifty miles north-east of the Lancashire industrial centres and the prevailing wind there is from the south west which, no doubt, regularly carries pollution over the Pennines. Corroborative evidence of this was given to me at a later date by skiers who told me that in winter they regularly encountered 'pockets of soot' in the snow drifts.

Apart from killing the vegetative lichens and blackening tree trunks and boughs, in some places, such as Chapeltown, the ground itself, along with the fallen twigs and dead leaves, becomes darkened (Plate 5.1). All these are the normal resting places for the majority of cryptic moths. In 150 years, backgrounds have entirely altered. At the same time considerable selection against the toxic effects of heavily polluted foliage, which the majority of larvae have to eat, must have been taking place. This, then, is the history of the changed environment (Plate 5.3, 1, 2, and 3) in which Industrial Melanism has arisen.

(iv) The history of the spread of industrial melanic species in Britain

In Britain there are approximately 780 different species of Macrolepidoptera, and of these over 100 are undergoing the same process of substituting for their highly complicated and specialized patterns, built up to perfection over thousands of years, a dark or black colouration. New melanic forms are being recorded each year, and we are able to watch a steady increase in the frequency in some of these. A list of some of the British species demonstrating this is given in Appendix B, and these can be compared with the same or nearly related species in other countries (see Chapter 6). The history of the spread of the melanic forms of *Biston betularia* in Britain, and the theory of how these forms adapted, is given in Chapter 9.

It must be admitted here that no other species has been able to provide so much data on adaptation and rapid spread as *Biston betularia* f. *carbonaria* which on no occasion can be considered a non-industrial melanic. This is because of the recorded past history of the species, its ubiquitous distribution, and the fact that large-scale experimentation has been carried out both in the field and in the laboratory.

Other species, however, have contributed different but equally important aspects to our knowledge of Industrial Melanism. The recent work of D. R. Lees (1971) on the distribution of *Phigalia pilosaria* (Selidosemidae (Plates 5.4, 5.5, and 5.9, right 3 and 4) has demonstrated this in many ways. *P. pilosaria* is a sexually dimorphic species: the female is brachypterous, the wings being but 1–2 mm in length, whilst the males are fully winged and fly freely. In contrast to *B. betularia* f. *carbonaria*, the melanics occur in completely rural districts such as the Highlands of Scotland (White, 1876, Kettlewell, 1957). Hence, this species, apart from being a non-industrial melanic as in the past, is also today a successful industrial one. A further important consideration is that *P. pilosaria* flies in the months of January, February, and March, and we have shown that in the laboratory it can survive a temperature of 0°C for up to two months. During cold periods (which are frequent at this time of year), imagines could live, in theory, for such a time, but in nature they are exposed to intensive predation as they sit on tree trunks. Lees in fact records a 30 per cent predation rate on the first day of release. The relative frequencies are therefore likely to be different at the beginning and the end of such a period, though their numbers are greatly reduced. This is then an entirely different situation from that of *Biston betularia* which flies in mid-summer and which in captivity can live only 16 days and in nature on average two to three days. *P. pilosaria* has several melanic forms and Lees (1971) has shown that two are controlled by a pair of alleles, with the more extreme melanic f. *monacharia* dominant to a paler one which is in turn dominant to typical.

Lees (1971) has obtained samples of *P. pilosaria* from 88 sites in England, Scotland, and Wales (Fig. 5.1) thus making possible a comparison of its melanic distribution with that of *B. betularia*. The two species show one basic similarity: highest frequencies occur in dense urban and industrial areas. There are, however, several points of difference:

(a) In rural central Scotland 25 per cent of *P. pilosaria* populations are melanic, yet the melanic *B. betularia* f. *carbonaria* is completely absent from this area.

(b) In parts of East Anglia melanic *B. betularia* constitute 75 per cent of the population; the equivalent value for *P. pilosaria* is less than 15 per cent.

(c) In rural south-west Wales where f. *carbonaria* seems to be absent and where f. *insularia* occurs at a frequency of less than 5 per cent (Lees, in prep.), *P. pilosaria* melanic frequencies as high as 25 per cent are found.

(d) Maximum melanic frequencies in highly polluted areas in *B.*

60   The phenomenon of Industrial Melanism

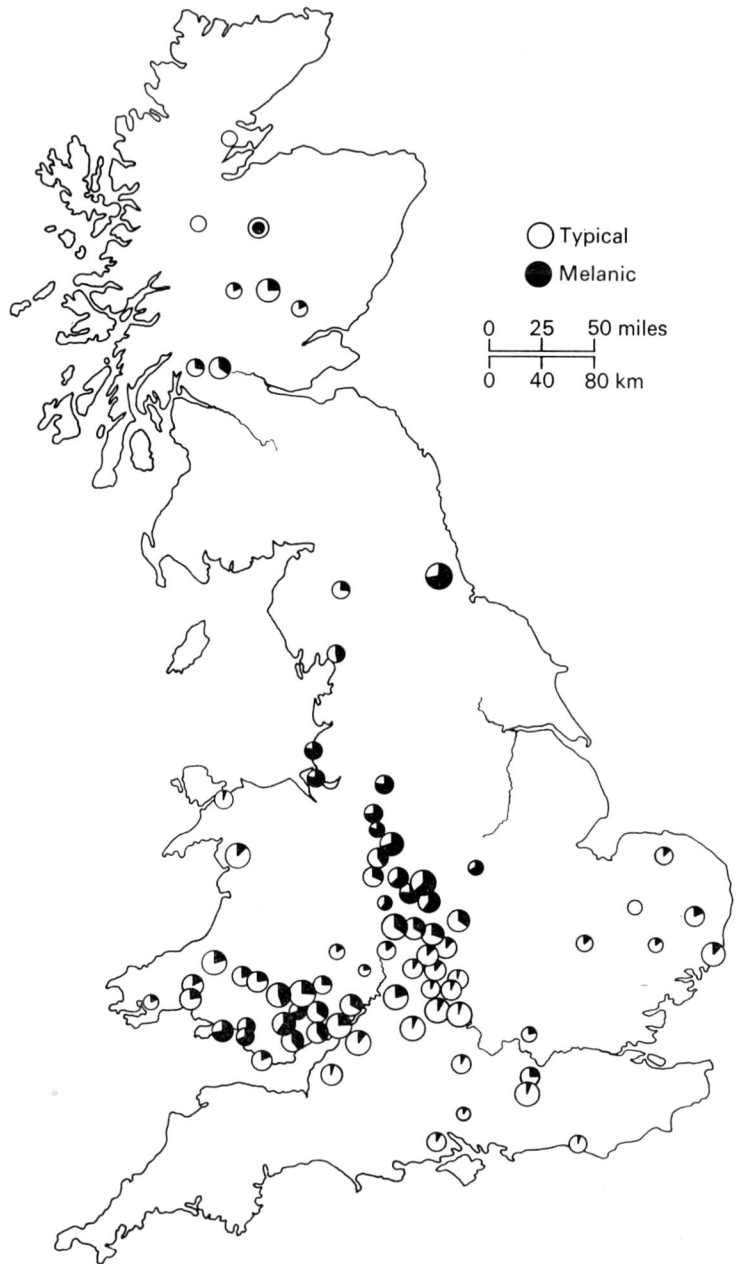

FIG. 5.1   Frequency Map of *Phigalia pilosaria* Fab. and its melanic forms from 88 localities in Britain (Lees 1971)

betularia are around 98 per cent; in *P. pedaria* no sample has yet been found where melanics exceed 75 per cent of the population.

These differences are not consistent from area to area and show no clear cut pattern, in that in some areas frequencies of melanic *B. betularia* are higher than of *P. pilosaria*, while in others the reverse is true. Lees points out that the simple differences between the species such as exist in the degree of sexual dimorphism, crypsis, or potential for migration cannot alone be invoked to explain the discrepancies in the distribution of the morphs. It is likely that the genes controlling the melanics of the two species have differing pleiotropic effects, the selective values of which may vary from area to area. It has been shown, for instance, that the homozygotes of both f. *monacharia* and another but paler melanic form of *P. pilosaria* are at a disadvantage compared with their heterozygotes, when reared under laboratory conditions. Pleiotropic effects of the genes controlling melanic *B. betularia* have also been demonstrated (Kettlewell, 1958c).

Creed, Duckett, and Lees (in prep.) have measured 17 physical and biotic parameters of the ecology of epiphytic plants on oak trees (*Quercus* spp.) at 103 sites in England and Wales, many of which were also the sampling points for *P. pilosaria*. In a multiple regression analysis of the geographic variation of *P. pilosaria* in relation to these factors Lees (loc. cit.) reports that 85 per cent of the variation in melanic frequency can be explained in terms of all seventeen variables, and 64 per cent on reduction to the single variable, oak-trunk reflectance. This parameter is influenced by many environmental factors; in towns the most important of these are the two main components of air pollution, smoke and sulphur dioxide, but in rural areas tree-trunk reflectance is influenced by many factors other than the concentrations of these pollutants.

In my own view an all-important factor is diminished light due to cloud-haze either from air pollution or from water vapour.

A further point must be considered. Lees has suggested that pleiotropic effects may be responsible for the frequency contrasts between *P. pilosaria* and *B. betularia*, and this must in part be true. Yet, it is impossible to compare predation which takes place in the winter months with that which occurs in the summer. On the one hand, *P. pedaria* avoids the predation of the summer migrants, including the Spotted Flycatcher, *Muscicapa striata*, and the Redstart, *Phoenicurus phoenicurus*, which I have shown to be so effective with *B. betularia*. On the other hand, the behaviour difference between winter and summer feeding (when birds have young) must affect the degree and quality of predation. It is in fact impossible to equate two situations in which different species of birds have profound effects on different species of Lepidoptera at different times of the year in different localities.

From the studies of the two species which have been the most extensively investigated, *B. betularia* and *P. pilosaria*, it is apparent that the distributions of their melanics are a result of many interacting selective factors. From this I think we can generalize by stating that direct comparisons between any two melanic polymorphisms are impossible, though many are influenced by a common factor, as for example, that the newly hatched larvae of both species under discussion become wind-borne and are distributed as aerial plankton (as in many other species of Lepidoptera).

The situation differs from that of *Gonodontis bidentata* and its black form *nigra* Prout (Plate 5.6, left 4 and 5), for here dispersal is minimal in all stages. Consequently, the melanic is confined to certain districts and absent in others where presumably the mutation has not taken place; thus f. *nigra* is absent from central Birmingham (where modified f. *typica* occur) but present 20 miles to the north-west on Cannock Chase, where it spends the day at rest on pine trunks, and where 50 per cent of the population are of this form. Bishop, Harper, and others, centred on the University of Liverpool, are at present working on this species and have detected highly significant frequency changes over short distances in samples from east and west Liverpool (Bishop and Harper, 1970). Here its larvae feed on privet hedges, even those bordering main traffic routes, so that minimal flight must be encouraged. Even here there are ample predators to ensure differential selection, for sparrows abound. The patchy distribution of melanic *G. bidentata* is therefore in contrast to *B. betularia* and *P. pilosaria* and may in part reflect centres of mutation rather than widespread dispersal.

There is only one clear example known to me, though I suspect others exist, where migration from the continent of Europe has been responsible for melanism in a species in Britain. *Tethea octogesima* Hübn. f. *fusca* Cockayne occurred commonly throughout Holland in the second quarter of this century, but with one doubtful record from south-west England, was unknown here till 1945, when it appeared simultaneously throughout south-east and eastern England. Since that time f. *fusca* has continued to increase and to spread (see Appendix B). The fact that another and more extreme melanic (Plate 5.7, right 2), f. *frankii*, which is common in south Netherlands, has doubtfully appeared in this country suggests that there may be a differential migration habit between it and f. *fusca*.

Plates 5.6, 5.7, and 5.8 show other examples of industrial melanics. Of particular interest is *Tethea or* Schiff. whose jet-black form f. *albingensis* Warnecke (Plate 5.7, right 3 and 4) has the central orbicular and reniform markings on the forewings conspicuously light. This was first recorded from the industrial centre of Hamburg about 1907. In 1917 a sample of wild larvae gave over 90 per cent of this form and at the present time it is wide-spread there and is found as far distant as Finland. The genus *Tethea*, whose species rest by day on trunks, boughs, and more particularly twigs, is prone to melanism. *T. duplaris* (Plate 5.7, right 5 and 6) demonstrates a further point, in that the dark form *obscura* Tutt has, from earliest records, occurred in rural northern England and Scotland. More recently a similar melanic form constituted up to 90 cent of the population in many central industrial areas. The common denominator here must surely be one of diminished light-reflectance resulting from cloud and increased humidity in the north and from industrial haze and darkening background in the industrial midlands.

An example in Britain of a melanic mutation reversing the fortunes of a species is *Miana* (syn. *Procus*) *literosa* Haw. (Plate 5.8, right 3 and 4) which rests on detritus on the ground by day. The light-coloured f. *typica* was common 50 years ago around Sheffield, a heavily polluted centre in Yorkshire. The species became extinct there, however, and was not seen for a number of years. More recently (*c.* 1940), it reappeared, but in its melanic phase f. *aethalodes* Rich. only. Such an example emphasizes the fact that had not the various melanic forms of *B. betularia* been available (Plate 5.6, left 1, 2, and 3) as a result of recent mutation, this species could have today been extinct throughout central, eastern, and south-eastern England.

I know of only one species in which a melanic form has been overwhelmed by the march of industry. *Boarmia roboraria* Schiff. (Plate 5.9, right 5 and 6) is the largest British Geometrid. It occurs locally in oak woods in central and southern England, where it rests by day on tree trunks where the smoke-grey forms *infuscata* Stdgr. and *varia* Cockayne have spread. In the Warwickshire woodlands round Coventry the black insect f. *melaina* was recorded about 1887 and again in 1893 and subsequently. On the evidence of one brood it is dominant to f. *typica*. Today this melanic form appears to be extinct, though the species still occurs at a low frequency there. Single examples, similar to f. *melaina*, have been recorded from near Reading and, more recently, in Surrey, to the south-west of London. There is no question of dispersal being responsible for this, for this form has not been recorded over the 70 miles which separate the localities. It must be assumed then, that the records of these f. *melaina* indicate local mutations and that these must be rare or alternatively physiologically disadvantageous. This is a salutory reminder that not all industrial melanic forms necessarily succeed and spread.

There are several species whose black forms have been recorded but recently; yet even after as short a period as 10 years or less, we have evidence of their spread. *Brachionycha sphinx* Hufn. (Dasypoliinae) (Plate 5.10) is a large Noctuid moth which hatches in the winter months and rests exposed on dead wood, palings, and fences (Plate 5.11). It depends utterly for survival on its high degree of crypsis, yet only in the last four years has a distinct melanic form occurred. Its recognition has been complicated because of a gradual darkening of f. *typica*, presumably polygenic and similar to the Birmingham *G. bidentata*. I recorded one of the first black *B. sphinx* (Plate 5.12) in 1968 at Steeple Barton, Oxfordshire—this was a single specimen in a sample of 43. In the previous eleven years, the samples averaged 100–200 a year, yet all had been f. *typica* (some greyer than others). In 1969, my sample of 157 included two clear-cut melanics, and in 1970, 12 in a sample of 142 (8·5 per cent); this form is dominant (H.B.D.K. 1972).

One of our largest Noctuid moths, *Catocala nupta* L. (the Red Underwing) rests on trunks, but more frequently today on wooden telegraph poles. In 1920, the first specimen with all-black forewings and body was bred from wild larvae taken at Dartford, south London (Lancum, 1920), but no others were recorded there subsequently. In 1962, I. Lorimer caught the second specimen of this form f. *nigra* Cockayne at Totteridge in north London (Cockayne, 1951b). Since then he has collected two more in samples averaging one hundred per annum. The form *nigra* is a distinctive insect comparable to the clear-cut melanic forms of *Catocala* spp. which occur in North America (for example *C. cerogama* Guenée). *C. nupta* f. *typica* used to be common in London and fifty years ago I found it regularly at rest on elms in Hyde Park. Since then it has become rare or absent there. The new mutation which must have occurred both in south and north London separately, is holding its own locally at Totteridge, and assuming that the melanic f. *nigra* is dominant (as in some *Catocala* species, for example *C. cerogama* (Plate 5.17)) it may have acquired a degree of physiological adaptation after ten years or more. It will be interesting to see the future frequencies of this large and beautiful moth. Other species which have recently and locally produced industrial melanic forms which are now increasing, are *Xylocampa areola* in Essex, *Notodonta anceps* (*trepida*) ab. *fusca* Cockayne. in Surrey, which is similar to the non-industrial melanic form in Westmorland, and *Drymonia ruficornis* Hufn., also in Surrey, where two melanics were recorded (R. F. Bretherton) out of a sample of 51 in 1970, after a total of 43 f. *typica* in the previous 6 years.

At the extreme end of the rarity range a black specimen of *Apamea sordens* Huf. (syn. *basilinea*) has recently been recorded from the industrial area of Leeds by Dr. L. S. Sutton. It is the only example known of this extreme melanic (Plate 5.17a).

In this brief résumé of the history of the spread of Industrial Melanism in Britain, I have referred to a few examples only in order to attract attention to the various biological causes which affect Industrial Melanism and its spread. Their occurrence in a particular locality must be due to either a local reservoir (non-industrial), dispersal, or to mutation. The subsequent success or failure of each is decided by the degree of local predation and the physiological efficiency of the new mutant. Bivoltine species (such as *Ectropis*) and others which have a long imaginal life span (for example *P. pilosaria*) are subject to a longer period of selective predation and in theory, the black forms if advantageous, have the opportunity of increasing more rapidly than a short-lived univoltine species, for example *Apatele alni* (Plate 5.8, left 3 and 4).

The rapidity of the spread of Industrial Melanism has served to show how one advantageous character can replace a previous one when the environment has altered. The earlier incredulity of biologists which greeted the reports of the rapidity of the spread of the new melanic forms, some even doubting their authenticity, was due to two failures on their part. First, they failed to realize that the ecology had changed with considerable rapidity due to the effects of industrialization, and secondly they failed to appreciate the degree of selective pressure which was involved. The rapid substitution of one *visible* character by another, black for light, was a previously unknown phenomenon. Much more common, but less frequently disclosed, must be instances of comparable selective pressures on *invisible* characters, and these for the most part must be concerned with physiological and biochemical processes.

(v) Behavioural and physiological differences

These include differences in background choice, viability, hatching time, and assembling scents.

Genes controlling melanism either in larvae or imagines, must at the same time exert other effects. In *Drosophila* every gene so far investigated, has shown pleiotropism. In the Lepidoptera by far the best documented species are *Ephestia kühniella* and *Bombyx mori*. As each is a specialization in its own right, I refer but briefly to them.

In the first named species 'a' alleles, apart from conferring black eye-colouration, also affect testes, larval pigmentation, growth rate, viability, and the whole of the courtship behaviour. These pleiotropic effects are, furthermore, influenced by temperature in each genotype. The more important references to this species are Kühn and Henke (1932), Kühn (1963), and Caspari (1948, and 1952).

The second species *Bombyx mori*, with a haploid chromosome number of 28, is the strain used in sericulture today. It has not been discovered in the wild, but the silk worm has been bred by the Japanese for several thousands of years in captivity. It may have arisen from *B. mandarina* found today on the east Asian mainland and with a haploid chromosome number of 27. The genetics of *B. mori* are even more fully documented that those of *E. kühniella*. For a detailed bibliography see Kikkawa (1953), Tanaka (1953), and Yokoyama (1959). The pupa, usually yellow-brown, has at least one melanic form which is recessive and whose expression is affected by temperature. In the imago there are three separate melanic forms, two of which are dominant, but may be controlled by the same gene in different gene-complexes (Tazima, 1964). The third melanic is a recessive. Of the 200 different loci recorded, many have been shown to have effects other than the visible ones such as colouration and pattern of larva, pupa, and imago. Apart from pleiotropism, these two species provide a host of other interesting data on the chemistry of pigments, mating preferences, linkages, heterosis vigour, the inheritance of voltine strains, variation in chromosome numbers, and of course multiple allelomorphism.

Now why should two species of Lepidoptera only, out of the thousands available, have contributed so much and the rest so little? The fact that they have been investigated so thoroughly is entirely fortuitous, and not primarily because of any particularly unusual character: *B. mori* because of the commercialism of silk; *E. kühniella* as a pest of stored produce. Both species have therefore been freed from the consequences of natural selection as usually found in nature and share with Man the experience of civilization with freedom from the earlier hazards. These two species demonstrate how pitifully small is our knowledge of the genetic potential of free-living Lepidoptera, and in particular our near-failure so far to demonstrate behavioural and physiological differences associated with the various melanic forms. Robinson (1971) referring to *B. mori*, in a most poignant sentence, states, 'the discovery of so many mutations in these species, affecting almost all aspects of physiology and morphology, may be taken as a guide to the sort of variation which is waiting to be discovered in other Lepidoptera species.' How right he is, and what opportunities exist for those large numbers of graduates who, having qualified for taking a higher degree in science are completely void of original ideas on which to work. The efforts that have been given to studying the other Lepidoptera have in fact so far been minimal. Because of the greater and more varied external selective pressures in nature, such studies should be far more rewarding than previous work, in particular in demonstrating the biochemical, behavioural, and physiological effects of major colour/pattern mutants. Because Industrial Melanism has taken place with such rapidity and is at present in the process of attaining balanced polymorphisms under modern conditions, pleiotropic effects may as yet not have been determined in the Industrial Melanisms. Nevertheless, one would expect to find them in that ancient group which we place under the heading of Non-industrial Melanism.

The ability to choose a correct background

Few field observers would dispute the statement that cryptic moths pass the hours of daylight on their correct resting-sites; hence the occurrence of selection for highly specialized patterns for camouflage on equally specialized backgrounds (Plate 5.14). Charles Darwin was deceived by such insects when collecting in Brazil as we can see from his statement in the Journal (Darwin, 1884):

The large and brilliantly coloured Lepidoptera bespeak the zone they inhabit, far more plainly than any other race of animals. I allude to only the butterflies; for the moths, contrary to what might have been expected from the rankness of the vegetation, certainly appeared in much fewer numbers than in our own temperate region.

Using mercury-vapour light traps in the same Corcavado forests which Darwin had visited in 1832, I was able to sample the moths here in 1958. I found the largest number of Lepidoptera I have experienced anywhere in the world, both of species numbers and total catch. Many of these depended entirely on crypsis for their survival. By day they were most difficult to find. Indeed, without the aid of modern sampling methods, one would have come to the same conclusion as had Darwin.

Yet there are still today a minority of biologists who, no doubt with insufficient field experience, question the selectivity of the choice of resting-sites of each species. The Senior Entomologist at a well-known University recently challenged my statement that 'many cryptic species rest on the trunks and boughs of trees'. Incredulous, I asked him why he held such doubts. He explained that he had a brilliant student who had chosen for her thesis 'the resting positions of Lepidoptera'. On the tree trunks in Berkshire where she worked she had found only two moths in six weeks. Professor X and his student, in the good company of Charles Darwin, have unwittingly borne tribute to the extreme efficiency of the cryptic colouration in the Lepidoptera. But such colouration is useless unless an equally correct behaviour pattern is associated with it. In particular, how do the Lepidoptera decide their background for survival during the daylight hours? Experience can be dismissed in this order as from the day of hatching any error of pattern or choice of background must end in death. Genetic inheritance alone can contribute to pattern and behaviour. This could be brought about in several ways: by scent attraction to specific trees or lichens (in *Biston betularia* f. *typica*), for example, in the same way as ovipositing females are attracted to food plants. Alternatively, it could be brought about by a mechanism for scent/colour/texture recognition.

Not until we study polymorphic species are we likely to get further evidence. For here, whilst many behavioural aspects between the different forms may be held in common, such as wing positioning, and orientation to the tree species on which they sit, the colour and pattern of one form is entirely different from that of another. Thus in *Biston betularia*, within a single brood, it is possible to have a proportion which are uniformly jet-black with no patterning (=f. *carbonaria*), some light with highly specialized lichen markings (=f. *typica*), and others of grey granular appearance (=f. *insularia*) whose optimum background for concealment depends on the presence of *Pleurococcus* covered trunks. Yet all these are sibs and by inheritance might have the same basic behaviour pattern in regard to resting-sites. Such polymorphic species provide excellent material for testing the mechanisms involved in the choice of backgrounds.

Background-choice experiments. It has not been sufficiently appreciated that the behaviour of moths in captivity is entirely different to that in the wild. For on the one hand it is limited and constrained while, on the other, by free flight at the correct moment, moths are able to take up their positioning by scent, vision, and also having regard for wind-direction, sun, and shade. It appears to me to be impossible to devise satisfactory experiments in the laboratory for testing their behaviour under normal conditions.

Laboratory experiments on background recognition. In 1954 I set up a series of experiments which took into account the following criteria: clear-cut background-reflectance differences (either uniform black or variegated grey), equal surface areas of each, identical texture of background material, a container of large cubic capacity, and overhead lighting from indirect sunlight.

Design of experiment. A large cider barrel (height 40 in., maximum diameter 28 in.) was lined with alternate black and white strips of cloth

TABLE 5.1
Background recognition in Biston betularia *(the barrel experiments)*

	Black (=f. *carbonaria*)	White (=f. *typica*)	Total
Black background	38	20	58
White background	21	39	60
Total	59	59	118

2 × 2 table gives $\chi^2_{(1)} = 10.9$, $P \simeq 0.001$.

or rough paper, all of identical texture (Plate 5.15). A sheet of glass was placed on top which was covered with white muslin. Each evening a maximum of six *B. betularia* of the same sex, and with f. *carbonaria* and f. *typica* in equality, were released inside. I came to the conclusion that even six individuals were too many to use at any one time because of disturbance effects. At dawn their resting positions were scored. All those individuals which overlapped the two contrasting backgrounds were eliminated. Similarly, those which rested on the top or the bottom were not included.

Results. Table 5.1 gives the findings of the 118 *B. betularia* scored out of 198 tested (40 per cent were disqualified). Of those which qualified, 65 per cent chose correct backgrounds ($\chi^2_{(1)} = 10.9$, for which $P \simeq 0.001$) and each form contributed equally (Kettlewell, 1955d).

I repeated these same experiments later in conjunction with P. M. Sheppard to whom I passed over my *B. betularia* specimens (unpublished). Unknown to him these insects fell into two groups. An earlier one which consisted entirely of bred examples and a later one which had been captured at mercury-vapour light; this was because *B. betularia* tend to hatch later in the wild than in captivity. He reported that he rather rapidly obtained near-significant figures for correct choice but that my more recent insects appeared to take up positions at random. I think the most likely explanation of this anomaly is that these later insects which, unknown to Sheppard, had passed several hours in the glare of a 120 W mercury-vapour bulb, had expended their visual purple and were virtually blind for the following 48 hours. This then is another hazard in undertaking such experiments.

More recently Sargent (1966, 1968, and 1969a and b) and Sargent and Keiper (1969) have carried out background-choice experiments on cryptic species using both Geometrid and Noctuid moths, but the majority of these were monomorphic. Some species only showed a correct choice ($P = >0.01$), others appeared to rest at random. Without having had the advantage of seeing their techniques, and therefore with some reservations I make the following criticisms: the containers were much too small (approx. 15 in. square by 35 in. high), the four angles of 90° were unsatisfactory, the insects were overcrowded (up to 10), and they appear to have tested individuals on the same night as they were captured, regardless of their being captured at 'light' or at 'bait'. More pertinent is it to point out that, whilst no one denies the ability of each species to recognize its special resting-site in the wild, the main issue is missed: this is to discover whether each form of a polymorphic species is able to choose its own correct background. As my own data are, up to date, inadequate I can do no more than state that they suggest that the morphs of *Biston betularia* are capable of correct individual choice. (See footnote on p. 88.)

TABLE 5.2

Resting Positions of f. typica *and f.* nigra *of* Ectropis consonaria *on light and dark surfaces*

Background	f. *typica*	f. *nigra*	Total
Light	22	5	27
Dark	4	15	19
			46

$\chi^2 = 14.202$, $P = 0.001$.

Background scoring in the field. Because of the seemingly impossible task of devising satisfactory laboratory experiments where moths have to be restrained inside a container, I attempted in 1957 to test background recognition under natural conditions. I chose the species *Ectropis consonaria* Hueb. (Plates 5.9, left 5 and 6) and the place, Stroud, Gloucestershire, where the frequencies of the melanic and typical forms were nearly equal (f. *nigra* 43 per cent, $n = 129$) (Kettlewell, 1958c). Here the imago spends the day at rest on the trunks of Beech (*Fagus*) (Plate 5.16). Initially the position of each individual was carefully noted and a second person captured the specimen. An area of approximately two inches square of the bark on which it had been resting was then cut out and marked beneath as 'M' or 'T' according to the phenotype. On return to the laboratory, the areas of bark were inspected by three separate individuals who had been given the number and proportion of the *E. consonaria* forms. Because of dehydration during transport to the laboratory there appeared to me to be less differences between them than when they were in their natural setting, and scoring showed no significant departure from random choice. I therefore repeated the experiment in 1958, but on this occasion we compared the correctness of the phenotype background according to human recognition. Of 46 *E. consonaria* assesssed in this way (27 f. *typica*, 19 f. *nigra*) (Table 5.2) the majority (37) had taken up correct positioning. However impartial one's approach, I am certain that bias must favour the theory one nurtures. Certainly in *E. consonaria*, which, like *C. repandata*, may have to reposition itself several times during the course of the day due to disturbance, more rigid analysis is indicated and this can now be undertaken using a light-reflectance photometer on background and morph, as successfully carried out by D. R. Lees on *Phigalia pilosaria* (Lees, 1971). This particular form of apparatus was not available to me in 1958 and I recommend its use, particularly with smooth-barked trees, and moths of uniform colouration. I must stress again that here we are concerned as to whether two morphs take up positions on different and correct backgrounds. We are not considering orientation positioning and other factors involved.

A theory of the mechanism of choice. In 1955, I put forward a theory which could account for the ability of each morph to choose its correct site, be it light or dark. I suggested that scent was likely to be the primary stimulant, in particular the scent of specific trees and those of the particular ecology favoured. At early daybreak, when the majority of cryptic moths fly (usually a second and quite distinct flight from an earlier one) a tree of the correct species is chosen at random and on alighting the moth runs over the trunk. Here it shifts its position maybe by several feet, so as to judge the best of the local advantages offered. This it does by vision. Having discovered its own optimum it now carries out a particular behaviour pattern which I believe has previously not been described. The insect turns on its own axis, maybe through the whole 360°, and during this act it from time to time 'clamps' its wings flush with the trunk. The final movement is a side to side wriggle when it succeeds in getting its thorax and abdomen aligned into a groove. I have made these observations from several hundred moths released at earliest daylight and subsequently followed to their destinations. I think that during this sophisticated act several different stimuli are being received and their correct answers determined. First, wind direction; no Lepidopteron chooses the windblown side of a trunk. Secondly, and more important, during the 'clamping' movements the insect is receiving light stimuli from its background which it can compare with the colour and pattern of its own circumocular tufts. If the two are out of phase I have suggested that a state of 'contrast/conflict' will occur (Kettlewell, 1955d) and that the insect will then move to another position or even take flight again. When once the final positioning for the day is decided, considerable disturbance may be necessary in order to stimulate the flight reflex. This is certainly so in *Biston betularia* where the usual behaviour is to feign death in a period of catalepsy. Sargent has worked on the North American *Catocala* and has shown how selective they are in choice of background and position. I have found all four of our British *Catocala* at rest in the country: *C. nupta* commonly on telegraph poles (p. 64), *C. fraxini* on Aspen, in Ham Street, Kent (three in all), and *C. promissa* and *C. sponsa* on Oaks in Kent on a rare occasion following flight at midday during a period of extreme heat. The various species of this genus are certainly most sensitive to their particular restingsites. How much more interesting it would be to test the melanic and non-melanic morphs of such a species as *C. cerogama* Guenée and its f. *ruperti* Franc (Plate 5.17), which occurs in the eastern United States.

The evidence of contrast/conflict in nature. The contribution of a correct background-assessment by each individual versus an inherited specific one is of the utmost importance to an understanding of colour polymorphism. For if specific choice, regardless of morph colour, decides the destination of each form, the behaviour pattern must be fixed for all without recognition of different backgrounds. Conversely, different backgrounds cannot be made use of by the particular mutants, and even if random choice was admitted a considerable body of available resting-sites would have to change before such a new form could expect any advantages. It is, in fact, unlikely that the evolution of melanic forms could have taken place without individual (as distinct from specific) background-recognition. I have suggested that this could have been brought about by 'contrast/conflict' and I have some evidence of this. In 1957 (Kettlewell, 1958d) I was working in the deciduous forests on the northern shore of Lake Eyrie. A large area of one such forest had experienced a severe fire about five years previously. The trunks and lower boughs of all the trees were bare and blackened over an area of several square miles. This was contiguous with a large extent of unburnt trees where I had been searching the lichened-covered trunks for several days and where I had found only nine *Catocala* (three species) during this time. I then chanced upon the burnt out portion of the forest. Here I found large numbers of specimens of fifteen different species in a short time. Most of these were visible many yards distant. What impressed me was the different behaviour of the moths in these two contrasting though adjacent woodlands. In the unburnt forest each insect, when discovered, could be approached and captured without eliciting an escape response. Without exception, the moths in the burnt area took flight on approach and this when I was several yards distant. On settling, they spent several minutes running over the surface with occasional wing clamping and frequently they took flight again spontaneously without further disturbance. This, I think, is an example of 'contrast/conflict' in the wild—a reaction of the individual to unsuitable conditions. What opportunities exist for similar observations to be made on *Catocala* species which have melanic polymorphisms?

Viability differences

There is a certain amount of confusion in comparing the differing viabilities between melanic forms and f. *typica*. There are three clearly distinct situations. First, one in which melanism is inherited as a recessive character. Secondly, that in which the dark forms are dominant and the expected 1:2:1 ratio from heterozygote pairings is disturbed, resulting in a deficiency of homozygous black individuals. Thirdly, there are instances where the expected 1:1 ratio in back-cross broods is distorted.

In regard to the viability of recessive melanics, the most frequent examples are amongst the aposematic Arctiidae, yet I can find little clear evidence that the rare melanic individuals, occasionally bred from wild collected larvae, are even inherited. Robinson (1971), who has studied the literature extensively, agrees with this. But here there is a practical difficulty, in that such insects are rarely fertile, and seldom pair in captivity. Also, similar melanic forms occur from time to time in broods subjected to temperature experiments (for instance in *Arctia caja*), in stock not known to contain such mutants. Nevertheless, it is likely that rare recessive melanic mutants exist due to inheritance alone, and that their heterozygotes must be constantly present in wild populations. Disregarding the effects of increased heterozygote vigour, they are likely to be commoner than realized, for, even assuming equal viability of the genotypes, the heterozygote of any recessive mutant whose visible homozygote is one in ten thousand individuals, will be nearly 2 per cent (see Hardy-Weinberg table).

Several lepidopterists have worked on the melanic and semi-melanic forms of *A. caja*; foremost amongst these are Wright and Smith (1956), who have worked on f. *fumosa*, with its variable heterozygote, f. *brunnescens*, at times indistinguishable from f. *typica*. This most interesting situation is discussed more fully in Chapter 16. Other melanic forms in *A. caja* give even less clear-cut results in regard to their origins. Stertz (1915) and Cuno (1932) bred such forms but their breeding results are inconclusive. Cockayne (1949a) has attempted to analyse these along with other melanic forms such as *clarki*, *nigropennalis*, *melanozosta*, and *obscura*, and favours an explanation involving genetic origins with lowered viabilities. In the light of our recent work I think this may be incorrect, for he has tended to ignore environmental effects on the genomes, in particular temperature-buffering.

Another aposematic species is *Panaxia dominula* with its homozygous melanic f. *bimacula* occurring at an extremely low frequency near Oxford, and with its recognizable heterozygote f. *medionigra* varying in frequency from 4 to 11 per cent. In the laboratory there is no evidence of lowered viability of the homozygote, and as stated in Chapter 16, f. *bimacula* can maintain itself in a pure form in artificial colonies. In one such experimental population (Elsfield), I released yellow (f. *lutea*) × red (f. *medionigra*, not heterozygous for f. *lutea*) in 1965. After five years, a sample of 56 gave the phenotype frequencies shown in Table 5.3.

TABLE 5.3
Phenotype frequencies of artificial colony of yellow lutea-medionigra × red medionigra *after five years.*

Year	Red f. *typica*	Yellow f. *typica*	Red *medionigra*	Yellow *medionigra*	Red *bimacula*	Yellow *bimacula*	Total
1970	5	12	11	25	1	2	56

This shows a marked deficiency of the homozygote *bimacula* of both colour forms but a significant excess of yellow mutants. This contrasts with two colonies of yellow *medionigra* founded six years ago in two differing types of ecology: a natural marsh maintaining wild comfrey plants (Southmoor), and a vallum in the grounds of a house where the comfrey had been introduced (Eynsham). In both these locations after five years, the 1:2:1 ratio has remained undisturbed (Eynsham, 1970, $n = 90$, and Southmoor, 1970, $n = 43$). As can be seen from Table 5.4, the viability, indeed the overall survival of the three genotypes appears to be equal; a quite surprising and unexpected result.

TABLE 5.4
Phenotype frequencies of an artificial colony in the wild (after five years)

	Yellow *typica*	Yellow *medionigra*	Yellow *bimacula*	Total
Eynsham	20	46	24	90
Southmoor	11	22	10	43

The aposematic Selidosemid *Abraxas grossulariata* L. has been extensively studied in Britain, U.S.A., and Russia. It has a number of melanic forms under the names of *varleyata, hazeleighensis, nigrosparsata, nigrotincta* (*nigra*), 'Poulton's sooty Oxford', and *aberdoniensis*; all are recessive. In the few broods in which the frequencies have been recorded, it appears that the number of melanic forms is deficient.

Ashwell (1953–4) tested the viability of a brood of 252 larvae containing genes for the melanic f. *varleyata*, the sex-linked f. *dohrni* (*lacticolor*), and the double recessive f. *exquisita*. That part of the brood which was overwintered under protected conditions indoors produced the expected proportions of the three forms along with f. *typica*. The survival rate of that portion subjected to harsh winter temperatures outdoors was 7 per cent, 15 per cent, and 0, respectively, compared with that of f. *typica*.

Apart from among aposematic species, Recessive Melanism also occurs, but rarely, in cryptic Lepidoptera. In back-cross broods of *Lycia hirtaria* f. *typica* × f. *nigra* the melanics have been below expectation in the majority, though this is not significant. Even in the recessive f. *olivacea* of *Lasiocampa quercus* ssp. *callunae*, which occurs at frequencies of up to 70 per cent in Caithness, the proportion of the melanic form tends to be lower in wild-caught imagines than in those bred in the laboratory from larvae collected from the same areas the previous year (see Chapter 14).

Another species with recessive melanics which has shown sub-viability is *Ennomos autumnaria* f. *schultzi*. Bretschneider (1936) states that from heterozygote pairings he obtained 508 f. *typica* to 112 f. *schultzi* (expected proportion 465:155), and that the melanic form has therefore less viability. This is probably correct, but the deviation from the expected proportion is not significant ($\chi^2_{(1)} = 3\cdot154$, $P > 0\cdot05$). A second, but less extreme melanic form in this species is f. *brunneata* named and discussed by Cockayne (1952*b*). Minnion (1957) recorded that whilst some broods maintained the normal 3:1 ratio, others showed a melanic deficiency of f. *brunneata*. A similar situation exists in the breeding of *Endromis versicolora* f. *lapponica* (Newman, L. H., 1943). The status of this form is discussed in

Chapter 13. A melanic form of *Selenia bilunaria* which showed melanic deficiency of the recessive mutant was recorded by Harrison (1935); in a back-cross brood the progeny were f. *typica* 92, melanic 1 (expected 1:1), and in another (heterozygote × heterozygote), they were in the ratio 85:1 (expected 3:1). One may anticipate then that even in protected laboratory broods, many other species containing recessive melanics will show viability differences. How much greater must be the survival differential in nature.

A second but comparable situation is frequently found in the homozygote portion of dominant melanics, for once again both genes carry the mutant form (see Chapter 1, *Heterosis*). Examples of this vary from the homozygote fraction being near lethal, as in *Biston strataria* Hufn. (Plate 5.18, 1 to 4) and f. *robiniaria*, to other species in which this class is present but deficient in numbers. A. pairing of *B. strataria* f. *robiniaria* x f. *robiniaria* produced 61 f. *robiniaria*: 32 f. *typica* (expected ratio 3:1). This seems to conform almost exactly to a 2:1 ratio, suggesting homozygote lethality (Cockayne, prsonal card index).

In *Phigalia pilosaria* at least two dominant melanic forms, f. *monacharia* (Plate 5.9, right 3 and 4) and '*intermediate*', exist and there is evidence that their melanic homozygotes are inviable. In three F_2 broods containing '*intermediate*' and f. *typica* the pooled data shows a significant deficiency of '*intermediate*' when tested against the expected 3:1 ratio but a close agreement with an expectation of 2:1 (Lees, 1971). It would therefore appear that 'intermediate' homozygotes are at a disadvantage to the heterozygotes, and that this disadvantage is greater than that of the typical form. It is not, however, possible to separate deleterious pleiotropic effects of the gene controlling the melanic form from close linkage to semi-lethal recessives.

Large F_2 broods containing the other melanic f. *monacharia* show no departure from an expectation of a 3:1 ratio, but small broods, where mortality in the larval stage has been greater, show a significant departure from 3:1, but an agreement with a 2:1 ratio. The simplest interpretation of these data is that homozygous f. *monacharia* is less viable than the heterozygotes when under some form of environmental stress, but not when such stress is lacking (Lees, 1971). In a similar manner, in *Biston betularia* many pairings from f. *insularia* × f. *insularia* produce a near 2:1 ratio. Also, as discussed later, nowhere in wild populations does the f. *carbonaria* frequency of the population of this species exceed 98 per cent (with one exception), even after 120 years of its presence as, for example, in Manchester. Nevertheless, I produce evidence in Chapter 14 which suggests that under exceptional circumstances even a homozygote recessive melanic form can be at an overall advantage.

Amathes glareosa has a variable melanic form *edda* which, until recently, was only recorded from Unst, the northernmost island of Shetland. Actually it occurs throughout this archipelago at varying frequencies, the island of Unst having the blackest individuals.

From breeding experiments I showed that f. *edda* was dominant to f. *typica*, and that the darker individuals from heterozygote pairings were the homozygotes. Unst is one of the main land-falls in the British Isles for birds migrating south in August, and here the scoring of 2539 f. *edda* out of over 12,500 caught suggested that homozygote *edda* formed 85 per cent of the population instead of the expected 68 per cent (see Chapter 11). This reflects the fact that the overall emphasis here may be that a perfect crypsis for the imago transcends physiological disadvantages; alternatively, that about 20 per cent of the heterozygote melanics on Unst have achieved full dominance for colour.

Back-cross broods of melanic species have been studied but few subjected to experimentation. Theory would suggest that the expected 1:1 ratio in those species which come into the class of Ancient Melanism would behave differently from their modern industrial counterparts. For the former have had thousands of years to adapt their physiological optima to their cryptic requirements. This implies that genes responsible for imaginal melanism must not produce disadvantageous effects in the larval or pupal stages. Furthermore, it could be argued that the recent industrial melanics have had insufficient time to produce such a balance. There is some, but minimal evidence that this is indeed true in both classes.

A number of different melanic forms of *Hemerophila abruptaria* (Plate 5.6 left 6 and 7) have been recorded (Harris, 1904, 1905, 1906, Hamling, 1905). All are dominant or semi-dominant. One of the earlier experimenters was Onslow (1921c), who worked on the melanic form of *Hemerophila abruptaria* f. *fuscata* Tutt which had dominance even by 1920. This form was found by him to have increased vigour under rigorous conditions. The combined progeny of three back-cross broods gave the offspring as shown in Table 5.5.

Brett (1935–6, 1937) worked on the same species but his data are complicated by a separate and less extreme melanic f. *brunneata* which he considered to be the more usual form at that date. It certainly could not have been a heterozygote of his darker f. *fuscata*. No viability differences were shown, probably because he did not subject his broods to stress. Cockayne (1949a) showed later that another doubtfully melanic but non-disruptive

TABLE 5.5
Survival rate of melanic and f. typica *under stress in the larval stage.*

Hemerophila abruptaria	f. typica	f. fuscata
Normal food	84	79
Semi-starved	19	37

$\chi^2_{(1)} = 5 \cdot 796$, $0 \cdot 05 > P > 0 \cdot 01$.

colour form of the species had a recognizable heterozygote f. *coarctata* whose homozygote f. *knightii* was distinct but sub-viable (f. *typica* 63, f. *coarctata* 108, f. *knightii* 33, with the expected 51:102:51 ratio).

Onslow (1919b) also worked on *Ectropis consonaria* Hübn. and its melanic f. *nigra* Bankes (Plate 5.9 left 5 and 6), but found no significant differential viability in his back-cross ratios. He showed however, that *Boarmia punctinalis* Scop. f. *humperti* (which Hasebroëk (1934) showed was dominant) was hardier than f. *typica*, and this was substantiated (though with no significance) by Walther (1927). Onslow also considered that f. *sericearia*, one of the melanic forms of *Cleora ribeata* Clerck (Plate 5.8, right 1 and 2) was more viable than f. *typica*. Williams (1933) described his breeding results with *Cleora rhomboidaria* and referred to the apparent hardiness of the melanic f. *rebeli* (Plate 5.6, right 3 and 4). Ford (1940a, 1945) tested the viability of *Cleora repandata* L. and a melanic form referred to, incorrectly I think, as f. *nigricata*. This name should be limited to the non-industrial form from Scotland. Because of its origin it should in fact be called f. *nigra* which is the present day industrial melanic. He divided back-cross broods into two groups, one of which had normal food, and the other of which was starved. In a total of 192 of the former, he bred 91 f. *typica* to 101 'f. *nigricata*', That portion of the brood subjected to starvation gave 31 f. *typica* to 52 'f. *nigricata*' in a total of 83 individuals (see below). There was a significant survival of the melanic form (see Table 5.6).

So far the evidence of differential viability occurring between melanic and *typica* forms in the various species is unconvincing, and the small quantity of data produced by myself on *Biston betularia* and its melanic form *carbonaria* have not taken us much farther. In the early 1950s I commenced to work on a small scale on the behavioural differences between these two forms. At this date, rather stupidly, I did not anticipate that there would be differences between f. *carbonaria* originating from an industrial area such as Birmingham and those resulting from out-crossing this form into f. *typica* from Devon and Cornwall where this form does not occur. That different gene-complexes had developed for each situation because

TABLE 5.6

Viability differences between Boarmia repandata
f. typica *and f.* nigra *(from the experiments of E. B. Ford).*

(a) Unstarved back-cross families of Boarmia repandata.
The first family represents the progeny of the wild female,
and contains the parents of the other two families.

Melanics	Pale forms	Totals
43	38	81
27	21	48
31	32	63
101	91	192

$\chi^2_{(1)} = 0.528, P = 0.05$.

(b) Starved back-cross families Boarmia repandata.

Melanics	Pale forms	Totals
3	0	3
18	12	30
10	6	16
4	3	7
17	10	27
52	31	83

$\chi^2_{(1)} = 5.8, 0.05 > P > 0.01$.

of differing ecological requirements had, I admit, not occurred to me then. Rather I was concentrating on the pleiotropic effects of the *carbonaria* gene regardless of its known genetic environment.

In six back-cross broods of industrial origin (Birmingham), in which there was an extremely high mortality rate, I bred 65 f. *typica* and 108 f. *carbonaria*, which had been fed on unwashed Oxford Sallow augmented by spraying small quantities of air pollution extracted from filters in London (Kettlewell, 1958a). The significant deficiency of f. *typica* (see Table 5.7) could be accounted for either by their intolerance to toxic food, or by polyhedrosis which infected my stock, or a combination of both. I compared these figures with the earliest back-cross broods I could find recorded between 1900 and 1905. The combined results of four broods bred by separate lepidopterists was 255 f. *typica* to 217 f. *carbonaria*. The stock originated from London, where f. *carbonaria* at that date was at a low frequency, probably around 40 per cent. Because of these results, I put forward the suggestion that it was possible that at that time the gene-complexes may have been at a physiological disadvantage for f. *carbonaria*,

Table 5.7

A comparison between early and recent back-cross broods of Biston betularia *segregating for f.* typica *and f.* carbonaria *showing a deficiency of the two forms in opposite directions.*

(a) Earliest broods recorded 1900–1905

typ.	carb.	Total	Breeder
123	109	232	Bacot
57	47	104	Main and Harrison
18	11	29	Fletcher
57	50	107	Harrison
255	217	472	carb.=46·6%

(b) Recent broods 1953–1956

typ.	carb.	Total	Breeder	Brood no.
14	22	36	H.B.D.K.	B/2/52
7	10	17	H.B.D.K.	19/53
1	5	6	H.B.D.K.	30/53
28	30	58	H.B.D.K.	15/54
1	2	3	H.B.D.K.	19/54
14	39	53	H.B.D.K.	11/54
65	108	173	carb.=62·4%	

(c) 2 × 2 Table

typ.	carb.	Total	Year
255	217	472	1900–1906
65	108	173	1953–1956
320	325	645	

(a) $\chi^2_{(1)} = 2\cdot 90$; $P = 0\cdot 089$
(b) $\chi^2_{(1)} = 10\cdot 20$; $P = 0\cdot 0014$
(c) $\chi^2_{(1)} = 13\cdot 27$; $P = 0\cdot 0003$

and that only after a considerable period of time did this form acquire enhanced vigour. In 1955 I repeated the experiments using a large brood (B/1/54) of f. *carbonaria* × Devon f. *typica* female (where f. *carbonaria* does not occur). The larvae were fed on unwashed Oxford Sallow (not augmented with London pollution). Of the 976 individuals bred, 486 were f. *carbonaria* and 490 f. *typica*.

Subsequently, Clarke and Sheppard (1963) produced figures from Cheshire–Lancashire stock which also showed equal survival of the two genotypes (151:150). Robinson (1971) quotes other back-cross figures taken from the literature.

Table 5.8

Phenotype frequencies from four other back-cross broods (from Robinson, 1971)

Breeder	Year	f. typica	f. carbonaria	Total
Miller	1910	72	72	144
Steinert	1892	75	90	165
Gerschler	1915	149	125	274
Walther	1927	48	56	104

PLATE 5.8 (p. 63). Examples of industrial melanic species (× 1).

Left
1. *Apatele aceris* L. f. *typica*.
2. *Apatele aceris* f. *candelisequa* Esp. (syn. *infuscata* Haw.)
3. *Apatele alni* L. f. *typica*.
4. *Apatele alni* f. *steinerti* Caspari.
5. *Apatele leporina* L. f. *typica*.
6. *Apatele leporina* f. *melanocephala* Mansbridge.

Right
1. *Cleora ribeata* Clerck f. *typica*.
2. *Cleora ribeata* Clerck f. *nigra* Ckyne.
3. *Miana literosa* Haw. f. *typica*.
4. *Miana literosa* Haw. f. *aethalodes* Richardson.
5. *Abrostola tripartita* Hufn. f. *typica*.
6. *Abrostola tripartita* f. *plumbea* Ckyne.

PLATE 5.9 (p. 63). Examples of industrial melanic species (× 1).

Left
1. *Apatele megacephala* Schiff. f. *typica*.
2. *Apatele megacephala* Schiff. f. *nigra* Shaw.
3. *Antitype chi* L. f. *typica*.
4. *Antitype chi* L. f. *olivacea* Stephens.
5. *Ectropis consonaria* Hubn. f. *typica*.
6. *Ectropis consonaria* Hubn. f. *nigra* Bankes.
7. *Erannis defoliaria* Cle. f. *typica*.
8. *Erannis defoliaria* Cle., a melanic form, possibly f. *obscurata*.

Right
1. *Cleora ribeata* Clerck f. *typica*.
2. *Cleora ribeata* Clerck f. *nigra* Cockayne.
3. *Phigalia pilosaria* Schiff. (syn. *pedaria* Fab.) f. *typica*.
4. *Phigalia pilosaria* Schiff., a melanic form, possibly f. *monacharia* Stdgr.
5. *Boarmia roboraria* Schiff. f. *typica*.
6. *Boarmia roboraria* Schiff. f. *melaina* Schiff.

PLATE 5.10 (p. 64). *Brachionycha sphinx* Hufn. ($\times 1\frac{1}{2}$) and its melanic form.

PLATE 5.11 (p. 64). *Brachionycha sphinx* Hufn. f. *typica* ($\times 1\frac{1}{2}$) at rest.

PLATE 5.12 (p. 64), *Brachionycha sphinx* Hufn. ($\times 1\frac{1}{2}$) melanic form at rest.

PLATE 5.13 (p. 70). *Ectropis consonaria* Hueb. f. *nigra* ($\times 1$) at rest on Beech trunk, Stroud, 1957.

PLATE 5.14 (p. 67). *Biston stratoria* Hufn. ($\times 1$) at rest on lichened trunk.

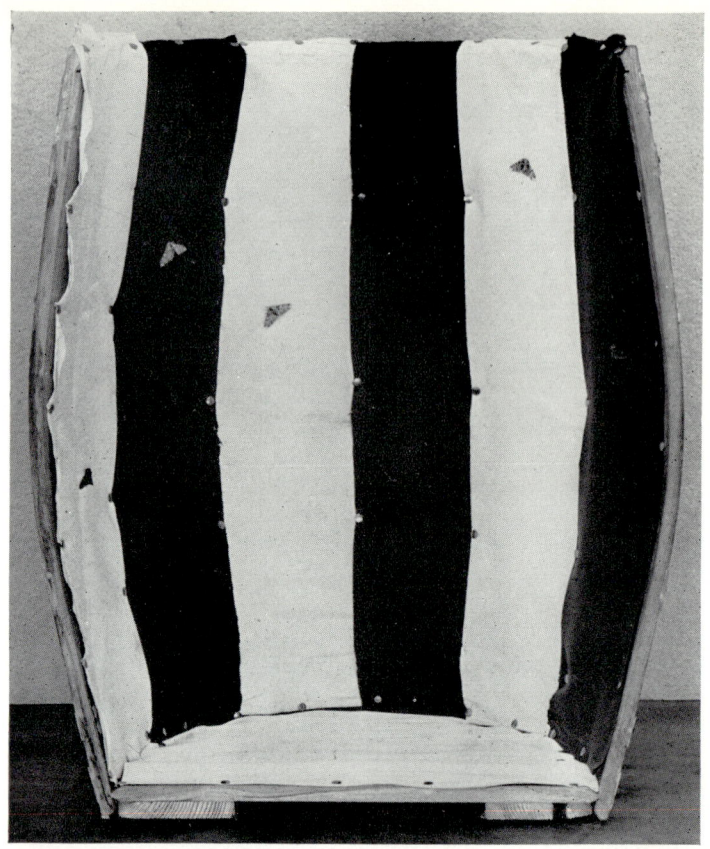

PLATE 5.15 (p. 69). Method of testing choice of background (barrel experiment).

PLATE 5.16 (p. 88). Experimental tree-trunk used for testing background choice (1972).

He concludes, and I think correctly, 'It is perhaps not legitimate to compare British and European populations too closely since different polygenic complexes may be at work and events may take slightly different courses.' The same applies for stock from different parts of Britain taken at different times in the past. Until broods from contrasting areas in Britain are bred on a large scale under stress (for example fed on leaves contaminated with the various types of pollution, starvation, polyhedrosis, etc.), we shall not know whether f. *carbonaria* survives better under certain conditions than f. *typica*. A reflection on this is that the rate of spread of the *carbonaria* form throughout polluted Britain could be accounted for entirely by the cryptic advantage of the imago alone.

In the course of this work an interesting side issue appeared, namely that there is a significant deficiency of f. *carbonaria* males in the majority of

TABLE 5.9
Sex-ratios of Biston betularia *f.* typica *in 18 broods.*

Brood	Male	Female
B/2/52	4	10
B/4/52	32	22
B/5/52	13	9
B/6/52	5	7
B/7/52	41	50
B/10/52	11	16
B/11/52	9	7
B/12/52	21	23
B/13/52	24	38
W/1/52	94	91
W/2/52	16	21
B/4/53	5	1
B/19/53	3	4
B/1/54	214	276
B/4/54	37	12
B/11/54	9	2
B/15/54	14	14
B/20/54	6	6
	558	609
Total		1167

broods ($P > 0.001$); this is also the case, but to a lesser extent, in f. *typica* ($P > 0.5 > 0.25$ (see Tables 5.9 and 5.10), and it is also observed in larger broods, which are usually those which did not suffer from polyhedrosis. Table 5.10 shows the sex ratio of 2886 f. *carbonaria* from 24 broods. The corresponding figures for 1167 f. *typica* recorded from 18 broods (Table 5.9)

TABLE 5.10
Sex-ratios of Biston betularia
f. carbonaria in 24 broods.

Brood	Male	Female
B/1/52	7	6
B/2/52	15	7
B/10/52	21	37
B/11/52	9	7
J/1/52	177	251
T/1/52	35	25
W/2/52	31	34
B/4/53	8	7
B/19/53	7	3
B/30/53	0	5
B/1/54	221	265
B/2/54	435	541
B/3/54	107	142
B/4/54	24	20
B/6/54	11	17
B/8/54	24	12
B/9/54	0	6
B/10/54	26	32
B/11/54	21	18
B/12/54	13	10
B/15/54	19	11
B/19/54	0	2
B/20/54	3	3
J/2/54	106	107
	1312	1574
Total		2886

were 558 males and 609 females. Clearly in laboratory bred stock there is a deficiency of males, particularly amongst f. *carbonaria*. This is well illustrated (see Table 5.11) in the largest brood I have ever bred, to which I have referred previously.

TABLE 5.11
Sex distribution of f. typica and f. carbonaria in a brood
(B/1/54, ex f. typica ♀ Devon × f. carbonaria ♂.

Brood	f. *carbonaria* male	female	f. *typica* male	female	Total
B/1/54	221	265	214	276	976

Here there were over 100 more females than males, yet on theory it is in broods with smaller numbers that differential survival is likely to be more marked, and in those in which a high mortality rate has taken place for one reason or another, as is usual in nature.

Behavioural differences

Pheromone. No other single character could contribute to the spread of a mutant form more rapidly than that of a male's scent-difference when it is more acceptable to a particular form of the female of the same species. That certain scent molecules are, because of their chemico-physical qualities, more successful than others, is demonstrated by the number of different species of Lepidoptera, frequently in widely separated genera, which use the same pheromone (Kettlewell, 1955*f* and 1956*g*.). It is interesting to see how the mating behaviour of two such species is geared

TABLE 5.12
Female assembling differences between f. typica and f. carbonaria in a population maintaining f. typica only. Biston betularia.

Date	f. *typica* males to f. *typica* females	f. *typica* males to f. *carbonaria* females
Warm (above 15°C)		
14.6.55	2	7
16.6.55	1	8
29.6.55	0	7
Total warm	3 (8%)	22 (92%)
Cold (below 10°C)		
17.6.55	10	5
20.6.55	4	4
28.6.55	10	5
30.6.55	6	1
Total cold	30 (67%)	15 (33%)
Cool (10°–15°C) 7 nights	21	38
Total all weathers	54	75

so that they do not overlap in their courtship times; for example, in the pair *Arctia villica* L. and *Parasemia plantaginis* L., the former is nocturnal (11 p.m.) and the latter diurnal (4–6 p.m.). Only by incarcerating females, thereby preventing copulation, can such instances be discovered.

Sheppard (1951) and Sheppard and Cook (1962) demonstrated convincingly that under caged conditions disassortative mating took place in *Panaxia dominula* between f. *typica* and f. *medionigra*. Having eliminated visual choice by painting out wing markings, such preferences must be attributable to scent/behaviour differences. In 32 of 46 instances in which a single male f. *typica* was placed with a female f. *typica* and a female f. *medionigra*, it paired with the f. *medionigra* female. In 32 of 51 tests with a female f. *typica* in the presence of male f. *typica* and male f. *medionigra*. disassortative mating again took place. By contrast when a male f. *medionigra* had the choice of either of the two female genotypes, random pairing took place. Similar results were obtained when female f. *medionigra* were tested with the two male genotypes. This suggests that the final decision is one of female choice, and that in each instance she prefers the male of the opposite form to that of herself. It also appears that f. *medionigra* of either sex has a mating advantage over f. *typica*.

Using entirely different methods, in 1954 I carried out a small series of tests using sib females of two forms of *Biston betularia* both of the same day of hatching. These were caged separately in muslin bags which were placed three to five yards apart at right angles to the direction of the wind. The experiments were carried out in a woodland in Dorset where the Peppered Moth population was 100 per cent f. *typica*. The number of males attracted to each genotype was recorded nightly. Sampling suggested that the nocturnal temperature influenced the results (Table 5.12), more males being attracted to f. *carbonaria* females on hot nights than on cold ones. We are planning to repeat these and other similar tests of mating preferences in the near future. It is indeed surprising that so little work has been undertaken so far in a field likely to be so rewarding.

Speed of larval feeding. The rate of larval growth is in some species strictly under the control of environmental conditions, in particular temperature and daylight period. In others it is controlled genetically, and different individuals may even pass a particular season of the year in different stages (for example *Pararge aegeria* L.). One of the major behavioural differences between *Biston betularia* in Britain and *Biston (Amphidasis) cognataria* in North America is that *B. betularia* is always univoltine and some individual larvae may continue feeding late into the autumn before attaining full growth. By contrast, in North America, *B. cognataria* may be bivoltine in parts of U.S.A. whilst, farther north, Canadian stock feed up with great rapidity and all pupate simultaneously. This no doubt reflects the rapid change from summer to winter which occurs in that continent, compared with its gradual onset here (in Britain). It was therefore interesting to see whether melanic genotypes behave differently as larvae from those of f. *typica*. In 1957 I attempted to test this by dividing pupae of a back-cross brood of B. betularia according to how early or how late each had pupated. The stock was derived from an industrial (Birmingham) source.

There was some evidence that those larvae which fed up rapidly and pupated first, produced a higher proportion of f. *typica*, whilst those that fed slowly and pupated late in the summer were mostly of the *carbonaria* form. I was able to check this by segregating all larvae that pupated at fortnightly intervals. Table 5.13 shows one such example. In regard to one brood (B/4/52), after the first two pupal containers (which produced 35 f. *typica* and 16 f. *carbonaria*) had been collected in August, mice ate the main one, and the third group, still feeding in October, produced five moths, all f. *carbonaria*. If a differential feeding rate exists between the two forms, it can be accounted for by two alternative hypotheses: first, that selection has favoured those f. *typica* which feed up rapidly, thereby avoiding the heavier pollution which takes place late in the year; or, secondly, f. *carbonaria* may be capable of getting rid of toxic substances. Slow feeding and a capacity for excreting noxious materials has been demonstrated in instances outside the Lepidoptera.

TABLE 5.13
Biston betularia. *Speed of larval growth. A comparison between f.* typica *and f.* carbonaria. *(Brood B/4/52)*

Pupation date		f. *typica*	f. *carbonaria*
Earliest	1 August	9	1
Early	8 August	26	15
Main fraction	September	?	?
Late	October	0	5
Laboratory controls	August–October	19	17

In 1954 I repeated pupal segregation in a large brood B/1/54, which I have previously referred to, and which was f. *carbonaria* (from Birmingham) × f. *typica* (from Devon). There was no corroboration of the earlier results obtained from pure Birmingham stock (120 Earliest f. *carbonaria* 92 f. *typica*, 103 late f. *carbonaria*, 111 f. *typica*). Only later did I realize that by introducing gene-complexes from Devon, I was indeed unlikely to obtain similar results.

In undertaking experiments of this kind it is important that one is sensitive to the origin of the stock used. Results so far suggest that differences are not pleiotropic, but the effect of secondary polygenic selection associated with each genotype.

In regard to sampling wild populations I can provide no evidence that an industrial melanic form changes its frequency during the flight period. However in some species which come under the heading of Ancient Melanism this does occur. For *Amathes glareosa* I give evidence (Chapters 11 and 12) that in Shetland (both on Unst and Fair Isle) f. *typica* increases in the population later in the season possibly after the main southern migration of birds is over. In the Black Wood of Rannoch in the Scottish Highlands where *Cleora repandata* occurs with its black form *nigricata* Fuchs, melanics rapidly increase their frequency in the course of their short season. This is well shown from the observations of Cadbury (see Chapter 10, Table 10.1, and Fig. 10.1) and confirmed by myself in 1968. Over a period of a few days the melanic frequency increased from 2 per cent to 10 per cent. In both these instances the shift in phenotype frequency could be the end result of differential visual predation and is not necessarily due to a differential emergence.

(vi) The genetics of Industrial Melanism

Over 90 per cent of industrial melanics are controlled by a single major gene, and the melanic character is inherited as a Mendelian dominant; but rarely today do intermediate forms occur. The genetics of many industrial melanics have been determined: for example, that of *Gonodontis bidentata* Clerck f. *nigra* Prout by Bowater (1915), (Plate 5.6, left 4 and 5), that of *Hemerophila abruptaria* Thunberg f. *fuscata* Tutt by Onslow (1921c) (Plate 5.6, left 6 and 7), and that of *Cleora rhomboidaria* Schiffermueller (=*gemmaria* Brahm) f. *rebeli* Aigner by Williams (1932–33) (Plate 5.6, right 3 and 4); all have complete dominance today. The genetics of the melanic forms of *Biston betularia* come under this heading, of single gene control, in which each form has dominance to f. *typica*. This is discussed in Chapter 7.

Other mechanisms, however, have been found to promote darkness. Multifactorial inheritance is found in the moth *Lymantria monacha*, the darkest forms of which are controlled by three genes, two autosomal and one six-linked (Plate 5.19). All are dominant, and each confers a degree of darkness. These genes do not occur simultaneously in Britain, though they do in Germany (Goldschmidt, 1921). The geometrid moth *Oporinia dilutata* Schiffermueller produces a series of varying melanics, and it is stated that these are controlled multifactorially (Harrison, 1920). Multiple allelomorphs are responsible for the melanic forms of the European moth *Aglia tau* L. (Standfuss, 1910a and b) as well as the Tortricid *Acleris*

(=*Acalla*) *comariana* Zeller (Fryer, 1928, 1931). More recently, Turner (1968) has limited melanism in this species to one form *fuscana* Sheldon, of recessive inheritance, though he concedes some doubts as to whether the locus of this is the same as that of the other morphs.

Female inheritance is found in the Geometrid moth *Phigalia pilosaria* Fab. (Bretschneider, 1939). In this species the female is apterous, and one type of inheritance is the normal Mendelian, in which both sexes can be black. There is another, however, in which black female *P. pilosaria* beget melanic females only, the males remaining f. *typica*. This character could therefore be carried either in the Y chromosome or in the cytoplasm; at present we do not known which. The genetics of *P. pilosaria* has been extensively investigated by D. Lees (1971). He has shown that genetically different melanic mutants occur. These are phenotypically distinct in the males and some have, in the past, in fact, received different names. He is at present testing whether they are allelic or not. They are likely to be so, and therefore be similar to the *insularia* complex in *Biston betularia*.

Polygenic inheritance has been overlooked in the past. Nevertheless, there is circumstantial evidence that it takes place. The Geometrid moth, *Gonodontis bidentata* Clerck has a dominant black form *nigra* (Bowater, 1914a). This species does not fly very actively and the genes therefore spread slowly. This situation is at present being studied by the group at Liverpool University (C. A. Clarke, P. M. Sheppard, and J. Bishop) in which district f. *nigra* is found locally at varying frequencies. At Cannock outside the city of Birmingham, over 50 per cent are of this melanic form; but rarely does the frequency exceed this figure (75 per cent in Leeds) and homozygous *nigra* are likely to be at a disadvantage in nature though Bowater's broods, bred indoors, did not show this (1914). In the centre of Birmingham, however, this melanic mutant does not occur. It is probable that the primary mutation has not taken place. Form *typica* is found here, but is very much greyer and darker than that found in the rural *G. bidentata*, and there is considerable variation among the individuals themselves (Kettlewell 1959d). The same situation is found in a large number of Noctuid moths, which have gradually been getting darker in the last 50 years. Polygenes and modifiers are constantly at work affecting the phenotypes of both melanic and typical individuals.

A comprehensive assessment of the genetics of Lepidoptera, including those of melanism, up to the year 1936 is given by Ford (1937). This was probably the first attempt, and a highly successful one, to bring together all that was known on the inheritance of mutant forms at that time. In particular he draws references from innumerable papers by E. A. Cockayne,

H. B. Williams, R. Goldschmidt, H. Onslow, J. L. F. Fryer, W. Bowater, J. W. H. Harrison, and others.

Apart from referring to species in the text whose genetics have been elucidated since then, I discuss separately (in Chapters 7 to 17) others of particular interest on which we have concentrated. The list includes the following:

Biston (Amphidasis) betularia L. and *cognataria* Gn.
Lasiocampa quercus L. and ssp. *callunae* Palmer.
Panaxia dominula L. and its subspecies.
Amathes glareosa Esp.
Papilio species.
Argynnis paphia ssp.
Polia nebulosa Hufn.
Spilosoma lutea Hufn.
Arctia caja L.
Aglia tau L.
Cycnia mendica Clerck.
Xylomyges conspicillaris L.

NOTE (see p. 70). August 1972. The experimental impasse in which I have previously been has resulted from having to test flying moths constrained in containers on the inside of which they have to take up day-time positioning. Using females only of *Biston betulari* which do not normally fly, we have now developed a different and more satisfactory design of experiment. A heavily lichened trunk or bough is denuded of its Bryophytes on one side, which is then darkened by painting with soot suspension. The trunk is placed upright with its axis in a north/south direction (Kettlewell and Conn, in press). Equal numbers of the two female morphs were released on to the mid-line, the whole trunk being covered by a muslin tent. Of 102 females so released, 35 (34 per cent) were not included as they took up position on the muslin. Of the 67 which were scored on the two halves of the trunk, all 31 f. *carbonaria* (100 per cent) were resting on the black background. Of the 36 f. *typica*, 27 (75 per cent) were resting on the lichened half. This is a highly significant result $\chi^2_{(1)} = 38 \cdot 99$ (significant at 0·1 per cent)) and demonstrated that female *betularia* are able to choose the background on which they sit, having regard to their own colouration. It would therefore appear that future laboratory experiments must be concentrated on certain species (these will usually be females) which do not fly but use their wings for concealment only.

6. The world distribution of Industrial Melanism

AT the present time, Industrial Melanism is found in many entirely unconnected places in the world, and, in each instance, it occurs in cryptic species only. In Britain, which was the first country to record it, it occurs today not only in all industrial areas, but has spread far beyond, usually in an easterly direction following the prevailing winds. In *Biston* (=*Amphidasis*) *betularia* L., the Peppered Moth, the best documented species, it has been shown that the clines to the west are sharp, while to the east they are long and gradual. Evidence has been provided that this distribution pattern is an indirect result of the prevailing south-westerly wind, which, day and night, carries for great distances the lighter portions of air pollution —smoke and SO_2—from industrial areas toward the east (Kettlewell, 1955a). I shall here give a résumé of melanism in Europe and also in North America.

(i) Europe

Similar situations to those in Britain exist in and around most industrial areas throughout the Northern Hemisphere. In Europe, Industrial Melanism was recorded in the Rhine Valley about the same time as in Britain, and a history of its spread can be obtained from Gerschler (1915), Hasebroëk (1934), Walther (1927), and Heydemann (1927–28). Lempke (1959, 1960) has recorded and named some of the many melanic forms that occur commonly in the Netherlands, and Warnecke (1913) those from around Hamburg, N. Germany. Many of these, though occurring in the same cryptic species as found in Britain, are phenotypically distinct from our own melanics. It would appear that in an area as small as Europe, though the same species are involved, different mutants have spread through each population. In the highly industrialized districts of eastern Europe, in Yugoslavia, Czechoslovakia, and Poland, Industrial Melanism is again widespread, and in 1965 I had an opportunity of seeing some of these in Czechoslovakia. More recently in 1972, Dr. R. Schummer of Humboldt University, Berlin, in East Germany has been kind enough to send me some current literature on melanism in Poland. Drozda (1970) has published an extensive list of species together with their named melanic mutants; there are also eight pages of plates showing melanism in 23 species. His paper provides evidence that this phenomenon is widespread in this area of eastern Europe.

That some of the black forms which have spread are genetically dissimilar from our British ones is even more evident here. A good example is the melanic form of *Hyloicus pinastri* L. (?ab. *brunnea* Spuler, ?ab. *nigrescens* Lpk.) which is common throughout Czechoslovakia. I have two examples in my possession and I am assured that this form is inherited as a simple dominant; they are indistinguishable, in fact, from our extremely rare ab. *unicolor* Tutt which in Britain is recessive and certainly cannot be considered an industrial melanic. Whether such instances reflect distinct mutants or different expressions in varying gene-complexes we, as yet, do not know.

In order to be able to compare the types of melanism, I have concentrated on two contrasting areas of Europe, both maintaining some species in common with Britain. First, Scandinavia, where Non-industrial Melanism, comparable to that found in northern and western Britain, overlaps with the earliest commencement of Industrial Melanism. Secondly, I give a résumé of melanism in the Netherlands, where, though only approximately 70 miles from the shores of Britain, and though similarly industrialized, melanism has spread in a different way—certainly in regard to the time factors and I think also in the selection of different mutant genes. For these records I am indebted to B. Lempke who has been good enough to provide them. These can be compared with melanic forms found in Britain, given in Appendix B. France and the Iberian Peninsula I have not included because of the paucity of records. For Germany, so rich in melanic forms, I am able to give certain references only, these owing to the help of Skat Hoffmeyer and Warnecke†, because the records are so thinly spread in so large a volume of literature. A study of melanism in the same species of Lepidoptera in these different areas would indeed be a worthwhile undertaking; also a comparison of these with their confrères in North America (see Synthesis, Chapter 19).

Scandinavia

It is impossible to separate Industrial Melanism from other types of melanism in this area of Europe. The reasons for this are first that, with the exception of a few centres (and excluding Denmark), there is not the same degree of pollution fall-out as occurs in many other European countries. Secondly, Non-industrial Melanism (see Chapter 10(**ii**)) is widespread

---

† R. Schummer has recently referred me to Von Ernst Urbahn's publication with photographs (Urbahn, 1972).

(Stockholm lies around latitude 60°N, the same as that of the Shetland Isles). Thirdly, in Finland and elsewhere in Scandinavia, indigenous pine forests occur where, in theory, melanism is always likely to occur. Nevertheless, though confused in Denmark, two zones are apparent in Sweden and Finland. The northern one between latitude 60°N and 70°N, where Ancient Melanism, in particular Northern Melanism, is widespread, and a more southern one, where Industrial Melanism is now commencing around urban centres.

Denmark. Melanism in Denmark is complicated because, as in Britain, Industrial Melanism is widespread, but merges into the various types of melanism of non-industrial origin. Hoffmeyer in his excellent publications (1948, 1949), records many industrial forms of Danish Lepidoptera but figures few of them. Unfortunately, I have no evidence of the frequencies, still less frequency-changes. The proximity of northern Germany where so many industrial melanic forms are available has no doubt contributed to their presence in Denmark.

Apart from the incidence of Industrial Melanism, this country shares many non-industrial forms in common with Britain. Thus the black form of *Amathes glareosa*, f. *edda*, was thought to be confined to the islands of Shetland and Orkney where we have investigated it extensively for many years (see Chapters 11 and 12). Recently, an apparently identical polymorphic situation has been found locally at Fanø. Also at least one recessive melanic form of *Lasiocampa quercus* occurs in Denmark as a polymorphism, probably not dependent on industrialization. There are in fact differences between it and the three separate (and genetically distinct) polymorphisms which occur in Britain (Chapter 14). Hoffmeyer records that melanism appears in all its stages, ova, larvae, cocoons, and imagines, a situation unknown in this country.

Sweden. Our knowledge of melanism in Sweden is at present quite inadequate. This is unfortunate because it holds, I believe, the answers to so many earlier but non-industrial melanisms. For this reason I was fortunate in obtaining the help of Dr. Guy Howard who sampled the Lepidoptera in Sweden in liaison with me from 1966 to 1968 using mercury-vapour light traps. He also made an extensive study of the Scandinavian literature and examined several of the larger collections. The main sources of information were Nordstrom's standard work (1941) and Hoffmeyer's volumes. Though the list is incomplete (Geometrids largely omitted), nevertheless, it adds considerably to our knowledge of melanism.

I have also gained further information from Dr. Douwes (personal communication) who, with Bjorn Petersen, is undertaking a survey of melanism at the University of Lund. I give a résumé of their records, in particular I compare certain species which differ from those in Britain in regard to their black forms.

Instances of melanism due to the effects of air pollution, no doubt carried from afar, are seen in such species as *Biston betularia* and *Dasychira pudibunda*. Douwes states, 'In both species the frequencies of dark forms is high in southernmost Sweden and eastern Denmark (around latitude 56°N) and decreases rapidly northwards.' This distribution coincides with the main concentration of industrial centres in Sweden and it is also the shortest distance from north Germany which is not only the source of so much air pollution but also a reservoir of industrial melanic forms. Hoffmeyer records the fact that *D. pudibunda* f. *concolor* (Plate 5.7, left 5 and 6) first occurred in Denmark at the end of the last century, but it appears to have been recorded from southern Sweden only in the last 20 years. Douwes states that a similar situation with a southern increase of melanic forms exists in the following species: *Lymantria monacha, Procus strigilis, P. latrunculus, Catocola fraxini, Oporinia autumnata, Dysstroma citrata, C. coenulata, Chloroclystis rectangulata, Boarmia repandata*, and *B. roboraria*. All these then are potential industrial melanics.

Such a north/south increase is in contrast to the majority of species in Sweden which more usually increase their melanic frequencies the farther north they are sampled and these must be classified under the heading of Northern Melanism (Chapter 10(ii)) as is found in northern England and Scotland.

The following Swedish Lepidoptera appear to have northern melanic forms in common with Britain. The three species of *Tethea, T. duplaris*, L., *T. or* Sch., and *T. octogesima* Hüb, but in each the melanic is somewhat darker in Sweden. *Achlya flavicornis* L., f. *finmarchica* Sch. is phenotypically similar to f. *scotica* Tutt, the Scottish race. Recently, in 1969, an all-black form of this species, f. *pseudoalbigensis* Franz (Plate 6.1), has appeared in Dutch Limburg, which must be considered an industrial melanic. It may be the same as that recorded from near Newcastle by Heslop Harrison (personal communication) many years ago.

There are a number of species with non-industrial melanic forms in Sweden which are absent from or exceedingly rare in Britain. Thus *Endromis versicolora* L., the Kentish Glory, whose males fly in sun in the spring, has a dark, smoky form in Lapland. In spite of constant inquiries I an unable to ascertain whether f. *lapponica* Bau. is a race, a polymorphism, or an occasional aberration. The same form has on one occasion been bred from inbred stock from Scotland (Newman, 1943). Further breeding showed that f. *lapponica* is recessive to f. *typica*.

In Britain, the four wings of *Cerura vinula* L., the Puss Moth, are pale silver-grey in both sexes with the forewings speckled and lined with darker markings (=f. *minax*). In southern Sweden females only have dark hindwings (=f. *vinula*), whilst f. *phantoma*, with all the wings blackened, occurs in northermost Sweden. Douwes (personal communication) states, 'all specimens I have seen taken north of latitude 65°N ($S = 7$) belong to f. *phantoma* and all specimens south of that latitude ($S = 50/100$) are normal'. (However, I would say that the females probably have black hindwings.) This species hatches in mid-summer and though normally nightflying must at that latitude at best take wing in twilight when light-coloured Lepidoptera are much more conspicuous than dark ones.

Federley (1943) has pointed out that in his opinion f. *phantoma* (syn. *arctica*) is likely to be inherited polygenically. An alternative and I think more compelling explanation could be that this is another example of incomplete dominance, as in *A. glareosa*, in which the expression of a major gene varies according to the local requirements. In *C. vinula* the situation is complicated in that Federley (1945, 1953) has shown that the diploid chromosome number varies from 42 (Finland) and 40 (Albania) to 62 (Morocco). If this is corroborated, the validity of the present taxonomic classification must be questioned (Robinson, 1971). *Phalera bucephala* L., the Buff Tip, is darker throughout Sweden and markedly so in central districts (=f. *tenebrata*). Other species with melanic forms in the north (in Lapland), though we know not whether they are polymorphisms or polygenic races, are: *Phoesia gnoma* Fab. f. *frigida*, *Notodonta ziczac* L. f. *lappona* Dix., and *Odontosia carmelita* Esp. f. *nocturnalis*. This last form is similar to the Scottish race, *nocturnalis* Stichel.

All the species so far referred to are cryptic, and with the exception of male *E. versicolora* they fly by night. Rather different is the case of the supposedly aposematic and certainly distasteful insect *Phragmatobia fuliginosa* L. This moth, in most of Europe including southern England, has reddish-brown forewings, with crimson abdomen and inner half of hindwings, and when flying by day, as it does, it is is conspicuous. Throughout Sweden and also in Scotland the dusky form *borealis* Stdg. replaces f. *typica*. I have only recently had the opportunity of seeing this northern race in action. During flight as it buzzes over heather it is extremely difficult to follow. When caught it goes into a state of catalepsy and exposes its red abdomen and at this somewhat belated hour shows its warning colouration.

This species demonstrates how both cryptic and warning colouration can be of advantage in the same individual in association with a particular behaviour pattern. In the north f. *borealis* is an inhabitant of heather moors and here gulls are one of the most voracious insectivores. Gulls appear to disregard warning colouration (see Chapter 16(ii)) and in their presence, crypsis must be the only successful mechanism of survival. It is interesting to note that *borealis* occurs as an occasional mutant on the southern heaths of Hampshire and Cornwall, where the Black-headed Gull, *Larus ridibundus*, is again common. It may be that similar advantages were responsible for two melanic specimens of *Arctia caja*, which is normally a rare mutant, that were bred from a sample of larvae taken on a small gull-infested island in a loch in central Sweden (G. Howard, personal communication).

One of the most surprising species to exhibit melanism in Scandinavia is *Simyra albovenosa* Goeze (Caradrinidae) (Plate 6.2). It is an Apatelid, yet the imago, on superficial examination, appears to be a typical Leucanid species; all rest by day on the stems of reeds (*Phragmites*) and grasses (*Gramineae*) and are pale-straw coloured. *S. albovenosa* is an excellent example of convergent adaptation. The near relative *S. henrici* Grote replaces it in North America. In Britain, only *S. albovenosa* f. *typica* occurs. In Scandinavia, around latitude 60°N, a dark or black form *murina* is present as a polymorphism, in particular on the archipelago of islands in the Baltic; the distribution of f. *murina* is from the northernmost tip of the Baltic, Salmia, through southern Finland to southern Sweden and Denmark. The species probably occurs on all the islands including the larger ones, for example Bornholm. De Worms (1958) records f. *murina* on a peninsula bordering the Baltic near Twärminne in southern Finland and states that this melanic, 'is almost the prevalent form'—presumably therefore around 50–60 per cent. The Hellman collection in Helsinki, and others, confirm that a high proportion of this form is taken in the northern Baltic, particularly in southern Finland, but it seems to be less frequent in the south (at Österslov, Ahus, and Krankesjön). It would be most instructive to sample the Baltic islands in order to find out whether a cline exists similar to that in *Amathes glareosa* in Shetland. It is likely that heavy predation by gulls is responsible for maintaining the polymorphism at this latitude, comparable to that in *A. glareosa*, though Howard saw no evidence of this.

Howard reports that there is considerable variation in the degree of melanism in f. *murina* and in fact suggests 'continuous variation'. More likely is it that the heterozygote is variable, but always distinguishable as

f. *murina*, in the same way as that of *A. glareosa* f. *edda* (Chapters 11 and 12).

In North America *S. henrici* Grote, in common with most *Apatele* species there, has a melanic form *fumosa* Morrison which is comparable with f. *murina*, but I have no knowledge of the natural history of this species.

Finland. Professor Esko Suomalainen has been good enough to give me the following information about melanism in Finland. He examined the material in the Helsinki Museum (entomological) and also 25 private collections, and lists eleven species with clear-cut melanism (personal communication)—presumably polymorphic:

Biston betularia L. f. *insularia* Th.-Mieg, 1965, Mäntyharju. 'The only melanic specimens have been taken in industrial centres there.'

Tethea or Schiff. f. *albingensis*. 'Five in all, earliest 1958. Helsinki (1) and Pvärminne.'

Apatele rumicis L. ab. *salicis* Curt. 'Twenty-four in all, earliest 1933. Widespread Karjalohja (4); Kuokkala (1).'

Procus latrunculus Schiff. ab. *intermedia* Horn. 'The dark form was very rare except in the 1930s. Later on it has grown much more general; in Porvoo, for instance, 30 per cent were dark in 1968. In the Helsinki region about 30 to 50 per cent of the *P. latrunculus* are nowadays dark.'

Gonodontis bidentata Clerck. ab. *nigra* Prout. 'Ten melanics since 1964.'

Cleora cinctaria Schiff. 'Five since 1948, also several dark specimens from Imatra.'

Cleora repandata L. ?f. *nigricata*, ?f. *nigra*. 'Six since 1943.'

Boarmia roboraria Schiff. f. *humperti*. 'Four since 1937, Helsinki (1), Karjalohja (1).'

Ectropis bistortata Goeze. 'Twenty-eight since 1925, now recorded yearly from many places, for example, Helsinki (8), Reposaari (11).'

Crocallis elinguaria L. ab. *fusca* Reutti. 'Helsinki (2), 1965 and 1967.'

Nonagria typhae Thnbg ab. *fraterna* Tr. 'Regularly found over a long period, it does not seem to have grown especially common.'

The following observations can be made:
(1) There is little sign of widespread Industrial Melanism in Finland. That most sensitive indicator, *Biston betularia*, shows little melanism as yet; there is some doubt as to whether f. *carbonaria* has, in fact, been recorded.
(2) No industrial melanics appear to have been collected from urban districts prior to 1930.

(3) Of the eleven species recorded by Suomalainen, the last two (*C. elinguaria* and *N. typhae*) are unlikely to be of industrial origin. Both occur in Britain as non-industrial melanics. The melanics of six of the remaining nine species are known to occur outside the influence of pollution fall-out. Only *B. betularia*, *T. or*, *G. bidentata*, *B. roboraria*, *C. repandata*, and *E. bistortata* (with eight melanic specimens taken in Helsinki) are likely to be true industrial melanics, and these at present occur at a low frequency.

The position of Industrial Melanism in Finland today is comparable to that in Britain in the mid-nineteenth century. It differs from Denmark and south Sweden in that, like Britain, it has sea barriers which separate it from northern Europe. The whole of the country lies north of latitude 60°N and Northern Melanism is widespread as also is Conifer Melanism in its forests; both sources are readily available for Industrial Melanism to exploit. I have dealt at considerable length with the little we know about melanism in Scandinavia in order to be able to compare the situation here with that in Britain and the Netherlands (see below). In many instances, though the same species are present in all three areas, melanic forms may be absent in one, but present in another. In other species different melanic mutants have spread through the separate populations. In Scandinavia the source of these must frequently have been a local reservoir of Ancient Melanism. In the Netherlands, as with many species in Britain, Industrial Melanism has had to await mutation. It is up to Scandinavia, in particular Finland, Sweden, and Norway, to record the early history of the spread which we have largely missed in Britain.

The Netherlands

I am grateful to Dr. B. J. Lempke for giving me a list of the more frequent species showing Industrial Melanism, and also for his table of the melanic frequencies of *Biston betularia* L. which appears at the end of the list.

Hyloicus pinastri L. f. *nigrescens* Lpk. Modern melanic, still rather rare, but in a comparatively short time dispersed over the whole country.

Dasychira pudibunda L. f. *concolor* Stgr. First specimen recorded in 1925, very rapidly dispersed over the whole country (with several variations), especially common with the ♂; much rarer in ♀.

Lymantria monacha L. Dark forms more and more common, pale typical specimens in many localities a minority.

Achlya flavicornis L. f. *pseudoalbigensis* Franzius. The first black specimen recorded in 1969 in the south of Dutch Limburg.

Tethea or Schiff. f. *albigensis* Warnecke. First recorded in 1938; now

known in many localities in the eastern half of the country and locally rather common.
Tethea octogesima Hübn. f. *frankii*. First recorded c. 1925, now very common in south Netherlands. Not yet known in north. f. *fusca* Cockayne widespread but not as abundant as f. *frankii* (this is the form which migrated to England in 1945).
Polia nebulosa Hufn. f. *robsoni* Collins. Only a few specimens.
Apatele megacephala Schiff. f. *nigra* Shaw. Only in a limited area, but regularly.
Apatele psi L. f. *suffusa* Tutt. Oldest specimen 1914; in the last few decades it has very rapidly dispersed over the country.
Craniophora ligustri Schiff. f. *nigra* Tutt. The prevalent form in the south of Dutch Limburg.
Xylophasia monoglypha Hufn. Dark forms dispersed over the whole country.
Procus strigilis Clerck f. *aethiops* Haw. the all-black form dispersed over the country.
Procus latrunculus Haw. f. *aeruginis* Tams. Common.
Colocasia coryli L. f. *melanotica* Haverkampf. First 1954, now several localities, in the eastern part, in central Limburg locally already rather common.
Allophyes oxyacanthae L. No melanics here.
Eupithecia albipunctata Haw. f. *angelicata* Barrett. Common.
Gonodontis bidentata Clerck f. *nigra* Prout. Only known in the last few years; still rare.
Semiothesia liturata Clerck f. *nigrofulvata* Collins. Known in many localities and locally common.
Erannis leucophaearia Schiff. Black forms common in the eastern part.
Biston strataria Hufn. f. *terrarius* Weymer. Very common. f. *robiniaria* Frings. Less dispersed, but dominate f. *terrarius* locally already (Plate 5.18a and 5.18b).
Cleora rhomboidaria Schiff. f. *rebeli* Aigner. Abafi. Only occasional black specimens.
Cleora repandata L. f. *nigricata* Fuchs. Not common, local.
Boarmia roboraria Schiff. f. *infuscata* Stgr. Common. f. *melaina* Schulze. All apparently of this form in Holland in Coll. Van Aartsen. (C. de W.).
Boarmia punctinalis Scop. f. *humperti* Humpert. Common. f. *nigra*. Warnecke. Less common, but known in many localities.
Ectropis luridata Bork. f. *cornelseni* Hoffman. Very local and still rare.
Phigalia pilosaria Fab. f. *monacharia* Stgr. First recorded in 1949 in the

south of Dutch Limburg. Now locally rather common there. In the rest of the country only local.

Biston betularia L. Table 6.1 gives a recent assessment of the frequencies of the three forms from 19 localities in the Netherlands.

TABLE 6.1

Biston betularia *samples from the Netherlands, 1969 Identified by B. Lempke.**

Locality	f. typica %	f. carbonaria %	f. insularia %	Total
Appelscha	30·00	27·27	42·42	33
Abdij sion	14·79	59·83	25·36	473
Deventer	12·50	50·00	37·50	32
Nunspeet	10·34	46·55	43·10	58
Amersfoort	11·11	51·85	37·03	27
Overveen	11·76	76·47	11·76	51
Bentveld	15·00	77·50	7·50	40
Aerdenhout	5·76	62·50	31·73	104
H. I. Ambacht	6·25	85·41	8·33	48
Ostvoorne	6·89	72·41	20·68	29
Melissant	0·00	70·00	30·00	20
Ostkapelle	9·09	39·39	51·51	33
Waalwijk	8·69	34·78	56·52	23
Nuenen	6·89	58·62	34·48	29
Swalmen	1·28	71·79	26·92	78
Annendaal	2·04	74·80	23·46	98
Stein	2·78	68·30	28·90	467
Schinfeld	1·63	63·93	34·42	61
Bergen op Zoom (30 Km N. of Antwerp)	3·87	73·50	22·63	181

* I am grateful to Dr. B. J. Lempke for permission to quote his work.

From this résumé several interesting points emerge. First, that species appear to behave differently in the Netherlands to how they have done in Britain; for example, a melanic of *Allophyes oxyacanthae* has not yet been found there, while here, f. *capucina* may contribute up to 75 per cent of a local population. A *Phigalia pilosaria* melanic was not found until 1949. In Britain it was common in many counties in the last century. The frequencies of some of the morphs are entirely different from those found in this country, for example, *Dasychira pudibunda* f. *concolor* (Plate 5.7, left 5 and 6) has become common throughout the whole region since 1925. Also, the melanic form of *Hyloicus pinastri* has spread rapidly. It is possible that this melanic is dominant to the wild type in the same way as is, allegedly, the Czechoslovakian one. It will be recalled that only two melanic *H. pinastri* have been taken in Britain, none recently, and that f.

unicolor is recessive (Chapter 13). Lastly, in regard to *Biston betularia*, the frequencies differ in one respect from those in Britain (Chapter 9): the f. *insularia* portion of the samples is much higher than that found in this country. All this suggests that different genes controlling melanism may be present in each country.

(ii) North America

Industrial Melanism is widespread throughout the North American continent. In the Pittsburgh area alone, it is probable that more than 100 species of Lepidoptera are in the process of becoming melanic. In Canada, Industrial Melanism is found around Montreal, Toronto, and probably elsewhere (Kettlewell, 1958c). Frequently it is found in species which are the western counterparts of those in Europe and they have similar melanic forms. Thus, for example, it occurs in *Amphidasis* (=*Biston*), *Phigalia*, and *Ectropis*. Also there are many genera in which species occur which have melanic forms which are not found in Europe, for example, *Graptolitha* and *Zale* which occur around industrial centres in the eastern States, as far north as Canada.

Owen (1961), 1962) has been one of the few biologists to make a study of the phenomenon in the United States. Under the heading 'The evolution of melanism in six species of North American Geometrid moths' he recorded his observations and I give a résumé of these:*

... In North America, the first melanic *B. cognataria* [f. *swettaria*] (Plate 9.1, right 3) was recorded in 1906 near Philadelphia; melanics now comprise over 90 per cent of the population in at least [some] parts of south-eastern Michigan, and they are probably more common than the pale form in other industrial areas. It appears that the spread of melanics began about fifty years later in North America than in Britain.

... In Livingstone and Washtenaw Counties, Michigan, it is probable that melanic *B. cognataria* increased from about 1 per cent to over 80 per cent of the population in less than 30 years.

... in southern Ontario ... Raizenne recorded five [f. *swettaria*] in a sample of 61 [*B. cognataria*] bred from the immature stages in 1939–48. Melanics were recorded for the first time in the Upper Peninsula of Michigan in 1960. They have been recorded in Champaign Country, Illinois, from 1935 onwards, and one was taken in New Castle County, Delaware, in 1928.

... In 1959–61, using a 100 W mercury-vapour lamp on every warm night throughout the summer, I collected a random sample of 576 specimens of *B. cognataria* at the Edwin S. George Reserve, Livingstone County, Michigan. This sample represents specimens from 5 successive generations. ... the melanics comprised 87·0 to 93·0 per cent of the population in this area. Collections made

* I am grateful to Professor D. F. Owen for permission to quote his work.

in the same area in the 1920s and 1930s did not include melanics, and it must be concluded that if they existed they must have been rare. . . . 7·6 per cent could be regarded as intermediate in colouration.

. . . Form *insularia* (from Europe) corresponds almost exactly in color and pattern to specimens of *B. cognataria* classified in category 2, and hence it is possible that a gene similar to that producing f. *insularia* is present in *B. cognataria*.

. . . hence it is possible that in North America some heterozygous melanics are incompletely dominant. An analysis of the genetics of melanism in *B. cognataria* is urgently needed.

Phigalia titea

The species occurs from Quebec south to Texas and west to Winconsin; there is a closely similar species in eastern Asia, which perhaps should be regarded as con-specific. The corresponding European species is *P. pilosaria*.

The melanic form, *deplorans*, of the male *P. titea*, is figured by Remington. The wings are not patterned and, with the exception of veins, which are very dark, they are thinly covered with black scales . . . There are two melanic forms of the European *P. pilosaria*. One form, *monacharia*, corresponds almost exactly with form *deplorans* of *P. titea*;

. . . No date of capture is given for the exceptionally early specimen described by Minot (1869) from West Roxbury, Mass. . . . Most of the first records are much later than those for *B. cognataria*, and there is no clear indication that the melanics first appeared near centres of industry. Despite large series in the museums, I have seen no melanics from the areas around Chicago, Pittsburgh, and Philadelphia; and it is possible that in other areas the melanics have only just started to increase in relative frequency. With the exception of the record by Minot, the earliest record I have been able to trace is for Ocean County, New Jersey, in 1915.

The frequency of melanic *P. titea* on the Edwin S. G. Reserve Michigan in 1960 and 1961 was 10·6 per cent and 7·9 per cent. About 170 miles north, at Frayling, Crawford County, Michigan, I found 1·4 per cent melanic in 1960. At Springfield, Clark County, Ohio, Laux recorded 8·3 per cent melanic in 1961. No melanics [were] found in a sample of 165 specimens collected in southern Ontario in 1939–48. Remington (1958) states that melanics have occurred recently at 'high frequency' around New Haven, Connecticut, but gives no figures. J. Muller states that 'about 10 per cent' are melanic around Lebanon, Hunterdon County, New Jersey. These are the only data I have on the relative frequency of melanic *P. titea*. Considering the scarcity of melanics in collections made more than 20 years ago, it seems possible that they do not yet comprise more than about 10 per cent of the population of the species in any area in north-eastern North America.

Phigalia olivacearia

The species is similar to *P. titea*. . . . It occurs from Mass. west to Illinois and south to Texas; little is known of its life history.

. . . The melanic of *P. olivacearia*, form *mephistaria*, was first described and figured from a male taken at West Roxbury, Mass. in 1912. It is similar to f.

deplorans of *P. titea.* The only melanic specimens I have seen are from southern Michigan.

Nacophora quernaria

A melanic of *N. quernaria*, form *atrescens*, was first described from London, Ontario (Julst, 1898). This specimen . . . is unlike any that I have taken in southern Michigan: it is predominantly black, but most of the lines on the wing are whitish.

The frequency of melanics on the E.S. George Reserve was 55·5 per cent in 1960 and 41·1 per cent in 1961.

. . . the earliest records of melanics are from industrial areas (Pittsburgh and New York City).

Epimecis hortaria

. . . A melanic form, *carbonaria* (Plate 6.3), was first described and figured from specimens collected near Philadelphia between 1909–11. In this form, the entire wings, with the exception of the white subterminal line, are much suffused with black. I have seen intermediates between f. *carbonaria* and the pale form but have not been able to examine a large random sample of the species; hence I shall defer discussion of intermediates . . .

. . . The earliest records (of melanic *E. hortaria*) were from around areas of heavy industry; thus, the species appears to be a true industrial melanic. . . . the earliest date . . . I have . . . found [is] 1909, [from] near Philadelphia.'

. . . In a small sample (8) collected on the Powder Nature Reserve, Westmorland County, Pennsylvania, in 1957, all were melanic. J. Muller reports about 90 per cent melanic at Lebanon, Hunterdon County, New Jersey. Judging by the large numbers in collections made during the past 20 years, melanic *E. hortaria* are now common in southern Pennsylvania, parts of New Jersey, and the New York City area . . .

Ectropis crepuscularia

. . . A melanic form, *fumataria*, was first described from West Roxbury, Mass (Minot 1869).

E. crepuscularia also occurs in Europe. The melanic form *delamerensis*, is figured by Ford (1955*a*); it is dominant to the pale form, and has spread in some industrial areas, and also in some rural areas.

. . . in 1961, the species was relatively common, and . . . just over one-third were melanic. . . . Judging by the number of specimens in collections made during the past 20 years, melanics are now common in the Pittsburgh and New York City areas.

. . . Most of the larger museums have long series of the species dating back to the nineteenth century. . . . The earliest record (of a melanic) was sometime before 1869. . . . The next record was from Pittsburgh in 1917. There were no records from the Chicago area until 1957, but one was recorded from Lee County, Illinois, in 1924. The first record for New York City was in 1931. The first definite record for southern Michigan was in 1959 in Livingstone County, but since in 1961 melanics comprised about a third of the population, it is likely that they occurred earlier than 1959.

From this account of Owen's it appears that except for one or two very early records of melanic forms (for example, *Phigalia titea* prior to 1869 and *Nacophora quernaria* in 1898) an industrial spread was not noted until much later—in the 1930s in *B. cognataria* and the 1950s and 1960s in *P. titea, Ectropis crepuscularia*, and in other species. Today melanism is widespread throughout the eastern States and elsewhere, though we have few records of moth samples. The opportunities for evaluating the selective advantages and disadvantages are immense. It is to be hoped that the unique opportunities which still exist today will not again be missed as they have been in Britain.

It is possible that the simultaneous appearance and the rapid spread of so many melanic forms on the North American continent may have resulted from the availability of such forms which already occurred as non-industrial polymorphisms in the nearby countryside. This would be different from the history of Industrial Melanism in Britain; this is discussed in the last chapter (see Synthesis, Chapter 19).

Much remains to be done in order to establish the relationship of the various *carbonaria* and *insularia* forms of *B. betularia* in Europe and even more in its co-species *B. cognataria*. For here, though the blackest form *swettaria* is easily recognizable, the melanics of a lower order may not be recognized because *cognataria* f. *typica* has a somewhat more smoky appearance than *B. betularia* f. *typica* (Plate 9, right column). Because of its multivoltine habits which one finds from central North America southwards, I would anticipate an even more rapid spread of melanic forms in industrial areas in that continent. Also it is of advantage that bivoltine species serve to hasten laboratory breeding experiments in regard to genetical answers.

Though Industrial Melanism has been recorded from around most centres of industry in the Palaearctic it is significant that in spite of constant inquiries I can find no record of its presence in the tropics. It may be that the advantages of camouflage are outweighed by the disadvantages of black pigmentation under conditions of great heat.

Part III. *Biston* (syn. *Amphidasis Pachys*) *betularia* L. (Selidosemidae). The original selection experiments (1952–5), The frequency surveys (1952–70)

7. The choice of material

(i) *Biston betularia* L. (syn. *Amphidasis Pachys*) (Selidosemidae) and other cryptic species, their advantages and disadvantages.

Up to 1952, no field experiments had been undertaken on any industrial melanic species whatsoever, in spite of the fact that biologists claimed that Industrial Melanism was the most striking evolutionary change ever to have been witnessed, and biometricans computed selective pressures for its rapid spread greatly in excess of any previously known. Furthermore entomologists could not accept the suggestion that its spread was in any way connected with industrialization. They pointed out, in fact, that melanic forms were found not only in and around urban centres, but in rural ones far beyond. The ornithologists could produce no shred of evidence that birds actively searched for and ate cryptic insects, still less that they did this selectively. The time was indeed ripe for experimentation in the field.

It was essential to choose the correct species for testing the cryptic advantages and disadvantages of contrasting light and dark forms. Such a species must ideally have the following attributes:

(1) Widespread distribution through both urban and rural areas.
(2) Occur at a high density (i.e. it must be common).
(3) Easy to capture (and recapture) at light or by some other method of sampling, such as 'assembling'.
(4) Adequate flight to ensure gene flow.
(5) Easy to breed in the laboratory on a large scale.
(6) Polyphagous feeding habits in the larval stage.

A limited number of species from the 100 or so common industrial melanics occurring in Britain qualified in these respects, the majority being unsatisfactory for one reason or another. From the commencement my choice was *Biston betularia*, the most famous example of Industrial Melanism and the first to be detected. Four others were listed as being likely to offer opportunities in the field—not as alternatives to *B. betularia*, but in addition. These were *Phigalia pilosaria* Fab. (syn. *pedaria*), *Gonodontis bidentata* Clerck, *Cleora repandata* L., and *Allophyes oxyacanthae* L. Each of these species presented different aspects in the study of Industrial Melanism. Thus *G. bidentata* is a slow flier and does not wander far. The females of *P. pilosaria* are apterous and therefore cannot fly. *C. repandata*, like *P.*

pilosaria and *A. oxyacanthae*, had melanic forms in completely unpolluted areas which were in some instances similar to the industrial melanic ones. Lastly, and an important point in planning one's work, three of the four species were on the wing at different times of the year; *P. pilosaria* in February and March, *G. bidentata* in May, and *A. oxyacanthae* in the autumn. Field work could therefore be spread over nine months of the year. Here we are concerned with *Biston betularia*, the Peppered Moth.

(ii) The genetics of *Biston betularia* L.

If one examines the extensive collection of *B. betularia* in the National Collection of British Lepidoptera (Plate 5.6, left 1, 2, and 3) in the British Museum it is possible to obtain a picture of specimens grading imperceptibly from light to black, suggesting that melanism in this species may be controlled polygenically. This most certainly is not the correct answer, although polygenes must contribute to the expression of each phenotype.

There are two main melanic complexes, 'f. *carbonaria*' Jordan representing the more extreme ones, and 'f. *insularia*' Thierry-Mieg, which is intermediate in appearance between f. *carbonaria* and f. *typica*. Furthermore, f. *typica* itself varies greatly from heavily speckled individuals (the usual form in the New Forest) to pale, almost white specimens with fine, granular black markings. I have outcrossed the heavily peppered New Forest type to normal Oxford wild types and there were no clear-cut segregations in any brood. This suggests polygenic control of this form.

The genetics of f. carbonaria

Until recently I had always considered that all f. *carbonaria* were controlled by the same pair of alleles which had dominance over f. *typica*. Certainly in broods which did not contain f. *insularia* clear-cut segregation always takes place and intermediates never occur. Yet within f. *carbonaria* there is considerable variation, as seen for example in the earliest specimens taken in the last century (Plate 7.1) (see Chapter 7 and Chapter 19); here many individuals had white wedges on the outer margin of both forewings and hindwings. Such specimens are seen but rarely today, though more frequently in rural than in urban areas.

J. B. S. Haldane has suggested (personal communication, 1959) that it is more likely that a number of different alleles, each varying phenotypically but only slightly, might be included in the *carbonaria* form. Moreover he extended this view as being applicable to the majority of melanic forms of other species. We have so far been unable to prove multiple allelomorphism in f. *carbonaria*, but this certainly is the situation in f. *insularia*, the less extreme melanic. (Clarke and Sheppard, 1964, Lees, 1968).

The genetics of the insularia *complex*
There are at least four separate forms referred to under this name, probably many more. Each is dominant to f. *typica* and though in 1955 I stated that f. *carbonaria* was epistatic to f. *insularia* we can now say that it is dominant.

In practice I score '*insularia*' into one of five categories according to the degree of darkness. *insularia*^{1} is a smoky f. *typica*; *insularia*^{5} looks like a f. *carbonaria* finely speckled with white.

In 1955 I (Kettlewell, 1955b) recorded that a 'medium' f. *insularia* of Oxford origin (now classified as *ins.*^{3}) appeared to be controlled by a gene at a locus different from that of f. *carbonaria*, and that this f. *insularia* was therefore non-allellic. This somewhat unexpected finding was the result of the segregation of all three forms, but in small numbers, from parents (f. *carbonaria* × f. *insularia*) crossed to wild types. This was in one brood only. Clarke and Sheppard (1963, 1964) have shown that a medium f. *insularia*, also from Oxford, was allelic to f. *carbonaria*, and also that the same was likely for f. *insularia* of Cheshire origin. More recently D. R. Lees (1968) has demonstrated that a dark f. *insularia* of Herefordshire origin was allelic to f. *carbonaria*; this was again shown conclusively in four broods.

In the light of these other breeding experiments, and of our present knowledge of the part played by allelomorphism in the long-term evolution of melanism, I think my earlier record was due to error. Although it may, in fact, be correct I now regret having made this statement in my 1955 paper (Kettlewell, 1955b). Its relevance to the theory of the origin of melanism had not at that time been appreciated by me. We must, I think, assume then that f. *insularia* is recessive to f. *carbonaria* but dominant to f. *typica*. The order of dominance is likely to be in a descending scale, from the darkest to the lightest and f. *typica*.

(iii) The life-history and behaviour of *Biston betularia* L.

Imagines
The Peppered Moth in Britain is univoltine and flies from late May until August. In common with many other cryptic species, it hatches from the pupa late in the day and dries its wings not necessarily on a tree trunk. By this means, it largely avoids the consequences of predation during the

first few hours of its imaginal life. Subsequently, both sexes spend the day at rest on the boughs and trunks of trees, and less frequently on walls and palings. They take up position at early daybreak (*c.* 4.00 a.m. in June). It appears to me that they choose a tree on which to rest at random but it is likely that scent plays a part in this. I have always found that deciduous trees were chosen.

I have described the pre-resting behaviour on page 71, and I believe that this is common to most cryptic species. Such moths, in fact, move but rarely during daylight, unless disturbed by the sun or by ants. During this time, they appear to be in a state of catalepsy and will drop into a box on being disturbed (except in hot weather).

Male *B. betularia* fly freely from dusk until midnight and again at dawn. Females fly rarely and can usually be found on the same trees, but higher up, as those onto which they had been previously released.

The females commence calling at dusk and this can be recognized by rhythmic movements in which the long ovipositor is extended and then retracted. The scent can be detected by males at a considerable distance and I have frequently assembled marked males, released down-wind at a distance of one to one and a half miles from a female whom they reached in under one hour (Kettlewell 1958*a*). This mating behaviour was first observed by John Ray on 29 May 1693, and recorded by him in his book 260 years ago (Ray, 1710) (Plate 7.2.) Copulation continues for up to 20 hours so that such pairs must spend the following day together when at rest. In f. *typica* × f. *carbonaria* pairings, this must necessitate that one of the two forms is on an incorrect background. As we have shown that the discovery of one individual on a tree trunk by a predator puts others at risk (page 116), it is likely that both partners are frequently eaten. Theoretically, then, one would have expected that pairings between unlike phenotypes would have been selected against. I have no evidence of this as yet. Only rarely is it necessary for a female to mate a second time.

A high proportion of females taken at fixed mercury-vapour lights are virgins. The reason for this is obscure, but it is likely that they have hatched in the immediate vicinity of the traps.

Ova

Females oviposit in the depths of bark crevices on deciduous trees. They are able to do this because of the length of their ovipositors ($\frac{1}{4}$ inch or more). The ova are at first pale green or yellow in colour but they darken prior to hatching. 'Cakes' of ova may contain up to 1,000 individuals. This may seem strange in a species whose larvae depend on crypsis and must

The choice of material 109

therefore essentially be widely dispersed. This all-important condition is however fulfilled in the larval stage.

Larvae

In ten to twenty days, according to the temperature, the eggs hatch. I believe that the next 24 hours is all-important both for individual survival, and for the spread of the species. The small larvae walk with great rapidity to the tips of twigs where they lower themselves on a fine thread of silk. They then become airborne and are carried by the wind in much the same way as young spiders are. As to how long they constitute 'aerial plankton' we have no knowledge. Many must be caught up on adjacent trees near by when, no doubt, they may repeat the procedure (Kettlewell, 1955c).

They feed on most trees and shrubs such as *Quercus* and *Salix* spp., also on *Betulus* and *Fagus*, more rarely on broom (*Cytisus* spp.), and have even been recorded from such plants as Michaelmas daisies (*Aster* spp.) and Lucerne (=Alfalfa).

The caterpillar is a typical 'looper' and at full-growth is stick-like in appearance, mimicking to perfection the twigs on which it rests by day. The larvae vary in colouration from green to purplish-black according to the food plant. The mechanisms controlling such colouration are discussed in Chapter 4.

They feed from June to October and in this they differ from North American *B. cognataria* which is referred to later (Chapter 19). They pupate in the earth and detritus beneath the trees in a frail cell and pass the winter thus.

Pupae

As has already been stated, the pupae hatch over four months, from May to August. The all-important factor in this is that an adequate number of males must be available to mate with females throughout this period. But we have found that when the pupae are kept at outdoor varying temperatures they most certainly do not hatch evenly throughout these four months. Hot spells are interspersed with cold ones throughout the season and the number of emergences bears a direct relationship to temperature. We have observed, however, that the males hatch first following each warm period with the maximum number of females occurring three to five days later. This pattern of behaviour is common to many other species. Such a happening was difficult to explain entirely on the grounds of genetic inheritance and it appears to me that it can only be brought about by a combined effect of inheritance and environment. Beneath the surface of the soil, a pupa is affected by many factors and foremost amongst these is the temperature of the air above and the pupa's distance from the surface. Pupae near the surface are subjected to heat (or cold) more rapidly than those at a depth of three or four inches. It could follow on theory that it would benefit female pupae to be deeper than male pupae which, if pupating superficially, would be affected earlier in each successive heat wave.

Since 1961 we have therefore tested pupal positioning with several species, in particular *Lycia hirtaria* which we have been breeding extensively since that date. The pupae of *B. betularia* appear to behave in a similar way, but I quote the work on *L. hirtaria*, about which I have obtained more extensive data. The following method has been used throughout. Fully grown larvae are provided with a 9 in. depth of fibre in which to pupate in large flower pots. We attempted to give uniform soil density. Subsequently, the surface of the fibre is lightly scratched and, keeping it level by discarding the loosened material, the pupae are extracted and arranged in their order of exhumation. Having discovered the total, each brood is divided into three categories representing a top, a middle, and a bottom layer. The middle one is disregarded (except its contribution to the sex ratio), the object being to differentiate more clearly the top pupae from those of the bottom. I examined 2630 *L. hirtaria* pupae from 39 broods in this way. Of these broods, some were too small to test, but eleven (comprising 1694 individuals) showed a departure from expectation in the distribution of the males and female pupae, ten having an excess of females in the bottom layer, and of males in the top; one had an excess of females in the top (Table 7.1). This table clearly shows an excess of male

TABLE 7.1
Data of eleven broods of Lycia hirtaria *which showed a departure from expectation in the distribution of male and female pupae.*

	♂♂	♀♀	Total
Top	511	347	858
Bottom	380	456	836
	891	803	1694

$\chi^2_{(1)} = 33.78$, $P > 0.001$.

pupae on the surface and of females below, though the χ^2 is invalid because the data is heterogeneous. The depth of the pupae could alone be responsible for females hatching later than the males, but we know that this is not the only reason. When, as in the laboratory, pupae of both sexes are subjected to the same temperature in a container, the earliest ones to hatch are the males. There must, therefore, be a genetical component as well as a behavioural–environmental one. This is often grossly upset in broods from crosses of different origin and even more so in interspecific offspring. Thus in broods of *Biston betularia* × *B. cognataria* (of Canadian origin) we frequently had small numbers of females hatch in February or March, the majority conforming to May and June. This, no doubt, is in some way associated with *B. cognataria* being bivoltine in the more southeastern areas of the U.S.A. (see life history of *B. cognataria*, p. 84). The hybrid offspring of *L. hirtaria* × *Nyssia zonaria* ssp. *atlantica* hatch in such a way that the two sexes never appear together; females appear two months earlier than the males. This type of behaviour is common in many hybrid crosses, for example, *Notodonta dromedarius* L. × *N. ziczac* L. where all the males hatch in the autumn of the same year and the females in May and June of the following. Our work on the pupae of *Biston betularia* and other related species has drawn attention to other interesting adaptations for survival. Though in *Biston betularia* I have never had a bivoltine strain from Britain, this is frequent in the eastern U.S.A. In *Lycia hirtaria* in Britain over 90 per cent of London pupae hatch the following spring. In a brood from Inverness-shire, Scotland, of 106 imagines, only 2 hatched in the first year, 18 in the second, 82 in the third, and 4 in the fourth. Such staggered hatchings must have been selected in northern species in order to avoid complete extermination in a particular year (for example, in the arctic spring of 1963). *Brachionycha nubeculosa* Esp. from Aviemore, Perthshire, may on occasions overwinter for as long as seven seasons. In British *B. betularia* I have never had a pupa spend more than a single winter in this state and here the species is univoltine.

Details of methods for the large scale breeding of this species (and others) for laboratory usage are given in the Appendix A.

8. The experiments on the differential survival of *Biston betularia* f. *typica* and its melanic forms

ACCOUNTS of three independent series of experiments are to be discussed here, and I am grateful to the editors of the journal *Heredity* for permitting me to quote these. I have recorded them in the order in which they were undertaken. I do this because in experiments of this nature, one is constantly having to change their design when an unexpected hazard or bias presents itself. Hence they form a natural sequence.

My basic approach throughout was to present the three forms of *Biston betularia* in near-equality on contrasting backgrounds, that is, on light (lichened) and dark (polluted) tree trunks (Plates 8.1, 8.2), and to observe whether predation, in particular a differential one, took place.

Data for the following aspects were obtained in each experiment:
A. Camouflage efficiency, having regard to the melanic and typical forms and their backgrounds, as adjudged by the human eye.
B. Direct observation of predation through binoculars, and the order in which predation took place. Also, the total number of each form which survived after a period of time.
C. Recapture of marked releases of the three forms in various types of trap.

The three series of experiments were the following: (i) *The preliminary selection experiments in an aviary*. Releases of the three forms on to introduced backgrounds in a large outdoor aviary in which there was a pair of nesting Great Tits (*Parus major*). (ii) The *selection experiments in polluted woodland*, in the centre of industrial England (Birmingham). (iii) The *selection experiments in an unpolluted woodland* (Deanend Wood, Dorset).

I decided that it was essential to commence my investigations on differential predation on the three forms of the Peppered Moth in an aviary for a number of reasons. In 1952 we had no knowledge of whether birds actually ate cryptic moths, still less as to whether they did this selectively, having regard to their colour and background. We had developed no techniques for scoring cryptic efficiency nor the selective advantages of different forms. Fisher and Ford (1947) had, however, recorded a successful method of individual marking, now referred to as 'the mark–release–recapture' technique (page 279) from which the daily death-rate, hatching-rate, and indeed, the total population could be deduced. This had been devised, in the first place for *Polyommatus icarus* Von Rott. (Lycaenidae), the Common Blue Butterfly, in the Isles of Scilly. Later it has been extensively used for the moth *Panaxia dominula*, a local colony insect. The method, however, had not been applied to continuous population covering miles of countryside.

(i) The preliminary selection experiments in an aviary

Both melanic and typical *B. betularia* were released in approximate equality on to light and dark trunks and boughs in a large outdoor aviary (18 yds × 6 yds), which contained a pair of Great Tits with young.

A. *Camouflage efficiency*. A method of scoring the moths on their backgrounds was devised by choosing one of six values for each insect at a distance of two yards after it had taken up its resting position (Kettlewell, 1955b). At this distance, in the first instance a decision as to whether the moth was conspicuous or inconspicuous had to be made; the degree of each was registered by one to three +s or −s. In these initial aviary experiments I attempted to provide equal surface areas of contrasting light and dark backgrounds.

B. *Direct observation on selective predation*. The Great Tits, for the first two hours of the first day, did not recognize either form of *B. betularia* as an article of diet. In the following hour, however, they took all five of those individuals which had been scored 'conspicuous' as well as two defined 'inconspicuous' in a total first release of 10 (5 f. *typica* and 5 f. *carbonaria*). The following day, the experiment was repeated but as well as using f. *typica* and f. *carbonaria* a second melanic, f. *insularia* was included, in all 18 individuals. In half an hour all except two had been taken and each of the survivors had been scored as 'inconspicuous'. These two, and other similar experiments suggested that it was necessary for Great Tits to have a period of contact with their prey before recognition. Also, that when a 'hunting image' had been established all the *B. betularia* were at risk—but selectively. The tits took them in order of conspicuousness as adjudged by the human eye. This was in accord with the earlier experiments of Sumner (1934) who showed that owls predated two forms of *Peromyscus* having regard to their cryptic efficiency.

These pilot experiments, therefore, were encouraging in that they showed that at least one species of bird ate large numbers of cryptic moths, including melanics, and that they did this selectively. It also emphasized

Differential survival of Biston betularia

an obvious fact that predation pressure must be greater during the breeding period of insectivorous birds which, in the Palaearctic, is from May to July.

The aviary experiments were an essential first step before testing bird behaviour to cryptic polymorphic moths in the wild. For these quite exceptional facilities, which were given me at the Research Station, Madingley, Cambridge I am indeed grateful to Professor R. A. Hinde.

(ii) The selection experiments in polluted woodland

Encouraged by the results of the aviary tests, I immediately moved from Cambridge to a wood near Birmingham so that I could carry out a full programme in the field. After our reconnaissance during the winter of 1952/53 I had chosen the Christopher Cadbury Bird Reserve near Rubery, Birmingham, for the following reasons. It was a large, mixed, deciduous wood, polluted by surrounding industries. Here the tree trunks maintained no vegetative lichens and they were black except for the green areas which were covered by *Pleurococcus*.

In spite of this it boasted a quite exceptional bird population, in terms of numbers both of species and of individuals. I was given facilities by the owners for housing the large number of *B. betularia* pupae as well as the considerable amount of apparatus necessary for a large scale experiment of this kind. Also, mains electricity was available so that I could place mercury-vapour sampling-traps around the releasing area.

Biston betularia was one of the commonest species here and in a sample of 621 local specimens taken in the summer of 1953 just under 90 per cent were melanic (f. *typica* 10·14 per cent, f. *carbonaria* 85·03 per cent, f. *insularia* 4·83 per cent).

We lived in a caravan-trailer *on the spot* and I consider this essential for 24 hours a day, when carrying out such field work. At this stage we had no idea as to how birds in the wild would behave with a cryptic polymorphic insect.

Data obtained from the Birmingham experiments

A. *Camouflage efficiency*. Of the 520 individuals of both forms which were scored for camouflage efficiency on oak trunks in this wood only 8 f. *carbonaria* (2·2 per cent) out of 366 were 'conspicuous'. By contrast, 137 (88·96 per cent) of the 154 f. *typica* fell into this category. Details of the various species of tree on to which *B. betularia* had been released have been recorded by me (Kettlewell, 1955b).

116 *Differential survival of* Biston betularia

From this it appeared that in this locality the *B. betularia* f. *typica* were, as recognized by us, at an overall disadvantage of 40 per cent.

B. *Direct observations.* After the first day of release it was noticed that some of the *B. betularia* were disappearing from one or two positions, so observation of these sites was maintained continuously from a distance through field glasses. H. M. Kettlewell first observed a bird fly up out of the bracken, snatch a *B. betularia* and return to the ground, the whole incident being over in a flash. Subsequently we were able to watch this bird, a Hedge Sparrow (*Prunella modularis*) regularly at work, and to score the order in which it took the phenotypes. At a somewhat later date moths began to disappear from another series of trees, and I and others witnessed a Robin (*Erithacus rubecula*) at work, particularly in the late afternoon. This bird was observed in a similar way to the Hedge Sparrow, flying on to the twigs and bracken near to the trees whence it viewed the trunks and branches, making occasional excursions to pick up *B. betularia*. This it frequently did on the wing, always returning to the ground to eat them, and there subsequently I found wings and remains as an additional check to what had happened. It came as a surprise to us to find the Robin and the Hedge Sparrow behaving in this way, and there were most certainly other birds at work, unseen by us. I give below (Table 8.1) a record of the few occasions when there was no doubt of the order in which these birds took their insects.

From this it would appear that when a conspicuous insect had been found, it at once put other insects in the immediate vicinity at a disadvantage because of the birds' active searchings. This is corroborated by the increased predation which took place on those trees which harboured a moth with the score of -3 compared with trees where no such conspicuous individuals existed. Nevertheless, on oaks there were nearly always one or more f. *carbonaria* left which had been overlooked.

Total figures for male releases show that while 62·57 per cent of f. *carbonaria* survived per day during observation (see Table 8.2) only 45·79 per cent of f. *typica* did so (Table 8.3). The form *insularia* had a 57·14 per cent survival rate for the comparatively few insects under observation. Furthermore, there is evidence that the birds took the individuals within each phenotype with regard to their degree of crypsis. Of the 508 males (of the three different phenotypes) released for observation over seven days, 210 had disappeared at the end of the day for one reason or another.

TABLE 8.1

Direct observations of the predation of B. betularia by two species of birds in a polluted woodland.

Robin	Oak	4 July 1953 Order of take	Robin	Oak	2 July 1953 Order of take	Hedge Sparrow	Oak	1 July 1953 Order of take
f. typica	−3	1	f. typica	−3	1	f. typica	−3	1
f. carbonaria	+1	2	f. typica	−3	2	f. typica	−3	2
f. typica	−3	3	f. carbonaria	+3	3	f. carbonaria	+3	3
f. typica	−2	4	f. typica	−3	4	f. typica	−1	4
f. carbonaria	+3	Not taken	f. carbonaria	+3	Not taken	f. carbonaria	+3	Not taken
f. carbonaria	+3	by 7 p.m.	f. carbonaria	+2	by 7 p.m.	f. carbonaria	+2	by 7 p.m.

TABLE 8.2
Observation and release (f. carbonaria) for 7 days (males only), Birmingham experiments 1953.

f. carbonaria

Score	+3	+2	+1	−1	−2	−3	Total
25.6.53	3	3	2	0	0	0	8
27.6.53	11	15	5	1	1	0	33
30.6.53	17	20	15	6	2	0	60
1.7.53	43	28	6	0	0	4	81
2.7.53	47	14	3	1	0	6	71
3.7.53	24	18	12	0	2	3	59
4.7.53	29	14	7	0	1	3	54
Total release before midday (seven days)	174	112	50	8	6	16	366
Total missing (seven days) up to 5 p.m.	65	35	20	3	3	11	137
Per cent missing	**37·35**	**31·25**	**40**	**37·5**	**50**	**68·75**	**37·43**
Per cent survival	**62·45**	**68·75**	**60**	**62·5**	**50**	**31·25**	**62·57**
Release escapes	—	—	—	—	—	—	44
Per cent escapes = activity	—	—	—	—	—	—	**10·75**

TABLE 8.3
Observation and release (f. typica) for 7 days (males only), Birmingham experiments, 1953.

f. typica

Score	+3	+2	+1	−1	−2	−3	Total
25.6.53	1	0	0	2	2	2	7
27.6.53	1	1	0	0	3	6	11
30.6.53	2	2	2	5	6	7	24
1.7.53	2	0	2	0	3	11	18
2.7.53	2	1	0	3	8	11	25
3.7.53	3	0	1	0	2	7	13
4.7.53	0	0	0	0	3	6	9
Total release before midday	11	4	5	10	27	50	107
Total missing up to 5 p.m.	4	2	3	5	13	31	58
Per cent missing	**36·36**	**50**	**60**	**50**	**48·15**	**62**	**54·21**
Per cent survival	**53·64**	**50**	**40**	**50**	**51·83**	**38**	**45·79**
Release escapes	—	—	—	—	—	—	9
Per cent escapes = activity	—	—	—	—	—	—	**6·67**

C. *Recapture figures.* It is evident from Table 8.4 that we recaptured more than twice as many f. *carbonaria* as f. *typica* relative to the number released. In fact, from 630 male *B. betularia* of the three forms which were

TABLE 8.4

Recapture figures for Biston betularia *(males only)*;
Rubery, near Birmingham, 1953.

Date (1953)	Releases				Catches				Recaptures			
	carb.	typ.	ins.	Totals	carb.	typ.	ins.	Totals	carb.	typ.	ins.	Totals
25.6	10	12	10	32	8	0	1	9	—	—	—	—
26.6	0	0	0	0	127	15	7	149	3	1	1	5
27.6	33	11	15	59	34	5	1	40	1	0	1	2
28.6	37	21	5	63	23	3	3	29	2	0	2	4
29.6	0	0	0	0	55	10	1	66	5	4	0	9
30.6	68	26	8	102	37	3	2	42	1	0	1	2
1.7	90	21	3	114	76	9	2	87	19	2	2	23
2.7	74	21	3	98	75	13	4	92	28	6	0	34
3.7	68	15	0	83	77	11	10	98	25	3	1	29
4.7	67	10	2	79	66	9	2	77	23	2	0	25
5.7	0	0	0	0	73	3	5	81	16	0	0	16
Totals	447	137	46	630	651	81	38	770	123	18	8	149

	Catches			
	carb.	typ.	ins.	Totals
Wild Birmingham population	528	63	30	621
Per cent phenotype	85.03	10.14	4.83	
Release after one day of self-determination	25	2	2	
Per cent phenotype	5.72	1.48	4.35	
Per cent return of releases	27.5	13.0	17.4	

marked and released, I recaptured 149 marked individuals. Of the 447 f. *carbonaria*, I got back 27.5 per cent; but of the 137 f. *typica* I recaptured only 13.0 per cent. This result could be accounted for in one or more of the following ways:

(a) The melanics were attracted to light more freely than f. *typica*.
(b) The f. *typica* had a shorter span of life physiologically than f. *carbonaria*.
(c) The f. *typica* wander or migrate more than the melanics.
(d) There was a differential predation between the phenotypes.

In regard to the first, we were able to show that an equal relative percentage of each form came both to light and assembling traps (Table 8.5). In a sample of 433 *B. betularia* taken at mercury-vapour light 83.6 per cent were f. *carbonaria* and 10.85 per cent f. *typica*. Of 328 individuals which assembled to virgin females, 85.68 per cent were f. *carbonaria* and 10.3 per cent f. *typica*. This excludes sampling errors due to behaviour differences

TABLE 8.5
*Comparison of the proportion (per cent) of phenotypes
which came to mercury-vapour light and to assembling (1953, 1955.)*

Year	Birmingham 1953 (for 10 nights)				Birmingham 1955 (for 6 nights)			
Phenotypes	*carb.*	*typ.*	*ins.*	Total	*carb.*	*typ.*	*ins.*	Total
Mercury-vapour light								
Totals	362	47	24	433	167	25	8	200
Percentage	83·60	10·85	5·55	100	86·44	10·17	3·39	100
Assembling								
Totals	281	34	13	328	255	30	10	295
Percentage	85·68	10·36	3·96	100	83·5	12·5	4	100

between the phenotypes. In regard to the life span, most wild caught insects were mark–released except for weak individuals, and when possible their numbers were subsidized by laboratory-bred specimens which were released at the same time. Many more f. *typica* than f. *carbonaria* were laboratory-bred specimens, and hence were in their first day of imaginal life and would, *a priori*, be at an advantage. The bias would therefore be in favour of f. *typica*. Nor in regard to the wild f. *typica* was there any evidence of a diminution of hatchings which would lead to the usage of older moths, with a consequent shorter expectation of life. Furthermore, the length of life of the two forms does not differ appreciably when bred and kept in the laboratory. The question of a different life span can therefore be excluded.

The possibility of a differential migration or dispersal rate taking place between f. *typica* and the melanics, was also considered. A comparison of the relative proportion of marked phenotypes taken at traps situated on the periphery of the release area to those caught at traps which were placed two hundred yards or more distant, shows no difference in the frequencies of the three forms.

One must, therefore, accept that the figures represent *selective differential predation*. Moreover, having in mind the state of the tree trunks, the results of the observation–release experiments (Tables 8.1, 8.2, and 8.3), and lastly, the frequent witnessing of the selective nature of the predation undertaken by two species of birds, it was fair to assume that this was the correct interpretation. The figures obtained reflected a 50 per cent advantage of the melanic form, f. *carbonaria*, which only a hundred years previously had been regarded as a rarity. Because such forms are dominant

M[e]ats clean,   Chap. xij.

you: ye shall not eat of their flesh, but you shall have their carcases in abomination.

12 Whatsoever hath no finnes nor scales in the waters, that *shall be* an abomination unto you.

13 ¶ And these *are they which* ye shall have in abomination among the fowls, they shall not be eaten, they *are* an abomination: the eagle, and the ossifrage, and the ospray,

14 And the vulture, and the kite, after his kinde:

15 Every raven after his kinde:

16 And the owl, and the night-hauk, and the cuckow, and the hauk after his kinde,

17 And the little owl, and the cormorant, and the great owl,

18 And the swan, and the pelican, and the

PLATE 3.1 (p. 33). Photograph of English Bible, 1672, Leviticus 11, v. 13 (Bodleian Library, Oxford).

Phalæna majuscula, corpore & alis ex cinereo albentibus maculis & lineolis nigris undique variis.

Huic speciei fere peculiare est corpus habere alis concolor & pariter maculatum. Caput noctuæ instar grande: Scapulæ crassæ. Facies nigricat. Antennæ in maribus plumosæ, ut plerisque Phalænis. Prona alarum pars supinæ concolor sed paucioribus nigris maculis infecta. In nonnullis alæ interiores supina parte lineas duas nigras transversas obtinent.

Maii 29. 1693. Ex Eruca Geometra bacilliformi exiit. Foemina erat quæ ex Aurelia nostra Pyxidi inclusa erupit: Verùm in camera seu conclavi ubi asservata est, fenestellis apertis, mares duo circumvolitantes ab uxore, forte fortuna in conclave intempesta nocte ingressa, captæ sunt, à foeminæ, ut nobis videtur, odore allectæ de foris in conclave intrantes.

PLATE 7.2 (p. 108). Photograph of page 177 of *History of Insects* by John Ray, 1710. The earliest record of the assembling of male *Biston betularia* to a female (1693).

PLATE 4.1 (p. 45). Larva of *Boarmia roboraria* Schiff. ($\times \frac{3}{4}$).

PLATE 4.2 (p. 46). The 'lichened' form of the larva of *Gonodontis bidentata* L. ($\times 1\frac{1}{4}$) (Scotland).

PLATE 4.3 (p. 46). The melanic form of the larva of *Gonodontis bidentata* L. ($\times 1\frac{1}{2}$).

PLATE 4.4 (p. 46). The 'intermediate' form of the larva of *Gonodontis bidentata* L. ($\times 1\frac{1}{2}$).

PLATE 5.1 (p. 56). Beech trunk, industrial Yorkshire, 1965.

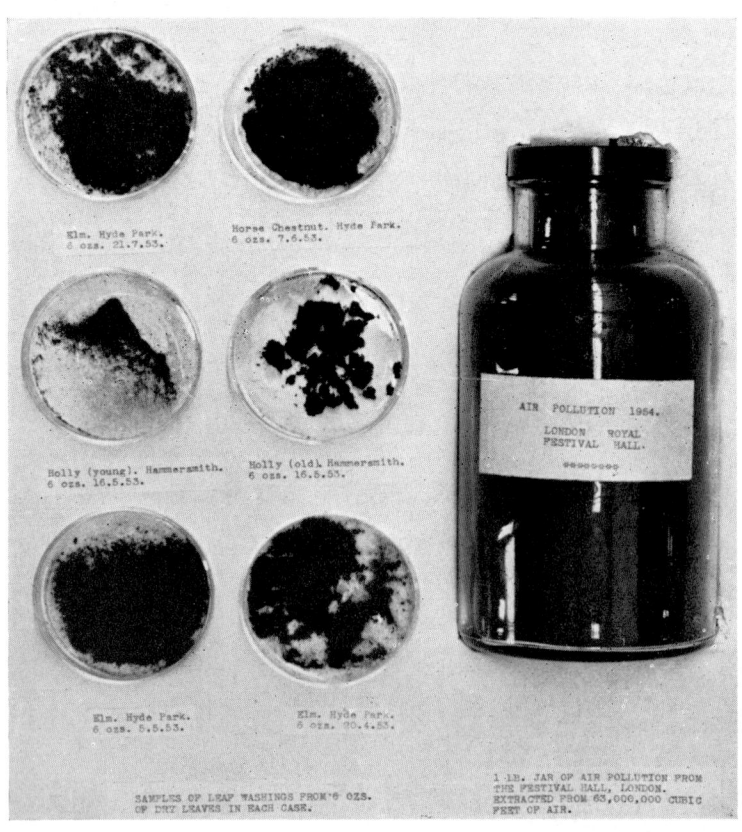

PLATE 5.2 (p. 56). Pollution obtained from washing 6 oz. leaves from different industrial areas, showing air pollution extracted from these.

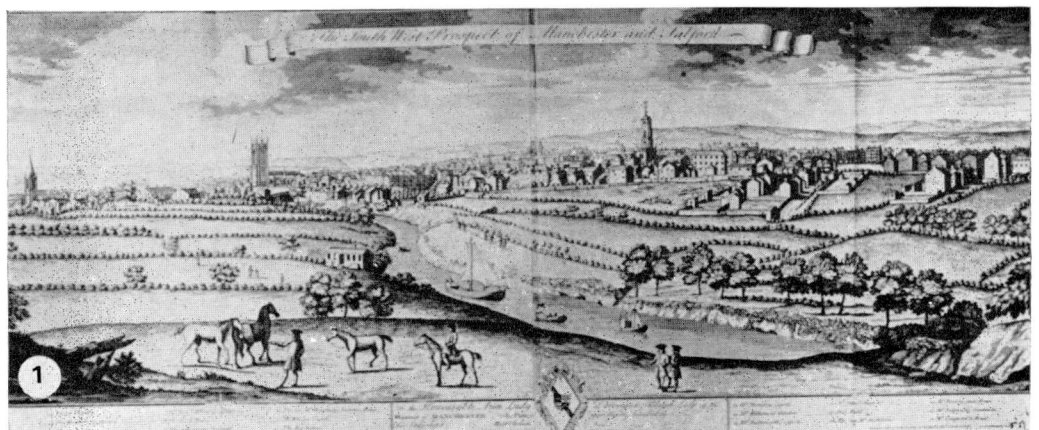

PLATE 5.3 (p. 58). 1. Manchester 1730. 2. Manchester 1860 (from the same site, note smoke plumes). 3. Manchester 1954, same district.

PLATE 5.4 (p. 59). Male *Phigalia pilosaria* Schiff. (× 1) and its black form at rest on lichened tree-trunk.

PLATE 5.5 (p. 59). Male *Phigalia pilosaria* Schiff. (× 1) and its black form at rest on heavily pollu[ted] Oak trunk.

PLATE 5.6 (p. 62). Examples of industrial melanic species ($\times$ 1).

Left
1. *Biston (Amphidasis) betularia* L. f. *typica*.
2. *Biston (Amphidasis) betularia* L. f. *carbonaria*.
3. *Biston (Amphidasis) betularia* L. f. *insularia* ($=$ins.[4])
4. *Gonodontis bidentata* Clerck f. *typica*.
5. *Gonodontis bidentata* Clerck f. *nigra*.
6. *Hemerophila abruptaria* Thngb. f. *typica*.
7. *Hemerophila abruptaria* Thngb. f. *fuscata*.

Right
1. *Colocasia coryli* f. *typica*.
2. *Colocasia coryli* f. *melanotica*.
3. *Cleora rhomboidaria* Schiff. f. *typica*.
4. *Cleora rhomboidaria* Schiff. f. *nigra*.
5. *Cleora repandata* L. f. *typica*.
6. *Cleora repandata* L. f. *nigra*.
7. *Semiothisa liturata* Clerck f. *typica*.
8. *Semiothisa liturata* Clerck f. *nigrofulvata*.

PLATE 5.7 (p. 62). Examples of industrial melanic species (× 1).

Left
1. *Stauropus fagi* L. f. *typica*.
2. *Stauropus fagi* f. *obscura* Rebel.
3. *Polia nebulosa* Hufn. f. *typica*.
4. *Polia nebulosa* Hufn. f. *thompsoni* Arkle
5. *Dasychira pudibunda* L. f. *typica*.
6. *Dasychira pudibunda* f. *concolor*.

Right
1. *Tethea octogesima* Hubn. (syn. *ocularis*) f. *typica*.
2. *Tethea octogesima* Hubn. (syn. *ocularis*) f. *frankii*.
3. *Tethea or* Schiff. f. *typica*.
4. *Tethea or* Schiff. f. *albingensis* Wncke.
5. *Tethea duplaris* L. f. *typica*.
6. *Tethea duplaris* f. *obscura* Tutt.
7. *Apatele alni* L. f. *typica*.
8. *Apatele alni* L. f. *steinerti* Caspari.

they could not have been maintained, undisclosed, within the populations in the same way as recessive heterozygotes. Their origin can only have come from either a new mutation or alternatively from some local reservoir where ecological conditions had in the past favoured such melanic forms.

The results of the 1953 experiments were of some interest because for the first time the main character change of Industrial Melanism—that of black or dark colouration replacing a previous light one—had been shown to offer a survival advantage when put to the test in countryside which had been subjected to heavy pollution for the previous one hundred years. Furthermore, this advantage was of the order of 50 per cent.

I recorded the results in *Heredity, Lond.*, in 1955 (Kettlewell 1955b) and elsewhere. The suggestion that birds could be the selective agents responsible for the phenomenon of Industrial Melanism produced a sceptical reaction amongst some biologists, in particular, the ornithologists, but even more so the lepidopterists. Also, ecological geneticists were surprised at the intensity of the selective advantage. The ornithologists could not believe that a pair of nesting Great Tits were capable of eating sixteen Peppered Moths in thirty minutes (Allen, 1955). Lepidopterists had never observed the bird predation of cryptic Lepidoptera in the wild (Harrison, 1920a, 1927a, 1928a, and 1956b), and geneticists had previously considered a selective advantage in the order of 0·1 per cent as being the normal.

I had been aware of this scepticism prior to the publication in 1955 and it appeared to me, therefore, that it was essential to have a visual record made in order to convince a group of doubting biologists.

Accordingly in 1955 I decided to repeat the Birmingham work and to make a film of the results as well as undertaking comparable experiments in unpolluted countryside.

I was indeed fortunate in being able to persuade Professor Niko Tinbergen, F.R.S., to join me in these ventures and to take advantage of his knowledge of bird behaviour and the techniques he had devised, for analysis. He also recorded the experiments on a 16 mm ciné film.

The repeat of the 1953 Birmingham experiments

The 1955 experiments differed from the earlier ones in that small numbers of marked males were released over a somewhat smaller area. Also, I considered my retrapping methods more effective. For this reason we had a higher proportion of recaptures of all forms than in 1953. A comparison of the results is, nevertheless, valid in regard to the *relative number* of each form which survived.

Mark–release–recapture results, 1955 (Table 8.6). A total of 227 *B. betularia* were released, 154 f. *carbonaria*, 64 f. *typica*, and 9 f. *insularia*. Of these, I recaptured a total of 100, 82 being f. *carbonaria*, 16 f. *typica*,

TABLE 8.6

Recapture figures for B. betularia (*males only*)
Rubery, Near Birmingham, 1955.

Date	Releases				Catches				Recaptures			
	carb.	typ.	ins.	Total	carb.	typ.	ins.	Total	carb.	typ.	ins.	Total
8.7	54	23	(5)	82	62	7	5	74	—	—	—	—
9.7	—	—	—	—	73	11	1	85	33	11	0	44
10.7	100	41	(4)	145	51	5	2	58	3	2	0	5
11.7	—	—	—	—	50	7	2	59	46	2	(2)	50
12.7	—	—	—	—	89	7	5	101	0	1	0	1
13.7	—	—	—	—	25	4	0	29	—	—	—	—
14.7	—	—	—	—	53	2	1	56	—	—	—	—
15.7	—	—	—	—	20	2	2	24	—	—	—	—
16.7	—	—	—	—	13	2	0	15	—	—	—	—
17.7	—	—	—	—	20	2	0	22	—	—	—	—
18.7	—	—	—	—	15	2	2	19	—	—	—	—
19.7	—	—	—	—	5	1	0	6	—	—	—	—
20.7	—	—	—	—	10	1	0	11	—	—	—	—
Totals	154	64	(9)	227	486	53	20	559	82	16	2	100

	1955				1953			
	carb.	typ.	ins.	Total	carb.	typ.	ins.	Total
Wild Birmingham population. per cent phenotype.	86·94	9·48	3·58	559	85·03	10·14	4·83	621
Per cent return of releases = recaptures	53·25	25	(22·2)	100	27·5	13·0	17·4	149

and 2 *insularia*. This represents a return of 53·25 per cent of the f. *carbonaria* release, 25 per cent of the f. *typica*, and 22·2 per cent of the f. *insularia*. Once again we recaptured twice as many of the black form as of the light, but this differed from the 1953 releases which were spread over 11 days, when we recaught 27·5 per cent of f. *carbonaria* and 13 per cent of f. *typica*. This no doubt reflected increased efficiency in our 1955 sampling methods, but it could also have been effected by decreased predation, diminished migration or a smaller natural population. The important point to note is that the *relative* return figures were of the same order as in my previous series of experiments.

Dr. Tinbergen succeeded in filming Redstarts (*Phoenicurus phoenicurus* L.) taking and eating our releases (Plate 8.3) and I am able to give (Table 8.7) his records of the predation which took place on two days whilst he kept observation from a hide. On each occasion the two forms were released in equality and subsequently replenished at intervals *after all of one form*

TABLE 8.7
Observation on predation of B. betularia, by Redstarts
for 2 days only (by Tinbergen from a hide), Birmingham, 1955.

	f. typica	f. carbonaria	Total
19.7 (a.m.)	12	3	15
20.7 (a.m.)	14	3	17
20.7 (p.m.)	17	9	26
Total	43	15	58

N.B. Replenished to equality as soon as all three of either phenotype had been taken.

had been taken. In this way, he recorded that 43 of the pale *typica* form were eaten to 15 of the black form *carbonaria* and, on the majority of occasions, two or more f. *typica* were eaten before a f. *carbonaria* was discovered. Once again we were able to show that the presence of a conspicuous light coloured f. *typica* immediately put the better hidden f. *carbonaria* at a disadvantage which they would not otherwise have incurred if they had been released on their own.

A total of 559 wild *B. betularia* were caught over a period of 13 nights, consisting of 486 f. *carbonaria*, 53 f. *typica*, and 20 f. *insularia*, giving the percentages respectively of 86·94, 9·48, and 3·58 (see Table 8.6). This compares with 621 *B. betularia* taken over 11 nights in 1953 with f. *carbonaria* 85·03 per cent, f. *typica* 10·14 per cent, and f. *insularia* 4·83 per cent; the 1953 introduction of f. *typica* certainly had no effect on the 1955 frequencies. Furthermore. Table 8.5 demonstrates that for the 6 nights on which light and assembling traps were both in operation the total catch (=wild population + releases) showed a similar proportion of each phenotype as coming to both methods of collecting. It again compares with the high degree of consistency with my 1953 figures.

This small repeat experiment fully corroborated the findings of the previous one, namely that the pale *typica* form, as found in unpolluted countryside, is at a cryptic disadvantage in an industrial area. The fact that birds eliminate these selectively, thereby affecting the evolution of the Peppered Moth, was again recorded.

There remained, however, for us to test the situation in which melanic forms were introduced into uncontaminated woodlands similar to those which prevailed 200 years ago and where such forms would only have occurred by recurrent mutation.

(iii) The selection experiments in an unpolluted woodland

The problem of finding an isolated deciduous wood where the trunks and boughs were heavily lichened proved much more difficult than that of choosing an optimum site affected by pollution fall-out. In fact, I examined large areas of countryside during 1954 but could find no lichened woods over the eastern half of England as far north as Yorkshire. With the exception of isolated Ash (*Fraxinus*) and Apple (*Pyrus*) trees and, in certain marshy areas in Norfolk, Alder (*Alnus*), woods with lichened covered trunks and boughs were non-existent. To the south of London it is true that green vegetative lichen growth can still occasionally be found on oaks in southern Sussex and Hampshire, as in the New Forest, and locally elsewhere. Not until I went as far west as Dorset could a heavily lichened woodland be found.

Here I chose Deanend Wood, an unspoilt relict part of an ancient deciduous forest. To the south-west there were no industrial areas whatsoever. Leaf washings taken in July suggested that there was little pollution fall-out.

I sampled *Biston betularia* here for one night in 1954 and of the 20 males I caught, 19 were f. *typica*, 1 f. *insularia*, but none were f. *carbonaria*. Subsequent work in 1955 (see below) confirmed the absence of this form there. Nevertheless, it can be assumed that such melanic mutations take place and have taken place from time to time throughout the range of the species. Accordingly, in 1955, in mid-June, I set up camp in these woods. Suitable sheds were found to house the three thousand *B. betularia* pupae. An electric generator with mercury-vapour traps was installed in the centre ride and the periphery of the wood was lined with cages destined to contain virgin females, so that at night the whole release area would be subject to a concentration of female assembling scent and this, no doubt, succeeded in holding a proportion of our male releases within the area of woodland.

The experiments were designed so that each aspect of the work could be compared with the previous Birmingham results, and the methods employed came under the following headings:

A. Scoring values as gauged by human standards.
B. Direct observation as to what happened to the individuals so scored.
C. Recapture figures which provided data over longer periods.

On every occasion, recapture results were assessed on male Peppered Moths only, because of the impossibility of recapturing females. However, these were used as releases for continuous observation from hides where, at the same time as a film record was made by Tinbergen, the order of predation of each insect was noted.

A. *Scoring values for crypsis*

The technique of scoring previously used was to assess, in the first place at a distance of two yards, whether a released moth was (*a*) conspicuous, or (*b*) inconspicuous. Subsequently, three categories were allowed for each, so that the score for (*a*) would be either -1, -2, or -3, and for (*b*) $+1$, $+2$, or $+3$, depending on the distance at which the insect faded into its background. This method of scoring had worked satisfactorily in the Birmingham experiments. It came as a surprise then to find that of 120 f. *typica* released in Deanend Wood during the first few days, all were classed as inconspicuous, with 85 per cent scoring $+3$. Conversely, all 75 f. *carbonaria* scored by us appeared conspicuous, and 80 per cent of these scored -3. In view of the random sample of the release points selected, it appeared unnecessary to continue the arduous procedure of scoring the cryptic value of each individual release, and it was accepted that in these woods, to the human eye, the majority of the light *typica* form were extremely well hidden, and the f. *carbonaria* were nearly always conspicuous (Plate 8.1).

B. *Direct observation*

A record of the number of individuals present or absent by late afternoon. Early on in the experiments an apparent deficiency of the light *typica* form was recorded when the afternoon check on releases took place. There were also many f. *carbonaria* missing. It became increasingly obvious that one was passing over the *typica* form on the lichened tree-trunks, and that they were practically impossible to see. To test this, I did a check immediately following a morning's release, in an area where my continual presence prevented predation. All the f. *carbonaria* were present, but over 30 per cent of the f. *typica* were unaccounted for. The cryptic efficiency of the f. *typica* on a lichened background is, in fact, greater than that of f. *carbonaria* on the blackened Birmingham tree trunks. For this reason, this type of recording was discontinued.

Observation from hides. In the course of filming, Tinbergen, who had to spend the greater part of each day in a hide, recorded the order in which predation occurred. These figures were added to my own observations. It must be emphasized that these records were of concentrations of female *B.*

betularia necessary for photography, but that they played no part in my release–recapture figures. On each occasion, an equal number of black and white forms were used at the commencement, but it was found impracticable to replace each phenotype after it had been taken. Their numbers were replenished after the last of one phenotype had been eaten, thus preventing simple statistical analysis. By doing this, it will be appreciated that the bias was, thereafter, in favour of that phenotype which had been eliminated previously, and it was found that on the majority of occasions the more conspicuous of the two forms were all taken before any of the others. This behaviour was common to five species of birds to a greater or lesser degree (Table 8.8, Plates 8.4, 8.5).

TABLE 8.8
Direct observation on predation by five species of birds. Deanend Wood, Dorset, 1955.

	Observer	f. *carbonaria*	f. *typica*
Spotted flycatcher	N.T.	46	8
(*Muscicapa striata* L.)	H.B.D.K.	35	1
Nuthatch	N.T.	22	8
(*Sitta europaea* L.)	H.B.D.K.	9	0 (first day)
	H.B.D.K.	9	3 (second day)
Yellowhammer	N.T.	8	0
(*Emberiza citrinella* L.)	H.B.D.K.	12	0
Robin	N.T.	12	2
(*Erithacus rubecula* L.)			
Song Thrush			
(*Turdus ericetocum* L.)	N.T.	11	4
Total predation observed (for days when records were kept)	—	164	26 Total 190

C. Recapture results

Releases were undertaken on fourteen occasions in all, making a total of 984 individuals, 473 being f. *carbonaria*, 496 f. *typica*, and 15 f. *insularia*. I recaptured 30 f. *carbonaria*, 62 f. *typica*, and 4 f. *insularia*. For all types of release we got back 12·5 per cent of the *typica* form, but only 6·3 per cent of the f. *carbonaria* (Table 8.9).

For various reasons, the releases on three of the days should be excluded from this total as they are unduly biased in one way or another. This gives a figure of 804 releases for the 11 correct days: 406 f. *carbonaria*, 393 f. *typica*, and 5 f. *insularia*. I got back 19 f. *carbonaria* and 54 f. *typica*, being 4·68 and 13·74 per cent respectively (see Table 8.9).

(iv) Release problems

Various types of release techniques were undertaken with the object of finding the best way of subjecting as many individuals as possible, under natural conditions, to maximum predation for the longest time. I propose, therefore, to discuss three in some detail, due to the unsatisfactory nature of the releases. These occurred on 13, 15, and 16 June, 1955 (Table 8.9).

TABLE 8.9
Recapture figures for B. betularia (males only).
Deanend Wood, Dorset, 1955.

Date (1953)	Releases				Catches				Recaptures			
	carb.	typ.		Total	carb.	typ.	ins.	Total	carb.	typ.	ins.	Total
(13.6)	(37)	(38)	(9)	(84)	—	—	—	—	—	—	—	—
14.6	8	17	2	27	4	17	2	23	(4)	(1)	(1)	(6)
(15.6)	(8)	(25)	(1)	(34)	2	27	5	34	2	9	(3)	14
(16.6)	(22)	(40)	(0)	(62)	1	34	2	37	(0)	(0)	(0)	(0)
17.6	—	—	—	—	7	58	3	68	(7)	(7)	(0)	(14)
18.6	42	65	(3)	110	0	30	1	31	0	1	0	1
19.6	39	72	0	111	2	26	1	29	2	6	0	8
20.6	24	57	0	81	1	44	2	47	1	3	0	4
21.6	42	29	0	71	1	13	2	16	1	4	0	5
22.6	—	—	—	—	5	13	1	19	5	4	0	9
23.6 } 24.6 }	No Releases				No Catches				No Recaptures			—
25.6	82	43	0	125	1	11	0	12	0	0	0	0
26.6	—	—	—	—	3	8	1	12	2	2	0	4
27.6	51	28	0	79	0	8	0	8	0	0	0	0
28.6	22	22	0	44	1	20	0	21	1	5	0	6
29.6	17	18	0	35	0	14	0	14	0	5	0	5
30.6	24	11	0	35	4	11	1	16	3	6	0	9
1.7	—	—	—	—	2	9	0	11	2	2	0	4
2.7	—	—	—	—	0	7	0	7	0	0	0	0
3.7	—	—	—	—	—	—	—	—	—	—	—	—
4.7	55	31	0	86	—	—	—	—	—	—	—	—
5.7	—	—	—	—	0	9	0	9	0	7	0	7
Totals	473	496	15	984	34	359	21	414	30	62	4	96

	carb.	typ.	ins.	Total
Wild Deanend population	(4)	297	17	318
Per cent phenotype	(Possible escapes)	94·60	5·4	
Release after 1 day of self-determination	2	4	(1)	
Per cent phenotype	—			
Per cent return of releases	6·34	12·50	(26·67)	

The release of 13 June, 1955. I used 84 *B. betularia* in the first experiment. These were released onto only twenty trees, the reason for this small number of release points being that I had been unable to locate and prepare the other trees in the area. The result was that I produced a high concentration of moths on comparatively few tree trunks (an average of 4 per tree), nor were the two forms on every occasion released in equality per tree. The late afternoon check showed that in nearly every case the moths were either all present or all absent per tree. Subsequently, by direct observation, Tinbergen and I found that a concentration of releases increased the predation risk for all present, even though the birds took the more conspicuous ones first. It will be noted that on this occasion we recaptured 4 f. *carbonaria* and 1 f. *typica* (expected—near equality). It is possible that, apart from a faulty release technique, a further factor may have played a part in the production of these figures. The birds in this wood were unlikely to have had any previous experience of the black Peppered Moths, and it is conceivable therefore that at first some of them did not recognize them as an article of diet. Tinbergen did, in fact, record a similar incident in one of the Birmingham 1955 experiments. A Wren (*Troglodytes troglodytes* L.) flew on to a tree trunk on which was released a conspicuous f. *typica* (Birmingham frequency 10 per cent) and after scrutinizing it closely it flew away without attacking it.

The release of 15 June, 1955. This was designed to test whether the presence of a f. *carbonaria* on the same tree trunk as a f. *typica* lowered the latter's expectation of survival. Apart from the 8 male f. *carbonaria* released, females had, on this occasion, to be used also (because of a temporary lack of f. *carbonaria* males), so that each of the male f. *typica* had a black moth on the same tree trunk. Two trees, however, were used as controls, and on these two f. *typica* were released on each, with no accompanying melanics. The late afternoon examination (6 p.m.) gave the surprising result that no *B. betularia* of any form were to be found on any of the release trees, with the exception that one of the control trees (without a f. *carbonaria* being present) had two f. *typica* on it. The following night, out of 37 *B. betularia* caught at light and assembling traps, none were from the releases of the 15 June. The next night, however, I recaught one f. *typica* (out of 68) from this release.

The important point is, however, that in order to obtain unbiased results a release of one moth per trunk must be the rule.

The release of 16 June, 1955. This release was undertaken between 5 p.m. and 6 p.m., long after the maximum predation time of birds, in order to test whether, having excluded predation, the design of experiment favoured unduly one phenotype more than another. Twenty-two f. *carbonaria* and 40 f. *typica* were released. Of these, 7 f. *carbonaria* and 6 f. *typica* were retrapped the same night (expected—4·61 to 8·39). It will be noted that this represents the return of nearly a quarter of our releases. I am unable to account for the increased proportion of f. *carbonaria* recaptures over f. *typica*, which occurred this night, but in view of the absence of predation it suggests that the recapture arrangements did not unduly favour the return of f. *typica* more than f. *carbonaria*.

It can be seen from this that the act of predation is no simple response to a single stimulus. It involves, apart from insect cryptic efficiency, such other considerations as insect density, bird conditioning, and searching intensity per trunk, stimulated by an immediate previous experience of finding a conspicuous insect. All other releases were in fact conducted in a uniform manner (with the exception of one other satisfactory method used). In these, on every occasion, there was one phenotype released per tree, and the experiment was conducted over a larger area of woodland. Furthermore, as in the Birmingham experiments, other species of moths which were inhabiting the wood were released within the area at the same time, to minimize the effect of conditioning. This, of course, involves a great deal of extra work, but it is necessary. One further point was noted, that the same trees must not be used as release points on succeeding days, as the birds become conditioned to them.

An alternative method of release, 18 June, 1955. Forty-two f. *carbonaria*, 65 f. *typica*, and 3 f. *insularia* were allowed to fly out of their separate boxes (which had been previously warmed on the engine of my car), just before sunrise, between 4 a.m. and 4.30 a.m. The majority flew and took up positions on the boughs and trunks of nearby trees. I, therefore, used many release points within the area. This method was not repeated because it is necessary to get each insect airborne over a very short period of time, which was difficult due to the coldness of the mornings, so that by the time the last few flew, birds were active and, in fact, a Spotted Flycatcher (*Muscicapa striata* L.) chased and caught two f. *typica*. A too early release, on the other hand, would involve a number of the moths coming to the various traps in action from the previous night. With the exception of the two f. *typica* taken by birds, it appeared that this release was satisfactory. One f. *carbonaria* and 6 f. *typica* were subsequently recovered.

Deanend results

A. *Human assessment.* One hundred per cent of f. *typica* were classified as 'inconspicuous' with 85 per cent scoring $+3$, the highest mark of six alternatives. By contrast, every f. *carbonaria* was scored 'conspicuous' with 75 per cent having -3, the lowest award.

B. *Direct observation.* In a release of female moths with f. *typica* and f. *carbonaria* in equality 190 were seen by two observers to be eaten by five species of bird. Of these 164 (86·3 per cent) were f. *carbonaria*. Direct observation suggested that this form was at a disadvantage of approximately 36 per cent.

C. *Recapture results.* Of all types of releases in near-equality 12·5 per cent of f. *typica* were recaptured to 6·3 per cent of f. *carbonaria*. Three of the 14 experiments, however, I deemed as biased for one reason or another; when this is taken into account and they are excluded, recaptures of f. *typica* were three times that of f. *carbonaria*.

A comparison of the Birmingham and Deanend Wood experiments

984 *B. betularia* of both forms in near-equality were released over 14 occasions at Deanend Wood. This can be compared with 857 similar releases at Birmingham. The results of these two contrasting sets of experiments are diametrically opposed, in scoring values for crypsis of the two forms in the two contrasting localities, in direct observation on bird predation, and in our recapture results.

The two woods were typical of the extremes of unpolluted and polluted areas respectively.

Throughout Britain, woods varying in the degree of pollution between these two are to be found. On every tree in every wood similar selective predation must be taking place, with the *insularia* forms finding their highest frequency in semi-polluted areas where *Pleurococcus* has replaced lichens. Such areas stretch from South Wales to Oxford.

The resulting frequency distribution of the different forms is discussed in the following chapter.

9. The distribution of *Biston betularia* L. and its melanic forms in Great Britain
(with special reference to frequency surveys between 1952 and 1970)

THE experiments quoted in the last chapter proved that intensive selective predation was taking place in two entirely different types of habitat, town and country. Surprisingly, the degree of predation in each locality was of the same order, though it was somewhat greater in the unpolluted deciduous woodlands in Dorset. This must surely reflect the ability of birds to adapt successfully to industrial conditions.

The frequency of the forms bears a direct relationship to the local backgrounds (although each form has other advantages and disadvantages). Furthermore, Clarke and Sheppard (1966) have shown that frequencies can vary over a distance of only four miles. Such frequency changes in Britain must therefore be a delicate indication of the amount of pollution fall-out in each locality. This has progressed insidiously in the last 150 years. One is able, therefore, not only to record the present spread of the melanic forms, but also to assess their rate of spread retrospectively, and from this their advantage.

This chapter therefore, attempts to give a retrospective assessment of the spread of the melanics; also a current record of the frequency of the forms which I have gathered from about 170 observers over a period of 19 years.

(i) Early spread of the melanic forms

There is every reason to believe that, prior to the middle of the last century, melanic forms were maintained in the population solely by recurrent mutation. For the next fifty years these melanics were much sought after by collectors, so that it is reasonable to suppose that their capture would be recorded assiduously when they first made their appearance. Table 9.1 gives a list of the earliest records of *Biston betularia*, f. *carbonaria* from 20 different localities, and it has been assumed that at these dates f. *carbonaria* would not, in the majority of instances, be at a local frequency higher than, say, 1 per cent.

Though sadly lacking essential data on sample size, certain information can be deduced from these earlier records. It appears that f. *carbonaria*

132 The distribution of Biston betularia

TABLE 9.1

Earliest records of f. carbonaria in Britain. Approximate selective advantages (S.A.) have been estimated for a period of 100 years on the assumption that the local frequency of f. carbonaria was not greater than 1 per cent when first recorded.

County Locality (recorder)	1st f. carbonaria recorded	1900–6 survey (when known)	S.A.	Locality (recorder)	1952–6 frequency sample size	S.A.	Locality (recorder)	S.A. over total period
BERKSHIRE (Barrett)	1885	—	—	—	11% (63)	—	Newbury (Saundby)	3%
CAMBRIDGESHIRE Cambridge (Farren)	1892	'now seen every year'	—	(Farren)	93% (88)	—	Cambridge (H.B.D.K.)	14%
Ely (Cross)	1895	—	—	—		—	,, ,,	15%
CHESHIRE Chester (Barrett)	Delamere 1860	83% sample 180	18%	Chester (Arkle)	94% (124)	5%	Chester (G. Smith)	10%
ESSEX Colchester (Harwood)	1892	—	—	—	86% (822)	11%	Bradwell (Dewick)	11%
Dovercourt (Mathew)	1902	'black not observed'	—	Stroud (Prideaux)	16% (185)	14% >6%	Nailsworth (Demuth)	14% >6%
GLOUCESTERSHIRE	—							
HAMPSHIRE New Forest (Barrett)	1897	—	—	—	10% (324)	4%	Eastleigh (Goater)	4%
IRELAND Castle Bellingham (Thornhill)	1894	—	—	—	7% (58)	3%	Belfast district (Wright)	3%
ISLE OF MAN (Cassall)	1904 2 sps	—	—	—	13% (69)	6%	Santon (Hedges)	6%
KENT	—	'black not observed'	—	Farnborough (Christy)	73% (224)	>12%	E. Malling (Groves)	>12%
LANCASHIRE Manchester (Edleston)	1848	'black prevalent. Type occurs'	—	Manchester (Clutton and Tait)	98% (350)	—	Manchester (Michaelis)	15%
LINCOLNSHIRE (Barrett)	c. 1860	'both light and black'	—	(Fowler)	91% (158)	—	Louth (H.B.D.K.)	9%
LONDON DISTRICT Woodford (Mira and Bacot)	1897	37% (sample 27)	52%	Woodford (Main and Harrison)	90% (327)	8%	Whetstone (Lovell)	14%

S.A. = Selective advantage of f. carbonaria as calculated from Haldane's (1924) table.

The distribution of Biston betularia   133

TABLE 9.1 (continued)

Location	Date	Description	% black (n)	S.A.	Locality (observer)	S.A.
MONMOUTH Newport (Wheeler)	—	—	—	—	—	—
NORFOLK King's Lynn (Atmore and Baker)	1892	'Prevalent' (=50%)	>60% (Atmore and Baker)	—	Fritton (H.B.D.K.)	10%
STAFFORDSHIRE Cannock Chase (Barrett)	1878	'all black' (=≥80%)	>25% (Frere)	<3%	Cannock Chase (Richardson)	11%
SUFFOLK Ipswich (Morley and Pyett)	1894	—	—	3%	Lowestoft (Burton)	10%
SURREY Croydon (Gower)	1906	'black not observed'	Reigate (Meyrick)	10%	Woking (C. de Worms and Trundell)	>13%
SUSSEX	—	'black not observed'	(Hewitt)	>13%	Ottershaw (Bretherton) Hastings (Astbury) Eastbourne (Ellison)	>13% >6% >7%
WARWICKSHIRE Birmingham	—	'50% black'	(Barrett)	>13% >6% >7%	Birmingham (H.B.D.K.)	5%
WESTMORLAND Kendal and Windermere (Moss)	1870	'black commoner than type' (=50%)	14% (Barrett)	5%	Kendal (Birkett)	—
WILTSHIRE Marlborough (Prentice)	1951	'black not observed'	—	Marlborough (Davis)	Marlborough College	>45%
YORKSHIRE Bradford (Butterfield)	'scarce' 1876	'black now prevalent' (=>50%)	>10% (Butterfield)	>45%	Bradford (Briggs)	10%
Huddersfield (Porritt)	1861	'now only black' =≥80%	≥15% (Porritt and Morley)	<10%	Sheffield (H.B.D.K.)	17%

S.A. = Selective advantage of f. *carbonaria* as calculated from Haldane's (1924) table.

was taken for the first time in many widely separated places between 1848 and 1900 (see Chapter 5). Whether this represents migration from one centre, or numerous discrete mutations, will be discussed later. Secondly, very shortly after the initial captures this melanic form increased rapidly in a brief space of time. Thirdly, until the end of the century, f. *carbonaria* was unknown in southern England. At this time it had a Lancashire, Midland, and eastern county distribution and was not taken in the London area till 1897. Fourthly, when these earliest records are considered with recent data, the selective advantages (Table 9.1) for the *carbonaria* form can, with a reasonable degree of accuracy, be assessed for each locality over the period involved. The ability to do this in one of the most "transient polymorphisms" ever known, makes these earliest records valuable.

(ii) The three surveys

1952–56 (Kettlewell, 1958*a*)
1957–64 (Kettlewell, 1965*d*)
1965–70 (Kettlewell, unpublished)

In view of the inadequate nature of the data on the frequencies of the melanic forms at any one period in the past, I decided in 1952 to obtain as much data on *B. betularia* as quickly as possible, with the object of providing figures for future reference.

Appendix C gives a list of frequencies of the three forms of *B. betularia* from many areas of Great Britain, both in the earliest survey and also two subsequent ones, and covers a period of 19 years. It must be stressed that the true frequency of the *insularia* forms is masked by f. *carbonaria* in whose presence it cannot be recognized. The form *carbonaria* is dominant to the majority of *insularia* forms; in one instance it may be epistatic. The higher the frequency of f. *carbonaria* therefore, the greater the error in assessing the f. *insularia* frequency. Even in large random samples from industrial areas the data are often inadequate.

A distribution map (Fig. 9.1) of the frequencies of the three forms shows:

(*a*) A correlation between the frequency of the melanic forms and the industrial areas of Britain.

(*b*) A high frequency of f. *carbonaria* throughout eastern England from north to south, though far removed from industrial centres. This has been brought about as an indirect effect of long-continued smoke fall-out carried by the prevailing south-westerly winds from central England.

(*c*) Western Britain, with the exception of Cheshire, Lancashire, and Westmorland (Kendal) is virtually melanic free.

The distribution of Biston betularia

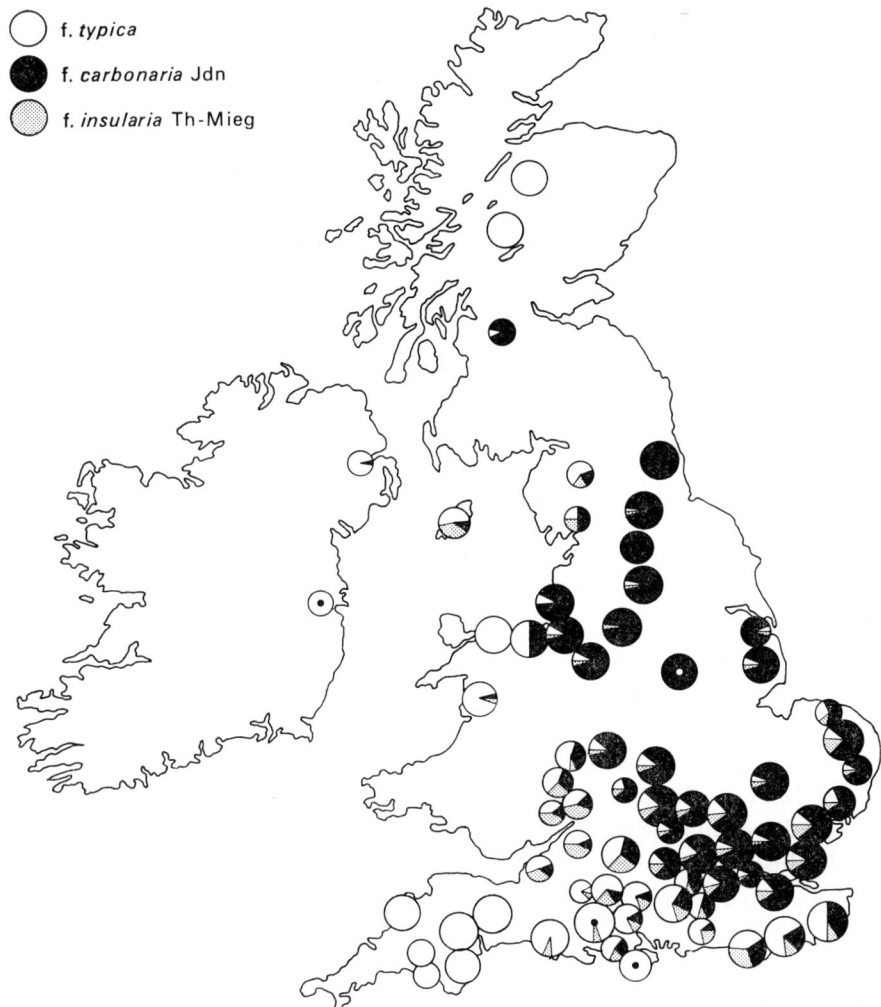

FIG. 9.1 A frequency map of *Biston betularia* and its two melanics, f. *carbonaria* and f. *insularia* comprising more than 30 000 records from 83 centres in Britain

(d) Northern Scotland has no melanic *B. betularia* (with the exception of f. *insularia*, which is extremely rare in the Western Isles of Scotland) but in the Glasgow district the f. *carbonaria* frequency is about 90 per cent.

(e) In Ireland, f. *carbonaria* has been recorded, first in 1894 from the Belfast district and also Dublin, but is still at a low frequency. It has been found nowhere in the west.

(f) The map shows multiple clines running out from the higher f. *carbonaria* frequencies in the centre of England. There may be a rapid decline from east to west as, for instance, from Birmingham to Wyre Forest, or from Cheshire to North Wales, which has been extensively investigated by Clarke and Sheppard (1963). Alternatively, the clines may be gradual as always found in an easterly direction.

(g) The highest frequencies of f. *insularia* phenotypes are, for the most part, found in the Gloucester and Severn Valley district of England. Fuller details of the 1952/56 survey and the two subsequent ones, 1957/64 and 1965/70, are given later in this chapter.

(iii) Analysis of f. *carbonaria* frequencies

The evidence for mutation rate versus dispersal

If one reviews the first county records, it is indeed difficult to resist the conclusion that the new mutant had radiated outwards from an original centre of mutation near Manchester. There is no doubt that following the original capture in 1848, the nearest counties to Lancashire were the next to record its appearance: Cheshire in 1860, Yorkshire in 1861, Staffordshire in 1878, and Westmorland in 1870. The London district, on the other hand, and southern England did not report a f. *carbonaria* until about twenty years later (1897). It is significant also that the eastern counties of Norfolk, Suffolk, and Cambridge all recorded their first f. *carbonaria* practically simultaneously between 1892 and 1895.

All this is consistent with the spread of a successful gene with no ecological barriers, from its centre of origin. Furthermore, on the Continent, the same spread was taking place, being recorded by A. Hofmann from Hanover in 1884, the Netherlands and Thuringia in 1888, 'and in the next few years in various parts of the Rhine Valley, indeed he [Hofmann] thinks that its progress was up the Rhine' (Barrett, 1895–1902). Recent frequency data for Europe are given in Chapter 6.

In a short series of evening releases undertaken in open country near Louth, Lincolnshire, in 1956 I attempted to find the distance which *B. betularia* flies per night. Using different releasing points, different markings, and several collecting centres of known distance apart, I was able to show that of 93 releases, 6 travelled $1\frac{1}{2}$ miles, and 9 over half a mile within 48 hours. At Fritton in Suffolk, of 78 releases over three occasions, 16 were recaptured three-quarters of a mile distant, 12 within 24 hours. Clarke and Sheppard (1966) have recorded that marked males on two occasions flew a distance of approximately three miles within a few hours. Yet the two local populations showed significant frequency differences. These therefore must be brought about by natural selection, and *not* by migration.

From these observations, it can be accepted that *B. betularia* frequently flies a mile per night, probably much farther. Nevertheless, dispersal from industrial centres, cannot be accepted as the actual cause of the widespread distribution of f. *carbonaria*. There is considerable evidence that recurrent mutation also takes place; f. *carbonaria* has constantly appeared in isolated centres separated from others by usually impassable barriers. Thus it was taken in Dublin about 1950, near Belfast in 1894, the Isle of Man in 1904, and in 1956 at Torquay, each locality being separated by fifty miles or more from the nearest possible contacts. It is, in fact, likely that the f. *carbonaria* allele has a high mutation rate. This is in contrast with certain melanics of other species, such as *Ectropis consonaria* f. *nigra* (see Chapter 5).

The rate of spread: the sigmoid curve (Fig. 9.2)

Haldane (1924) pointed out that, if in Manchester in 1848 the *B. betularia* population was 99 per cent of the *typica* form, and if by 1898 it was not less than 95 per cent f. *carbonaria*, this represented an approximately 50 per cent advantage of the black form over the light during this period.* Magnitudes, somewhat less than this, are borne out from the earliest records (Table 9.1).

On this basis it is possible to plot a curve of the theoretical rate of spread and to consider the consequences of a single mutation occurring in a highly polluted area in Britain. I have had to assume certain criteria: a mutation rate of one in a million; that the new melanic mutant has a fifty per cent advantage over f. *typica*; and that the heterozygote has a constant advantage over both homozygotes which, in fact, is unlikely to happen. Figure 9.2 shows a diagram of the rate of spread of a dominant melanic mutant, such as f. *carbonaria*, under these specifications, and the graph can be divided into three parts: (*a*) a period of adjustment, (*b*) a period of

---

* It has recently been pointed out to me by Dr. E. R. Creed that this has, in the past, been incorrectly stated as a 33 per cent advantage, both by myself and others. Haldane's actual words were, 'therefore k (fitness) = 0·332 at least, i.e. at least three dominants (=f. *carbonaria*) must survive for every two recessives, and probably more.' Creed says, correctly, that the survival of three f. *carbonaria* to two f. *typica* per generation must reflect a 50 per cent advantage (150 *carbonaria* to 100 *typica*). Alternatively, this can be construed as a 33 per cent disadvantage of f. *typica* (66 f. *typica* to 100 f. *carbonaria*).

In my own references to selective advantages I use the first interpretation, so that Haldane's figure should be quoted as a 50 per cent advantage.

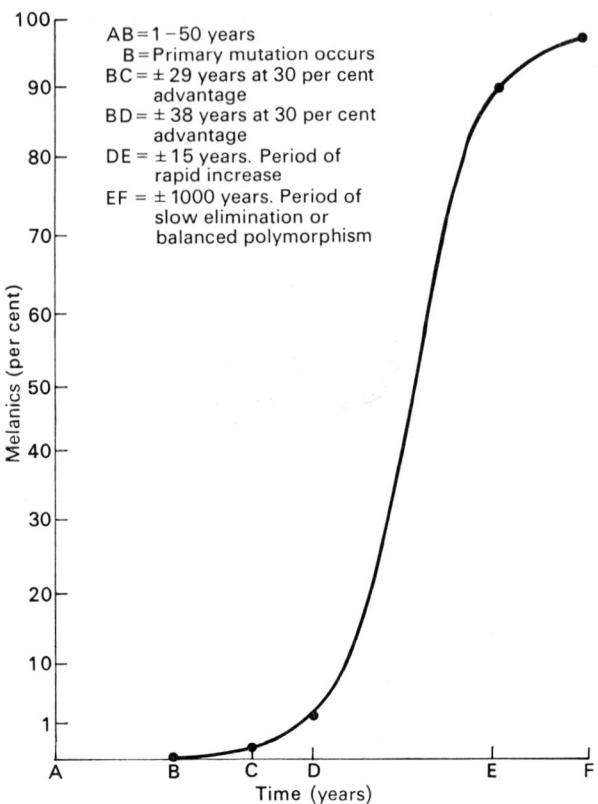

FIG. 9.2 Diagram showing rate of increase of a melanic (dominant) mutant with a mutation rate of one in a million, assuming a constant advantage of the heterozygote throughout (which in practice will not occur), with a 50 per cent selective advantage over its typical form

rapid spread, and (c) a period of slow elimination of f. *typica* or, alternatively, a balanced polymorphism.

(*a*) *The period of adjustment.* We must, in the first place, consider the time lag between the original mutation (B) and its arrival at the one per cent level (C). At frequencies below this, it is unlikely to have been recognized in the population because of the poorer techniques in sampling used at that date. As we have fixed certain criteria it is possible to theorize on the rate of spread. It would take about 29 years to reach the 0·1 per cent level, and 38 years to reach the 1 per cent level. Observed data are in agreement with this type of curve. The period under present consideration is represented by BC on the abscissa, and if in fact there was only one successful mutation in Manchester, and if by 1848 f. *carbonaria* had reached

1 per cent of the population, we can say retrospectively that this original mutation must have taken place about 1810, which is approximately fifty years after the commencement of the Industrial Revolution.

Cryptic adjustment. During this initial period (BC), the black mutants, though having great cryptic advantage, nevertheless would find themselves in a gene-complex entirely fitted for the specialized pattern of the *typica* form. It is of great interest that many of the earliest examples of f. *carbonaria* which I have obtained by searching innumerable old collections are different from the majority of heterozygous f. *carbonaria* bred today. Plate 7.1 shows two rows of f. *carbonaria*, the right-hand one of which is comprised of insects which were caught in the last century. It will be noted that they bear no resemblance to the forms of f. *insularia* which have an entirely different genetic origin. In contrast, the left-hand column shows five modern heterozygous f. *carbonaria* which differ considerably from those collected 100 years ago. Many of the earliest specimens show white markings on all the wings in both sexes greatly in excess of what are generally found today. The present-day heterozygote only occasionally shows minor degrees of white markings on the wings. On the other hand the majority of f. *carbonaria* have four white dots around the head and at the base of the wings. In the Sheffield area and, to a lesser degree, in other centres (Chester, for example) these dots are also disappearing in present-day specimens, leaving a completely black insect.

There can be two alternative explanations for this: first, that during this period the gene-complex has been adjusting itself by incorporating modifying genes in order to attain its greatest advantage; alternatively, different alleles could have spread through the population and replaced the old ones (see Chapter 19).

Such profound changes must lead to conflict as between colour and physiological advantages and disadvantages. This is because each gene must exert an effect upon both. Yet it appears that a satisfactory balance has been achieved but only after one hundred years of intensive selection.

Physiological adjustment. Back-cross broods of present-day Birmingham origin have consistently shown a small excess of melanics, whereas earlier broods at the begining of the century reflect an opposite tendency. Table 5.7 compares my own results of back-cross broods in the 1950s, fed on unwashed Oxford Sallow (*Salix* sp.) which corroborates this fact. It will be seen that there is a significant excess of melanics. Yet this does not happen in broods in which f. *carbonaria* is out-crossed to West Country stock (Devon) where this form does not occur, for here I have bred the two forms in equality (Brood B/1/54. Table 5.11). In contrast to this, Doncaster (1906a) records figures which deviate in the opposite direction for back-cross broods occurring between 1900 and 1906 (Table 5.7). It is probable that these also were fed on unwashed foliage. Both sets of data are homogeneous but are significantly different. ($\chi^2_{(1)} = 13\cdot27$.) It appears, therefore, that in its early phases the melanic mutant was not at a physiological advantage to the f. *typica*, and that only after a considerable period of adjustment did it become so. It is possible then that certain genes, previously adapted for keeping the complicated f. *typica* pattern in check, freed from cryptic responsibilities, are now able to exploit themselves, or their alleles, in order to bring about optimum physiological advantage. This period then (BC) must be regarded as a time during which the gene-complex is adjusting itself in various ways to the new mutant.

(b) *Period of rapid spread*. In theory, when once adjusted, the new mutant is free to spread, provided the selective pressures are maintained. In the earlier periods, because back-cross matings will leave a higher percentage of melanic offspring, this increase will proceed with greater rapidity up to a time when 75 per cent of the population is of the *carbonaria* form, with the highest number (50 per cent) of heterozygotes available. Thereafter it will tend to slow down at a speed directly proportionate to the increasing f. *carbonaria* frequency, when the number of heterozygotes in the population will commence to drop, but the homozygotes increase, as pointed out by Fisher (1937). Evidence of this obtained from the three surveys is given later in this chapter. The period of rapid spread is shown on the graph (Fig. 9.2) as CD, during which, under certain conditions, it is probable that f. *carbonaria* can change from being 10 per cent of the local population to 70 per cent within a period of ten to fifteen years, which reflects accordingly a selective advantage of from 30 to 50 per cent.

(c) *Period of slow elimination of f. typica or, alternatively, a balanced polymorphism*. The data in Appendix C show that in every industrial area f. *carbonaria* is now at a frequency of at least 85 per cent, but in no large sample (with one exception) is the value 100 per cent. Even after one hundred years, Manchester and Sheffield still have 1–2 per cent of non-*carbonaria* forms, and Lincolnshire, 9 per cent. Furthermore, the single sample in which actual figures are available in the earliest records is for Chester (Table 9.1), which shows that f. *carbonaria* rose in fifty years (1906–56) from 83 to 94 per cent only, giving a selective advantage of 5 per cent, after having achieved the earlier frequency from 1 per cent in forty-five years (=selective advantage 18 per cent). All the available evidence goes to show that f. *typica* continues in the population however great are the apparent cryptic selective pressures against it. This period is referred to as DE in the diagram (Fig. 9.2), and may represent many hundreds of years, depending on unpredictable variables. The evidence available suggests that by now, after experiencing one hundred generations, the new mutant has succeeded in achieving a gene-complex suitable for its optimum expression, and that the heterozygote has an advantage over both the homozygotes. This balanced polymorphism has been brought about by natural selection. Modifying genes, disadvantageous in the homozygous state, but advantageous as heterozygotes, will have become more closely linked to the new mutant. One would probably find that in the wild there is a figure below the expected Hardy–Weinberg ratio, because of the elimination of the homozygous f. *carbonaria*. Indeed the recent work of Clarke and Sheppard (1966) brings evidence to support this. This will inevitably be followed by a balanced polymorphism. Unfortunately, we as yet have no method of distinguishing heterozygous from homozygous f. *carbonaria* in wild samples.

(iv) Analysis of 'f. *insularia*' frequencies, and its distribution, past and present

The early history of 'f. *insularia*' is unsatisfactory in every respect. It is generally agreed that it appeared about the same time as f. *carbonaria*, but in certain districts, particularly in the south, before it. Doncaster (1906b) tried to differentiate 'two distinct forms', a light and a dark, and in this he was certainly correct. For the same reason I have placed 'f. *insularia*' in recent years within inverted commas, for this name covers a number of genetically distinct morphs, better referred to as 'the *insularia* complex'. A further complication is that the earlier breeders were confused on finding that back-cross f. *carbonaria* × f. *typica* broods, on some occasions, 'produced intermediates' (=*insularia* complex), whilst on others Mendelian segregation took place (Doncaster, 1906b). I must emphasize the fact that a cross of *pure* f. *carbonaria* × f. *typica* of British origin can never produce intermediate *insularia*-like forms.

History of spread

The form *insularia* occurred in Ireland in 1894, in Scotland (Kincardine) some time prior to 1906, and elsewhere in completely rural areas, but always as a rarity. Only towards the end of the last century did it appear in industrial areas in central England, and at a somewhat later date in London and the south. It was recorded on the Continent (Belgium) in 1886. All observers agree that it is an industrial melanic, and that it gets rarer following the upsurge of f. *carbonaria*. Nevertheless, f. *insularia* must be considered also as a non-industrial melanic occurring at a low frequency.

It is not surprising that f. *insularia*, being intermediate in appearance between f. *typica* and f. *carbonaria*, is found most frequently at the present time on the periphery of industrial areas, outside centres with a high frequency of f. *carbonaria*, in whose presence it is impossible to detect phenotypically. Background scoring efficiency (Kettlewell, 1955d), for the comparatively few releases of this form undertaken by me, gives f. *insularia* a position intermediate between f. *typica* and f. *carbonaria* in both industrial and rural environments. It is, in fact, admirably suited for resting on boughs covered with *Pleurococcus* and not lichens. Nevertheless, it must be pointed out that, in contrast with f. *carbonaria*, the phenotype frequency of f. *insularia* has never been found higher than 50 per cent in Britain (this is in contrast to the Netherlands).

The form *insularia* has the great drawback that it varies from those individuals which are indistinguishable from f. *typica* to those which are as dark as f. *carbonaria*.

I have therefore more recently attempted to classify individual f. *insularia* into one of five classes, from *ins.*^{1}, the lightest, to *ins.*^{5}, the darkest (Plate 9.1), though these do not necessarily imply different genotypes (see p. 107).

There are at least two separate genetic forms of f. *insularia* which are allelic to f. *carbonaria* (Clarke and Sheppard, 1964, Lees, 1968), probably many more, and these must in part contribute to the variable expression of f. *insularia*. Also in some broods from heterozygous pairings the homozygous individuals are dark (*ins.*^{4} and *ins.*^{5}) (Cadbury, personal communication) though these are probably rare in nature. A minority of the darkest form (*ins.*^{5}) are difficult to distinguish from f. *carbonaria*. At the opposite end of the scale rarely pale f. *insularia* may be mis-classified as dark f. *typica*. Because of the small numbers involved in these two categories, the frequency figures for f. *typica* and f. *carbonaria* are largely unaffected and can be accepted as being accurate (see Table 9.2).

Nevertheless, in spite of identification difficulties, the *insularia* complex offers many points of considerable interest. Haldane pointed out (personal communication) that a triangular graph of the three phenotypes (Fig. 9.3) shows that high values of f. *insularia* are associated with f. *carbonaria* frequencies of from 10 to 30 per cent. Districts showing this, for the most part, are centred around the Severn watershed and Gloucestershire which,

TABLE 9.2
A comparison of B. betularia *phenotypes from the same locality by four observers.*

Percentage phenotypes

typ.	carb.	ins.	Total	Name	ins. estimated frequency
18 per cent	77 per cent	5 per cent	484	J.B.P.	22·32
14 ,,	80 ,,	6 ,,	1435	R.F.B.	31·60
18 ,,	78 ,,	4 ,,	1191	C.deW.	18·80
15 ,,	74 ,,	11 ,,	424	E.T.	40·54

without doubt, receive slight but constant pollution fall-out from Bristol and South Wales. Also, the Isle of Man, which is subject to occasional but definite smoke drift from the industrial areas of Lancashire, when the

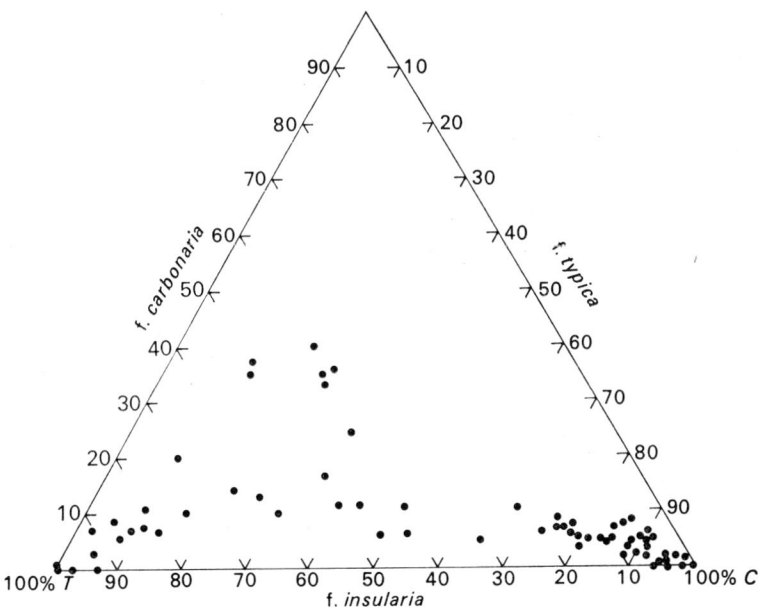

FIG. 9.3 Graph showing frequencies of the three phenotypes of *Biston betularia* in Britain

wind is in an easterly direction, has allegedly the high f. *insularia* frequency of 36 per cent, though, having recently (1972) seen these samples, I must admit that I would have classified them as dark f. *typica*.

Unfortunately the earlier records of f. *insularia* are, for the most part, lost, and what I have been able to extract is largely circumstantial. For example, the Giles Collection of Lepidoptera in the Folkestone Museum was amassed between the years 1880 and 1890. It was a local collection and in it is a series of 18 *B. betularia*, 12 f. *typica* and 6 f. *insularia*. They are by appearance not bred, in fact many of both forms are worn, and it is likely that this is a small random sample, and reflects correctly a frequency of about 33 per cent f. *insularia* at this date. The form *carbonaria* did not appear in Folkestone until a very much later date, about 1927 (Morley, personal communication). As will be seen from Appendix C, a more recent f. *insularia* frequency is about 11 per cent, with f. *carbonaria* standing at 42 per cent, which gives an estimated f. *insularia* frequency of 17 per cent. It has, in fact, dropped 16 per cent in thirty years, during which time the frequency of f. *carbonaria* has risen from approximately 1 to 42 per cent. The true frequency of f. *typica* would, therefore, have increased during this period from 66 to 83 per cent. It appears then that f. *carbonaria* has expanded at the expense of f. *insularia* and not f. *typica* and one explanation is that f. *insularia* in the presence of heterozygous f. *carbonaria* may interact with it in a disadvantageous manner. More likely is it that, as some f. *insularia* are allelic to f. *carbonaria*, this blackest form has replaced its former allele.

At the upper end of the f. *carbonaria* frequency, there appear to be only one set of early figures available. Arkle (Doncaster, 1906b) took a sample of 180 *B. betularia* at electric light in Chester, and this can be compared with a sample of 124 taken during 1956 by S. Gordon Smith. In a period of over fifty years, the estimated frequency of f. *insularia* has dropped by only 7·5 per cent in the population in spite of its cryptic disadvantage. Unfortunately, I have no more recent samples from here. Clarke and Sheppard (1966) have discussed the frequency of this form around Liverpool, although it also occurs on the Wirral Peninsula, Cheshire, and southward to north Wales. Using dead deep-freeze specimens they also ascertained its cryptic value in a number of differing localities. They confirmed that the frequency of this form in Cheshire has not changed greatly since 1956.

The f. *insularia* records obtained in the last few years, both in my own surveys for Britain and those of Clarke and Sheppard for a large area in the west are more helpful. Nowhere have we so far obtained a phenotype frequency at a level higher than 50 per cent. This can be interpreted in any one of several ways. Either f. *carbonaria* has swept through the population, due to the selective advantage of a more extreme mutant, enabling it to replace f. *insularia*; or, alternatively, homozygous f. *insularia* are at a considerable disadvantage to the heterozygotes. A more compelling explanation is that f. *insularia* is allelic to f. *carbonaria* so that fewer f.

insularia phenotypes appear in the population when the f. *carbonaria* mutant is at an advantage. Fig. 9.4 confirms, in fact, that f. *carbonaria* increases at the expense of f. *insularia* and only to a lesser degree of f. *typica*.

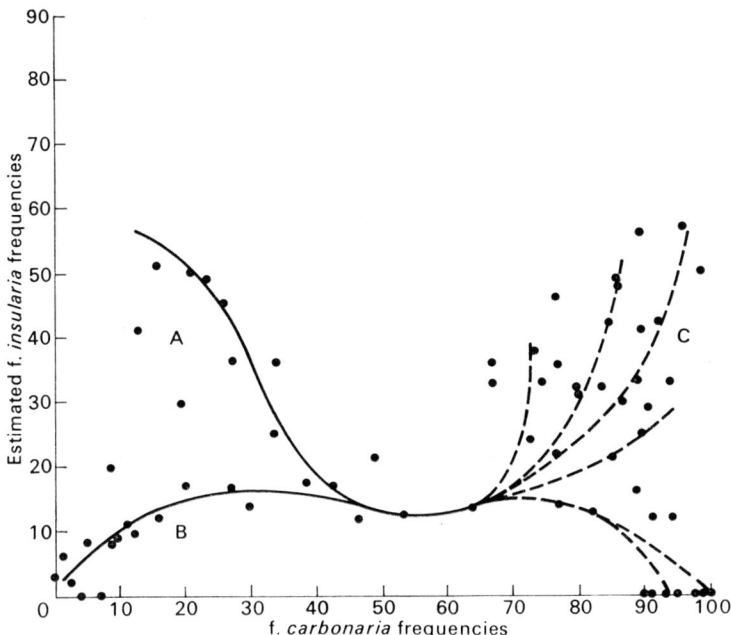

FIG. 9.4 Graph showing frequencies of *insularia* for the majority of districts in Britain (plotted against *carbonaria* frequencies)

Figure 9.4 shows the result of graphing the estimated gene-frequencies from Appendix C in which the totals sampled were larger than fifty individuals. The data suggest the following deductions:
  (1) When f. *carbonaria* and f. *insularia* simultaneously enter a population consisting of f. *typica* only, f. *insularia* will rise to about 15 per cent, but no higher as long as the f. *carbonaria* frequency remains under 10 per cent (District B).
  (2) When f. *carbonaria* enters a population which has a high frequency of f. *insularia* (40 per cent), it is likely that this will drop to this same level of 15 per cent (District A).
  (3) f. *carbonaria* frequencies of between 40 and 65 per cent are associated with the lowest gene frequencies of f. *insularia*, and this coincides with the most rapid spread of f. *carbonaria* (Fig. 9.2).

(4) Thereafter, a rising f. *carbonaria* frequency is associated with f. *insularia* gene-frequencies varying from high to low (District C).

It is of interest then to find that each of the three main groupings, designated with a letter, correspond geographically with three different areas of Britain in the majority of recordings. Thus group A, with a high frequency of f. *insularia*, is largely constituted by the Severn Valley, Gloucestershire, and Oxfordshire; group B by the counties of Berkshire, Wiltshire, Hampshire, and Dorset; while group C represents the whole of the eastern half of England from Yorkshire to London, an area which corresponds with the earlier map of f. *carbonaria* as drawn up by Adkin (1925).

For the purpose of analysis, it is convenient to divide the graph (Fig. 9.4) into two portions: below 50 per cent f. *carbonaria* frequency and above.

Below a 50 per cent f. *carbonaria* frequency, it is evident that no matter at what level the f. *insularia* has been maintained in the population, the entry of f. *carbonaria* into it will canalize the f. *insularia* for a period of time at an approximate 15 per cent level. It is possible on the other hand that this may represent an equilibrium frequency at the level of which f. *insularia*, in the presence of prolonged but slight pollution fall-out, is capable of adjusting itself in the population in a state of balanced polymorphism. A more likely explanation, however, is that it is in a transient state, in which f. *carbonaria*, the more extreme melanic, is replacing a less suitable allele in the presence of ever-increasing air pollution.

Above a 50 per cent f. *carbonaria* frequency (Fig. 9.4), the estimated f. *insularia* values appear to vary from 0 to 65 per cent, the majority being high except in populations practically entirely overrun by f. *carbonaria*, where the f. *insularia* samples are always inadequate, and where homozygous *carbonaria* must be present thereby excluding the allelic *insularia*. Standard statistical analysis shows a highly significant degree of heterogeneity ($\chi^2_{(38)} = 99 \cdot 86$) in the proportions of f. *insularia* in such populations, but it is evident that f. *insularia* continues to maintain itself within a f. *carbonaria* population even after losing its cryptic advantages and, in fact, may even expand. On the other hand, slow elimination is indicated by the subsequent history of the only recorded early frequency known to me (previously referred to), that in Chester; where the estimated frequency of f. *insularia* has dropped 7·5 per cent in 55 years, but the sample is inadequate to justify the assumption of slow elimination. The same tendency is shown in the latest analysis on comparing the 1952/56 survey with the 1965/70 period, thus covering nineteen years. This is discussed in the following pages when I give a résumé of the changes.

(v) Frequency changes in the two forms between 1952 and 1970

This nineteen-year period can be divided into three surveys in each of which the samples have been aggregated if there was no heterogeneity. The periods are 1952–56 (Kettlewell 1958a), 1957–64 (Kettlewell 1965d), and 1965–70 (not previously published). It must be appreciated that only a limited number of places have been sampled consecutively throughout this time and that because of this many comparisons are not always possible. Also, many of the recorders who contributed during the first period, some of whom were not particularly interested in *Biston betularia*, had by 1970 dropped out. Yet their joint efforts by that date had logged over 40 000 records.

When we compare the 1952–56 samples of *B. betularia* and its morphs with the two more recent surveys, 1956–70, the analysis can be divided into three main headings:

(a) Places in which f. *insularia* has *increased*.
(b) Places in which f. *carbonaria* has *increased*.
(c) Places in which f. *typica* has increased and f. *carbonaria* has declined (=smokeless zones).

(a) Increase in frequency of 'f. insularia'

Because individual insects in the *insularia* complex can be subject to misscoring, particularly as between different recorders, I am convinced that a strict comparison can only be made in those localities where the same observer has collected samples throughout the period. For example, I have obtained details of 1437 *B. betularia* from the Marlborough district of Wiltshire between 1952 and 1967, but nine different people have contributed to this (Appendix C.) Whilst a decrease in f. *typica* is common to all records, the frequencies of the melanics show gross heterogeneity. I will confine my remarks on f. *insularia* therefore to the few places where the same person has sampled throughout.

Fortunately I live in an area where two *insularia* forms (=phenotypes) are increasing, Steeple Barton in Oxfordshire. Since 1952 I have taken daily samples each year (with the exception of 1965) and the frequency has risen from 16·61 per cent to 24·33 per cent ($n = 1121$, $\chi^2_{(1)} = 5\cdot324$). At the same time f. *typica* has dropped from 49 per cent to 39 per cent. Steeple Barton is a rural area in central England. There are no local industries, but a large American air-base is four miles distant with constant fall-out from heavy jet planes. Vegetative bryophytes are scarce except on the trunks of apple and ash. Steeple Barton has much in common with other districts where f. *insularia* is increasing, namely a low pollution fall-out over a wooded countryside.

Newbury is about 50 miles due west of London with little local fall-out, but no doubt receives long distance fall-out from the south-west. Twenty years ago I recall that this district was one of the nearest to the metropolis with heavily lichened oak trunks; today it is less so. R. Saundby has recorded samples since 1952 when f. *insularia* was at a frequency of 9·5 per cent and f. *typica* 79 per cent. The most recent survey gives comparable figures of 19·9 per cent and 67 per cent respectively (total $n = 601$, $\chi^2_{(1)} = 3 \cdot 046, P > 0 \cdot 05$).

Similarly in the New Forest at Minstead 12 miles south-west of Southampton, L. W. Siggs records an increase of f. *insularia* from 2·32 per cent ($n = 518$) in the 1957–64 survey to 7·53 per cent in a more recent one ($n = 491$) ($\chi^2_{(1)} = 13 \cdot 752, P < 0 \cdot 001$), here strangely at the expense of f. *carbonaria* and, to a lesser degree, of f. *typica*.

It is indeed unfortunate that during the years 1952 to 1970 I have been unable to find a single place where f. *insularia* is increasing in the complete absence of f. *carbonaria*. Such situations were common no doubt a century ago (as for example Folkestone) but such opportunities have now been lost in Britain for ever.

(b) Increase in frequency of f. carbonaria

The earlier sampling from industrial districts showed a rapid increase of f. *carbonaria* from all such areas when once this mutant reached 1 per cent. Yet nowhere, except Leeds, has this form increased over 98 per cent and usually its frequency is somewhat less. Yet from here in a recent (1968–9) sample of 190 all have been of the f. *carbonaria* form. This must reflect a high proportion of homozygous f. *carbonaria* in the population.

There are several places showing an increase of this form as between the earliest sample (1952–6) and the second (1957–64) but only a few are continuing in this direction in the more recent (1965–70) records. This is in part due to my failure to get samples from but a few of the heavily polluted parts of England. Thus, of 3095 *B. betularia* taken by the late D. Wright at Borden, Hampshire, between 1952 and 1956, 20·29 were f. *carbonaria*. In 1966 sample of 353 the frequency had increased to 34·28 per cent ($\chi^2_{(1)} = 35 \cdot 637, P < 0 \cdot 001$). At Woking, Surrey, the 1952–56 recording of 1615 *B. betularia* by C. de Worms showed a frequency of 76·66 per cent, the 1957–64, 82·64 per cent ($n = 334$), and a smaller one of 150 in 1967, 86·8 per cent ($P > 0 \cdot 001$). Other places in Surrey show small increases, for example Witley (J.K.M.) and probably Bramley (R.F.B.)

though here the individual frequencies show rather violent fluctuations per year, not due to sample size. Though these Surrey localities are 40–60 miles south-west of London, the country is rapidly becoming a commuter area for London with consequent increase in the number of domestic dwellings.

It would appear that in most industrial areas the upper end of the sigmoid curve (Fig. 9.2) is now straightening out.

(c) Increase in frequency of f. typica *and decline of f.* carbonaria; *'smokeless zones'*

Following the Beaver report on air pollution (1954) and the public outcry which resulted, the British Parliament decreed that certain districts in industrial areas, hitherto most heavily polluted, should become 'smokeless zones'. This has resulted for the first time since the commencement of the Industrial Revolution in a detectable reversal of the previous ever-increasing pollution fall-out and its effects on nature. Unfortunately this is a present limited to a few local areas only, and these are likely to revert in the near future due to the fiasco of smokeless-fuel planning.

The evidence of a reversal

Since 1962 the frequency of f. *typica* has increased in several areas: Liverpool, Manchester, Brentwood Essex (K.M.G.); also Bishop's Stortford and Tring in Hertfordshire (C.C. and L.G.), and Totteridge in the London area (R.I.L.).

Clarke and Sheppard (1966) have, in particular, investigated this around smokeless zones in the Liverpool district. Firstly they did this by sampling; secondly by 'release experiments' of dead insects onto local tree trunks.

At Caldy, Cheshire, which is ten miles from Liverpool, the frequency of f. *carbonaria* dropped from 93·3 per cent in 1959 (93·1 per cent in 1962) to 90·2 per cent in 1965, $(n = 4229)$ $P < 0·01$. Secondly, Clarke and Sheppard undertook extensive field experiments which were designed to test the cryptic advantages or disadvantages by presenting deep-freeze specimens of the two forms in their normal day-time position on trunks. By these experiments and by making certain assumptions they showed that, whereas previously f. *typica* had had approximately 50 per cent cryptic disadvantage at Caldy, since 1962 this had fallen to 20 per cent. They account for the recent increase in f. *typica*, even though having a 20 per cent cryptic disadvantage, as being due to even greater physiological disadvantages of the homozygous f. *carbonaria*. (The theory of heterosis has been discussed in Chapter 1.)

The results of these experiments also showed that similar selection pressures were involved to those estimated by Haldane (in 1924), deduced from the rate of spread of f. *carbonaria* between 1848 and 1898 in the Manchester area. Also, my own releases of just under 900 marked living insects in the industrial areas of Birmingham demonstrated that a 50 per cent advantage of f. *carbonaria* was actually at that moment (Kettlewell, 1955b) being maintained.

The observations of Clarke and Sheppard have clearly shown an increase in the frequency of the pale f. *typica* since 1962. At the same time their release experiments suggest cryptic disadvantage of f. *carbonaria* about 10 per cent in excess of that estimated from the local phenotype frequencies.

I am convinced that the usage of *dead* insects in this type of experiment, superficially such a quick-cut method of putting two forms to the test, is not completely satisfactory because dead specimens cannot compare with the perfect crypsis obtained by living insects. The capacity of these to choose correct backgrounds and to establish positioning of wings and antennae is all-important. Human manipulation of these on dead specimens cannot substitute the behaviour of the living insects. For this reason, I believe that higher predation took place among their dead insects, not necessarily relative to the two forms. This does not detract from the importance of their observations and experiments which have shown, for the first time, a reversal in the trend with a diminished cryptic disadvantage of f. *typica*.

More recently, Cook, Askew, and Bishop (1970) have resampled the Manchester area. The original samples had been taken between 1952 and 1964 at Didsbury, four miles from the centre of Manchester, by H. N. Michaelis. Of 760 individuals 11 were recorded as f. *insularia* and none were f. *typica* (Kettlewell 1958a, 1965d). The recent survey was taken along a transect from north to south Manchester between Prestwick and Jodrell Bank which includes Didsbury. Of 972 *B. betularia* collected at mercury-vapour and assembling traps between 1966 and 1969, 25 were recorded as being f. *typica* (=2·57 per cent), and 12 as f. *insularia* (=1·24 per cent). Allowing for the inevitable discrepancies of scoring f. *typica* as distinct from f. *insularia* (and this is likely to occur when different samples are assessed by different observers—see p. 143) the frequency of f. *carbonaria* has dropped from 98·5 per cent to 96·2 per cent. They (Askew, Cook, and Bishop, 1971) conclude by recording 364 *B. betularia* from around Didsbury, the site of Michaelis' original collecting. Of these 8 were f. *typica*, and 2 f. *insularia*. This shows a highly significant ($P = 0.0001$) increase of f. *typica* from the earlier samples.

Though the samples from Hertfordshire are smaller, the swing to f. *typica* is even greater at both Bishop's Stortford and Tring. In each, the same observer has recorded a drop in both f. *carbonaria* and f. *insularia*. The 1952–6 frequencies for the former place were f. *typica* 4·7 per cent, f. *carbonaria* 89·18 per cent, and f. *insularia* 6·06 per cent ($n = 231$, C. Craufurd). The 1957–64 frequencies are f. *typica* 8·7 per cent, f. *carbonaria* 87·55 per cent, and f. *insularia* 3·69 per cent ($n = 217$). Similarly at Tring, the earlier sample ($n = 214$, L. Goodson) had a frequency of f. *typica* of 16·89 per cent and the later one ($n = 165$) 22·42 per cent, and again at the expense of both f. *carbonaria* and f. *insularia*.

In the London area, the 1967 sample from Totteridge ($n = 453$, R. I. Lorimer) varies from that in 1970 ($n = 206$), where f. *typica* increased from 3·75 per cent to 6·31 per cent, though this is not significant.

In all these places it appears that lowered pollution fall-out may have led to somewhat paler backgrounds, though there is no sign of the return of the bryophytes, and on these f. *typica* largely depends for concealment.

It will be interesting to see, now that the Coal Board have run out of smokeless fuel (1971) whether this reverse will once again lead to an upsurge in the frequency of f. *carbonaria* during the 1970s.

A summary of the 1952–1970 survey

During the initial stages of pollution contamination it appears that the various forms of f. *insularia* spread through the population of f. *typica*, from the lightest, *ins.*[1] (as in the New Forest 20 years ago), to the darkest, *ins.*[5] (as found at Marlborough at the present time and elsewhere). Some of these are allelic to f. *carbonaria* which at a later date replace the *insularia* forms and (though to a lesser extent only) f. *typica* which may even undergo an increase in frequency. It is likely that they will retreat in a reverse order (f. *carbonaria* → f. *insularia*[5] → f. *insularia*[1] → f. *typica*), should the trends of air pollution be reversed in the future.

The overall picture of the recent 19 years' survey is that f. *insularia* is increasing in a few somewhat specialized areas, and f. *carbonaria* only in peripheral districts of recent urbanization. More helpful is it (and I think somewhat surprising) that the smokeless zones are having an effect. For here this most sensitive indicator, *Biston betularia*, is for the first time in 100 years reversing its trend to melanism even though no vegetative lichens have reappeared, nor are they likely to in the foreseeable future.

Part IV. Non-industrial Melanisms

10. Geographic, Relict, or Ancient Melanism: A survey and classification

Definition. Melanic polymorphisms occurring today in several quite different types of habitat, which are similar to those which have existed in earlier times; inheritance of melanic forms is usually dominant. In certain instances this Ancient Melanism has present-day links with Industrial Melanism where the two ecologies overlap.

Introduction

Of the nearly 800 species of Macrolepidoptera occurring in Britain, approximately 80 (10 per cent) maintain melanic forms under certain specialized ecological conditions, which are far removed from the effects of industrialization. These melanisms are discussed later in this chapter under seven separate groups. The experimental work we have carried out on one of these, group (ii), is discussed in the next two chapters. The seven groups are:
  (i) Rural or Background Choice Melanism.
  (ii) Northern (latitude 60°) Melanism.
  (iii) Western coastline Melanism.
  (iv) Ancient conifer Melanism (Relict Melanism).
  (v) Melanism associated with fire-resistant trees.
  (vi) Pluvial Melanism.
  (vii) Thermal Melanism.

In Chapter 5 (**iv**), under the heading of 'The history of the spread of Industrial Melanism in Britain', it was pointed out that, at the beginning of this century, and indeed until the late 1920s, there was a school of thought which held that industrial melanic forms of Lepidoptera were the *direct* consequence of the environmental conditions imposed on them; today we know that this is not so.

Charles Darwin must have pondered on this problem. Though his early days and training made him primarily a coleopterist in his entomological pursuits, he, nevertheless, was knowledgeable of the British Lepidoptera, though it is evident that in Brazil he seriously underestimated the capacity of these insects for crypsis. When he moved to Downe House in 1842, although there could have been no Industrial Melanism at a recognizable frequency, many species of common moths must have occurred there, maintaining melanic polymorphisms, as they had for centuries. *Xylophasia monoglypha* L. (Plate 10.1, left 1 and 2), for example, must have flown into his drawing room at night and it is likely that 5 to 10 per cent of these were of its dark forms. Yet nowhere can I find a reference to melanic Lepidoptera. This is even more remarkable when one considers the emergence of Industrial Melanism elsewhere during his lifetime. When the earliest black Peppered Moth, *Biston betularia*, was taken in Manchester in 1848, Darwin was aged 41. By 1882, when he died, the frequency there of f. *carbonaria*, the melanic form, must have been of the order of 60 per cent. Darwin must have read records of the increasing numbers of such forms occurring farther north.

For example, in the *Entomologist* for 1864–5 (p. 150) under the heading of *Amphydasis betularia*, R. S. Edleston states that 'some sixteen years ago the "negro" aberration of this common species was almost unknown'. He then records that in 1864 the majority of wild males which assembled to his virgin females at Bowden, Manchester, were of the 'negro' aberration. 'If this goes on for a few years the original type of *A. betularia* will be extinct in this locality.'

I have no knowledge whether Darwin read the *Entomologist* though this is likely because he was a keen collector of Lepidoptera in his youth. It is surprising that this great observer leaves no comment on melanism in moths. Under the heading of *Darwin's missing evidence*, I made this point many years ago (Kettlewell, 1959a).

Today, nearly 100 years later, we still have little published on the connection between the Ancient Melanisms of the past and the Industrial Melanism of the twentieth century. The emphasis in this chapter, indeed in this book, must be towards providing such evidence.

There are, in fact, a number of species which occur today in remote rural regions which maintain melanic forms phenotypically identical to those occurring in industrial England. As yet in no instance have we been able to prove that an identical allele is common to both populations.

There are two separate situations in which Ancient Melanism may have links with the Industrial Melanism of today. Firstly, there are several instances, which are found in group (i) (Rural Melanism), in which a black form, occurring at a low frequency throughout the southern England populations, increases its frequency when it enters urban districts, and it may even be replaced at a later date by a darker morph. Such an example is *Allophyes oxyacanthae* (Plate 10, right 1 and 2), where f. *capucina* is found at low, but varying, frequencies throughout the countryside (see Appendix

B). Yet in urban areas such as Wellingborough, in a sample of 75, 53 per cent were melanic (Gent) and at Bradford over 90 per cent (Briggs).

A second link, but differing in that it is not a direct one, is seen in many species whose melanic forms until recently were confined to the isolated and specialized ecologies about to be discussed (groups (ii)–(vii)). Though there can have been no gene flow because of impassable barriers, black forms, phenotypically similar to the ancient melanics, have occurred recently in industrial areas of Britain. The black form of *Cleora repandata* L., f. *nigricata* Fuchs (Plate 5.6, right 5 and 6), which has a frequency of up to 10 per cent in the Black Wood of Rannoch (relict Caledonian pine forest), is similar to f. *nigra* Tutt which today forms over 90 per cent of many populations in central England. Clearly these industrial melanics had their origin from fresh mutations and not by migration from outside. The genes controlling them are unlikely to be identical and even if occurring at the same locus, they are likely to be different alleles.

The importance of being able to find an association between the Industrial Melanism of today and the Ancient Melanism of the past is that light may be thrown on the origin of the industrial melanics.

These two major classes have several common denominators. In each, the black forms occur among cryptic species only and the same species may contribute similar melanic forms to each. In both, the inheritance of these forms is dominant (or near-dominant), though we know the genetics of but few in the group under discussion.

Ancient Melanisms are frequently found as disruptive polymorphisms; but in some the heterozygotes are recognizably visible as intermediates. Rarely (as I think has happened in some species of North American *Catocala*) the melanic mutant forms 100 per cent of a population. Such situations are found, usually locally, in specialized ecological conditions. They may even be confined to a small island, a particular mountain top, or even to a few acres of relict forest. It is these we will consider in this and the following two chapters.

(i) Rural or Background Choice Melanism

A classification of melanism in the Lepidoptera is however never clear-cut; nor is it likely to be if one considers the number of components which affect an ecology suitable for maintaining such forms. The angle of the sun, the length of daylight, temperature, humidity, peat or chalk, the proportion of deciduous or coniferous trees present—all contribute. Furthermore, a point which I have not seen stressed is the effect of intense *limited* local predation. Thus, during autumn or spring migration, birds on islands tend to gather at the north or south end according to the direction of the migration, and the intervening areas may not be severely predated. Similarly, a colony of Black-headed Gulls, *Larus ridibundus*, or even a single pair of Merlin, *Falco eleonorae*, can selectively annihilate a moth as large as the Oak Eggar, *Lasiocampa quercus*, in a circumscribed area around a nesting site.

Rural Melanism cannot always be clearly differentiated from Northern nor, today, from Industrial Melanism. In many species the farther north we proceed the higher the frequency of a melanic morph. Thus, in the genus *Tethea*, occasional dark forms of *T. duplaris* are found in rural southern England, but in northern England and Scotland the majority of individuals are of a similar blackish or purplish-grey colour (Plate 5.7, right 5 and 6). In *Achlya flavicornis* L., a closely related species common throughout Britain, the Scottish race, f. *scotica* Tutt., is darker and larger, and the same applies in northern Scandinavia to the dark f. *finmarchia* (Plate 6.1). The Geometrid moth *Lampropteryx suffumata* has, in southern England, a disruptive black and white pattern; it can be beaten out of vegetation such as hedgerows on downland by day. In northern England and Scotland, the melanic f. *piccata* Stephens replaces this form. In Westmorland woods both forms occur.

It could be argued therefore that a few species which exhibit Non-industrial Melanism in southern England reflect an extension southwards of Northern Melanism. Alternatively, it is possible that the differences between northern and southern British populations represent an Ice Age difference, the northern forms having been preadapted for dark colouration at an earlier period, and light southern ones representing a Continental spread northwards.

Shade and shadow

If we disregard geologically different backgrounds, such as peat and chalk, and consider the angle of incidence of the sun and its brightness in relation to latitude, there is a general rule, which can be applied to many cryptic creatures which depend on disruptive camouflage by day: that is, the nearer the Equator and a vertical sun, the greater a clear-cut contrast is permitted. Hence, the patterning of the Zebra, *Equus burchelli*, the Tiger, *Panthera tigris*, and the young of many Hogs, *Sus scrofa*, in the Tropics. Conversely, at high latitudes shade and shadow are more diffuse. In the British Isles the latitude from Shetland to the Isle of Wight varies only by $10°$ ($60°-50°$). Yet, I think over this short distance of 1000 miles the angle of the sun varies sufficiently for natural selection to have favoured, within species, a disruptive pattern in the south and a diffuse, cryptic one in the north. The Noctuid moth *Orthosia gothica* L., which hatches in the early spring, rests by day on the ground among dead leaves. In southern England, the whole population of this very common moth has a chequered black and brown patterning. In Scotland, from Perthshire northwards, such contrast markings are usually absent and a more uniform patterned population is represented by f. *gothicina*. These forms are distributed in a similar way, according to latitude, in Scandinavia also. Of 505 individuals collected from four non-heterogeneous samples around Stockholm (latitude 60°N) in 1967, 8·2 per cent were of this form. (G. Howard, personal communication.)

An extreme example of the substitution of a melanic cryptic pattern in high latitudes for a disruptive one in the south is found in the Noctuid moth *Amathes glareosa* Esp. and its melanic f. *edda* Stdgr., which is confined to the Shetland Isles, the Orkneys, and Denmark. The distribution of the melanic phase in Shetland is discussed later in the next two chapters together with our experimental work.

The advantages of polymorphism

Unfortunately, due to lack of observations and records, we have little knowledge as to where many night-flying moths rest during daylight hours. Even in Britain, where we probably have the highest number of naturalists in the world proportionate to our population, I would hazard a guess that only two out of every three species of Lepidoptera have been found at rest in nature. We have but little evidence that different forms of a species choose different (and correct) backgrounds on which to pass the day camouflaged. Nevertheless, it is likely that two or more forms are maintained in a population by choice of alternative backgrounds. For this reason I refer to this type of polymorphism as Rural or Background-Choice Melanism. The ability of each form to appreciate its correct resting site (along with varying degrees of physiological fitness) would indeed contribute to the success or failure of each in a population (see Chapter 5, p. 67).

In Britain most species in the genera *Nonagria* and *Tapinostola*, which feed on *Phragmites* or grasses, have dark forms in rural areas. These may be clear-cut (disruptive polymorphism) as in *N. dissoluta* or graded as in *N. geminipuncta*.

The normal pale form of *N. typhae* has been found on the light dead stems of *Typha* but the rarer black form *fraterna* Borich, has not, to my knowledge, been detected; suitable resting sites exist, however. In some localities the dead stems of such plants as Water Dock (*Rumex* species) present dark surfaces, and these grow alongside *Typha*, the food plant of the larvae. The Water Hen, *Gallinula chloropus*, which is a voracious insectivore when such food is available, must ensure a high predation rate amongst reed and bulrush resters.

Background Choice Melanism in *Nonagria* is widespread in Britain. It may occur in eastern or western fens but the frequency of the phenotypes differs in each locality. Thus, in *N. dissoluta* (Plate 10.1, right 3 and 4 (which, incidentally, is nominotypically the melanic form and which was originally described as a separate species from f. *arundineta*, the usual light form), the frequency of f. *arundineta* was 100 per cent in a sample of 16 from Cheltenham (Jackson, personal communication, 1959), 'rare' at Mudeford, Christchurch Harbour, Hampshire, 20 miles to the west of Southampton (Carr, personal communication, 1956), but up to 40 per cent in a large sample at Titchfield, 10 miles to the east (R. S. Jackson). Plate 10.1, shows photographs of other British species which, in my opinion, fall into the category of Rural Melanism. Many of them are found widespread throughout England. As far as we know, their melanic forms occur at a frequency well above mutation rate at the time the earliest records were made. In many instances it appears that their habitats have altered but little in the last 100 years. In others it is evident that the frequencies of the melanic forms have increased when their original habitat came to be in juxtaposition with an industrial area.

Predator–prey relationship

In the last few years a considerable amount of work has been carried out on behavioural elements in the predator–prey relationship (de Ruiter, 1953, Tinbergen, L., 1960, Clarke, B. 1962).

The concept of hunting by visual image is fairly well established. Any species which can present an alternative and very different form may gain advantage over monomorphic species. Disruptive selection must usually be the rule because intermediate forms would, from time to time, be recognized and eaten by birds working to each image. A melanic morph, alternative to a light one (assuming for the present that each is able to choose its correct background, and that both backgrounds are available) must be the simplest way of spreading predation risks resulting in halving them. Other species resembling either morph but occurring at a lower density will be profoundly affected by the abundance or dearth of either form. A consideration of such a situation enables us to appreciate one of the major forces at work in interspecific competition.

Let us now extend the principle of alternative images to a number more than two morphs. What would now be the consequences? In order to demonstrate these, we will consider what happens during two entirely different and contrasting periods of the year.

First, during the summer months, the greatest weight and number of insects are eaten by birds at a time when their population, enhanced by their young, demands maximum food supply. This, of course, coincides with maximum insect output both in the number of species and the total number of individuals available.

Secondly, polymorphism is common during the winter months (October to March in the Palaearctic) when approximately only one in one hundred of the available Lepidoptera species are in the imaginal state. In Britain all of these depend for their survival on crypsis; certainly none display warning colouration.

In each of these situations *it is amongst commoner species* that multiple polymorphism is found and in each series a melanic morph is available. It is known that in some species such a situation is controlled by multiple alleles; in others (the majority) we have no evidence. Each form must vary in frequency from year to year and this will depend on environmental changes. Some may be entirely eliminated (except by mutation) and it is possible that there exist today examples of melanism which represent the end of a chain of allelomorphs whose co-morphs are now extinct in the population. Thus, in some districts, the melanic forms of *Celaena secalis* L., an abundant species in summer and a near-ground sitter by day, are now the commonest phenotypes. Similarly, *Triphaena pronuba* L., which occurs at the same time and in the same places, another ground sitter, has today a high proportion of black-forewinged individuals in some industrial areas. Elsewhere each of these species maintains at least four distinct forms in their populations. Any bird predator, in order to produce maximum yield, would therefore have to hunt to eight images for these two species alone.

Polymorphism, during the winter months, reflects another aspect of predator–prey adaptation. In the summer months, each parent bird needs to obtain as much insect protein as it can as quickly as possible for its offspring. In winter, because insects obtainable for food are rare, it is likely that comparable pressures are maintained, even though fewer birds are searching for fewer moths, which must be chosen from the few available species. There are, in Britain, under a dozen species of Macrolepidoptera which hatch in the three coldest months of the year (December–February). Yet amongst these are found three with polymorphisms, each with at least three morphs and each morph showing considerable variability. They are *Phigalia pilosasia*, *Erranis leucophaeria*, and *E. defoliaria* (Plate 5.9, left 7 and 8). In each, the female is apterous, small and difficult to find, and the males rest by day exposed on trunks, palings or dead leaves. On such varied backgrounds a number of non-specialized patterns are exploited; they are similar in each species. Each morph falls under the heading of 'diffuse', 'disruptive', or 'melanic', and in each there are several forms. The frequency of each varies in each locality. Thus the melanic forms of *E. defoliaria* have increased in the last half-century in Epping Forest.

In Britain the greatest degree of variability found during the winter months possibly occurs among hibernating species which spend the winter exposed at rest amongst dead leaves. *Sarcothripus revayana* has a multitude of phenotypes and its melanic f. *nigricans* Sheldon is one of the commoner ones. Two other abundant species, *Peronea crystana* and *P. hastiana*, which pass the winter as imagines in Blackthorn thickets, each have about 10 different forms.

In other parts of the world, where predation pressures are greater, a single species may have as many as 15 separate forms. It is strange indeed that such opportunities for assessing the selective visual advantages have not been investigated. Unfortunately, many of the obvious species to choose for field and laboratory observation are the most difficult to breed in captivity.

Ground colour due to geological differences = geographical races
Previously we have been discussing variability largely as a polymorphic situation, most cases of which are controlled by a one-gene difference. In some species a gradation of forms occurs from light to dark, such as in *Nonagria geminipuncta* Hatchett (Plate 10.1, right 5 and 6), and these involve either a polygenic (multifactorial) inheritance or more rarely heterozygous variability and therefore should by definition not be considered as constituting polymorphisms. As has been pointed out, alternative forms enable a species to take up the different but usually contrasting resting sites which are available in order to pass the day undetected. This can only be of advantage when such alternative backgrounds are available side by side in a particular locality.

Many ground resting species have a discontinuous distribution and have had to adapt to living on areas of black peat in some localities, but on white chalk in others. A good example of this is the Geometrid *Gnophos obscurata* these situations which occur round latitude 60°N., the highest proportion of black Lepidoptera, both species and frequency per species, is found.

For this reason I chose the Shetland Isles, through which this latitude passes, for intensive study, and I and my co-workers have recently spent part of each summer there for several years. Here we were mainly concerned with *Amathes glareosa* Esp. and its melanic f. *edda*; and the results of our investigations on this species are given in Chapters 11 and 12.

Apart from in Shetland, similar situations exist around latitude 60° N. in Scandinavia; in particular, on the islands in the Baltic off Stockholm. From an examination of the Lepidoptera collections in Canada, it appears that melanism may be frequent around this latitude there also. There is evidence that apart from a northern increase there is a west–east cline in certain places in some species.

Northern Melanism, for example, as in Scotland cannot be clearly differentiated from Western or Conifer Melanism, each of which has arisen from a particular ecological contribution, be it high latitude, increased humidity, or the proportion of dark trunk conifers. Such contributions must each be considered separately, the one from the other, and the concept that a pine forest situated on the west coast of a northern isle would, on theory, have the maximum number of melanic moths, is a lesson in false deduction. I make this point because this suggestion has, on occasions, been put forward in discussion; nor can I find any evidence that it actually happens (for example, in pine plantations on western islands). Because of the greatly increased humidity on western coasts, most pine species have, under these conditions, a heavy growth of light-coloured lichens and this can give little advantage to melanic trunk-sitting species. In Northern Melanism, I refer mostly to those species exhibiting it which inhabit the large areas of low peat-covered hills and islands which occur around latitude 60°N. The Shetland Isles are representative of many others which occur in the Northern Hemisphere around this latitude, though the temperature–stabilization effect of the Gulf Stream must be taken into consideration in each.

*Melanism in the Shetland Isles**

It appeared to me that the Shetland Isles, where melanic forms were known to be common, would be the ideal place for a fuller investigation into Northern Melanism. Here there could be no question of pollution fall-out,

---

* I quote, with minor insertions, from my paper in *Heredity, Lond.* (1961c), **16** Part 4, pp. 395–9, by kind permission of the editors.

and here for a thousand years the countryside has remained comparatively undisturbed by Man, as it had even before that, except for the elimination of trees due to cutting for fuel and the introduction of domestic animals. During the last Ice Age (40 000 B.C. to 18 000 B.C.), it is likely that much of Shetland was not covered by ice, and it is not certain that it was during the last but one Ice Age (Strathmore), approximately 100 000 years ago. If this is so, it is possible that some of the fauna may be very much older than previously believed.

We have undertaken five expeditions to the Shetland Isles in the last few years. Of the 78 species of moths recorded from these islands, 16 are migratory and are therefore unlikely to have acquired local forms. Migration may put a species at a disadvantage on the periphery of its range, in that an advantageous gene-complex, built up over previous years to adjust to local conditions, may be destroyed by an influx of fresh individuals from another environment (Haldane, 1956). Of the 62 native species unaffected by migrant individuals, 27 have produced local forms, of which twenty-one (33 per cent) are melanic. These may be classified into three groups:

(1) Species in which 100 per cent of their population is melanic. They have discarded their relevant normal alleles found elsewhere.
(2) Species in which the two forms co-exist as a balanced disruptive polymorphism.
(3) Species in which there is found a continucus range of forms from black to pale; these presumably reflect either a multifactorial situation, or one in which variable heterozygotes occur.

Apart from the dark background of peat and a high rainfall, the most obvious environmental change at these latitudes is the greatly diminished number of hours of darkness during summer and, what is probably more important, the long periods of twilight. These coincide with the warmest months in the year, when the majority of moth species appear as adults. This necessitates their taking flight in daylight. Under these conditions, light-coloured insects, which farther south have the protection of darkness, are more conspicuous than dark ones. In Shetland bird predators are extremely common, particularly in the north, and furthermore they feed on moths both on the wing and on the ground. Farther north, however (above latitude 61°N.), melanic forms are less frequent. In the Faeroes, of the 23 indigenous species (Wolff, 1929) only 4 (23 per cent) show melanism, and in Iceland 2 (10 per cent) out of 21. This may reflect a relaxation of bird predation farther north, though this is unlikely to be the only reason.

Examples of melanic Lepidoptera found in Shetland are shown on Plate 10.2 in the right hand column; the corresponding and more usual

Geographic, Relict, or Ancient Melanism

British mainland forms are on the left. The top pair depict the English and Shetland forms of *Diarsia festiva* Schiff. respectively, the latter being referred to as ssp. *thulei* Stdgr. This race is distinct in appearance from f. *typica* of the British mainland, being darker and smaller. We have bred the larvae from ova, and we find that they bear little resemblance to those of *D. festiva* f. *typica*. One of our team, C. J. Cadbury, succeeded in crossing the two races. The majority of larvae and the single male moth bred from them were intermediate in appearance between the parental races. It is likely that several genes are involved in the differences; ssp. *thulei* seems, in fact, to be well on the way to speciation. It is a good example of group 1 of our classification. Other species which are in a similar position are *Eupithecia venosata* Fabr. and *Hadena conspersa* Esp. (Plate 10.2, 3 and 4). Both these feed exclusively in the Shetlands on *Silene maritima*, the Sea Campion, and are therefore limited to the shoreline. Though in Shetland considerable variability occurs in both these species, examples are never as light as the f. *typica* found elsewhere.

Species showing a clear-cut polymorphism (group 2) with distinct phenotypes suggesting disruptive selection are uncommon. However, one of the most successful of species in the Shetlands, *Amathes glareosa* (Esper, 1788), which is to be discussed in considerable detail in the next two chapters, falls into this category (Plate 11.1). It will be noted that f. *typica* has little resemblance to the black f. *edda* (Staudinger, 1891). *Lygris populata* L. should be relegated to the same group with its black f. *masanaria* Freyer (Plate 10.2, 5 and 6), but in this species the melanic is variable; *Amathes xanthographa* Fab. (Plate 10.2, 7 and 8) should also be included in group 2.

Group 3, in which an infinite range from pale to dark forms occurs, contains the majority of species. Amongst these are the following: *Entephria caesiata* Schiff. (Plate 10.2, 9 and 10), *Xanthorhoë fluctuata* L. (Plate 10.2, 11 and 12), and *Hepialus humuli* L. (Plate 17.2) (see Chapters 16 and 17). The Shetland melanic form of *E. caesiata*, ab. *atrata* Lange, is phenotypically identical with the industrial melanic form found near Paisley in Scotland, and specimens from both localities are referred to by this name.

This grouping of Shetland melanism under three headings must therefore be considered arbitrary. Though others might place certain species in a different category, it is a convenient method of classification as a first step to understanding the origin of melanic forms in Shetland. The classification into three groups reflects phenotypic expression, and may bear no relationship to their genetic control.

Previous Shetland data. Not until the end of the last century was it realized that the Shetland insect fauna was very different from that found on the British mainland. At that time the interest was, as it still is to a large extent, in obtaining specimens for collections. Shetland offered few comforts then for collectors and was largely inaccessible. It is not surprising, therefore, that the few lepidopterists who visited these islands were professional collectors, sent there to bring back as many aberrant specimens as possible, and paid by wealthy collectors who were not prepared to undergo the hardships themselves. A more important consideration is that the emphasis was therefore on extreme examples and that, as is the rule in all collections, the insects which were kept bore no relationship to their frequency in the populations from which they had been extracted. All normal specimens, similar to those found on the British mainland, were discarded. Violent frequency changes appeared in their yearly samples, and these were interpreted as reflecting the direct effect of seasonal differences on the insects themselves.

One of the first collectors to work in Shetland was McArthur, who visited the Mainland in 1880 and 1881 and subsequently Unst in 1883 (Weir, 1880, 1881, and 1884). He recorded a large number of melanic forms. The current explanation at that time was that melanism was the direct effect of humidity, and so it is not surprising to find that South (1893), who had never been to Shetland, decided that McArthur's melanic forms came from 'boggy meadows', whereas the more normal individuals were collected on drier cliffs and hillsides. More specifically he refers to *H. humuli* as behaving in this way. Similarly he appears to have thought that the dark form of *A. glareosa*, f. *edda* was confined to 'low-lying wet moors near Baltasound'. Other collectors about this same period were L. A. E. Sabine and Salvage, many of whose specimens can be found in the National Collection of British Lepidoptera (R.-C.-K.) at the British Museum of Natural History, London; however, they published no records, presumably wishing to retain a financial monopoly.

Shetland has been visited on many occasions by collectors during the present century (Hare, 1957, Harper, 1958) but with the same object of procuring specimens for their own private collections only. On no occasion have frequencies of the forms been recorded. The reasons for melanism in the Shetland Lepidoptera had remained obscure.

For this reason I decided to try and evaluate the various factors which could contribute to the prevalence of melanism in these islands, and this we commenced in 1959 and have continued since. The frequency and experimental data are given in full in Chapters 11 and 12.

(iii) Western Coastline Melanism

The melanism which is found on the western-facing shoreline of northern Europe appears to be independent of latitude; in fact, there are frequent indications of west–east clines. Thus, dark forms of many species form 100 per cent of the population along the western coast of Ireland. Further east, dark forms become less extreme but in no instance do I know of a clear-cut polymorphism. Presumably these melanic forms are by inheritance controlled multifactorially.

The Geometrid moth *Euphyia bilineata* L. is usually of a bright yellow colour with faint scalloping. On the Tearach, a small island situated about 10 miles offshore from Valencia, all the specimens are black and of large size (= f. *atlantica* Kane) (Plate 10.5) (Kane, 1896). Similar forms, but not so extreme, have more recently been found on the Blasket Islands (Huggins, 1956) which lie a few miles eastward. Though I failed to find f. *atlantica* on the Tearach, I found every gradation, from bright-yellow to brownish forms, along the shore of the Irish mainland. Five miles inland, all of the one dozen specimens I caught were yellow. It would appear that there must be considerable selective pressures controlling these forms, particularly near the coastline. Unfortunately the genetics of the dark forms are unknown; I have consistently failed to breed this species.

The Noctuid moth *Dianthaecia carpophaga* is found throughout Britain and Ireland. In eastern Britain (Suffolk) light yellowish specimens predominate and in south-east England (Dungeness) there is a high proportion of white ones. Along the whole of the western coastline of Ireland a black form constitutes 100 per cent of the population, f. *capsophila*. In Wales a form intermediate between *D. carpophaga* f. *typica* and f. *capsophila* is found. In this species there is a definite cline from west to east: the darkest specimens being found once again in the Blasket Islands and the lightest along the eastern and south-eastern coastline of England.

Full speciation has, in fact, taken place in a number of instances—species which have a distribution limited to the west and south-west shores of Britain. All of them have dark colouration. *Hadena caesia* Schiff. (Caradrinidae), another 'Campion species', which feeds on the flowers of *Silene maritima*, is confined to the west and south-west of Ireland, the Isle of Man, and Canna. *Hadena barettii* Dbl. (Caradrinidae), whose larvae feed on the roots of the same plant and also, strangely, Rock Spurry, *Spergularia marginata*, is found on the shores only of north and south Devon and Cornwall (=south-west England), western Ireland, and Wales. I once found larvae, however, in a quarry two miles inland near Boscastle. Both these species pass the day as moths resting on rocks well above heavy-spray level, where their dark grey or brown colouration makes them inconspicuous. Both feed at early dusk on the *Silene* flowers, when, no doubt, they are predated by gulls, though I have no evidence of this. The Noctuid moth *Antitype xanthomista* Hübn. (Caradrinidae), whose larvae feed on Thrift, is another dark, cryptic, rock-sitting insect and is restricted to the rocky coasts of north Devon and Cornwall, the Isle of Man, and recently it has been recorded by Demuth from southern Ireland.

In south-west England, along the shores of north Devon and Cornwall, melanism is again found among shore-breeding Lepidoptera. For example, specimens of *Hadena conspersa* normally have a contrasting black and white colouration. Here they are uniform dark blue-grey, similar to, but not as extreme as, individuals from Shetland (Plate 10.2, 3 and 4). This form appears to be limited to coastlines because of its preferred food plant, *Silene maritima*. Immediately inland, however, *H. conspersa* feeds on other *Silene* species and here it has the normal black and white patterning.

We do not as yet know the reasons for 'Western Coastline Melanism' but it appears to exhibit its maximum effect on or around the actual beach-line. All the species hatch around mid-summer and fly at dusk, when they feed on the nectar of the white *Silene* flowers, and they are then likely to be subject to predation. They should form a ready source of food for gull species and such passerines which live there, such as the Shore Lark *Otocorys alpestris* L. By day most of the moth species rest on rocks and on the ground, and the background here is frequently darker because of increased humidity due to salt spray and increased cloud cover. In addition, it is possible that these forms have a tolerance to the absorption of salt physiologically but as yet we have no evidence of this. On the other hand, we have some information that industrial melanic forms are hardier under adverse conditions than f. *typica*.

Western Coastline Melanism is probably found in many regions of the Palaearctic, where it is likely to merge into more general melanism found around latitude 60°N. It undoubtedly reflects Gloger's Rule in the Lepidoptera, which states that creatures inhabiting western shores are darker than those found farther east: furthermore, it postulates that this is due to increased humidity.

(iv) Ancient Conifer Melanism (Relict Melanism)

Melanism is frequently found in regions of indigenous coniferous forest. This is certainly so in the relict Caledonian pine forests, and also in the pine barrens of New Jersey, and, no doubt, elsewhere in North America.

It is likely that 10 000 years B.P., before the succession of conifers by deciduous tree species, large areas of northern Europe, including Britain, were covered with pine species. Their trunks are very different from those of deciduous trees in two ways: they are usually void of lichens except under conditions of high humidity, and they present areas of dark coloured surface particularly in wet weather. It is in my opinion likely that melanic forms of many cryptic species were common at this period. It is possible that many of these fed then on *Pinus* species even though there was an underlying carpet of *Vaccinium* (Bilberry) and *Erica* (Heather). Pine is still today an alternative food plant to deciduous trees for many species which exhibit melanic forms: *Gonodontis bidentata*, *Boarmia abietaria*, and *Cleora rhomboidaria* and even *Biston betularia* has been recorded as feeding on it; also *Lymantria monacha*, *Oporinia autumnata*, and *Ectropis consonaria*.

Past and present

There remain in Britain today only a few small areas of relict indigenous pine forests and these are largely confined to Scotland. They are very different from the pine plantations so widespread under modern silviculture. The original pine forests were interspersed with Birch (*Betulus*) and permitted a thick carpet of vegetation beneath. Pine seedlings are only able to regenerate naturally in fact in patches where the carpet has been broken up by animals such as deer; hence the patchy distribution of the pines. The modern pine plantations so beloved by the Forestry Commission today are planted in neat rows with geometric precision, which has the effect of sterilizing the flora beneath completely. Though the same food plant, pine, is present, these plantations bear little resemblance to the earlier pine forest ecology. Nearly all those Lepidoptera species which feed on ground-level plants are absent in our modern plantations (except for Blackberry (*Rubus*) feeding species).

I decided to sample the Lepidoptera of a relict Caledonian pine forest and I chose the Black Wood of Rannoch, Perthshire, which covers about a mile square of the southern shore of the Loch. It probably bears a resemblance in miniature to the forests which were present 10 000 years ago, from which it was derived and has survived but little altered.

Here I have sampled the Lepidoptera during the months of June, July, and August, on seven separate visits during the last 15 years. Of approximately 45 species taken, nine had melanic forms. Though most of these rested by day on pine trunks and on the bare ground beneath, some of them fed on plants other than pine. I chose one species, *Cleora repandata*

L., for study. Here it exists in a state of balanced polymorphism, the black and the light forms being easily distinguishable.

In a sample of 428 recorded by me in the second half of July in 1958, 10 per cent were of the melanic form, *nigricata* Fuchs (Plate 10.6), a figure which is 5 per cent higher than one recorded from the same wood in 1942 (Williams, 1949), but these were bred from larvae collected ($n = 93$) at Rannoch by Crewdson and the figure therefore reflects the frequency over the whole season.

An interesting point has arisen more recently. In 1963, Cadbury visited the Black Wood at the commencement of the hatching period in early July,

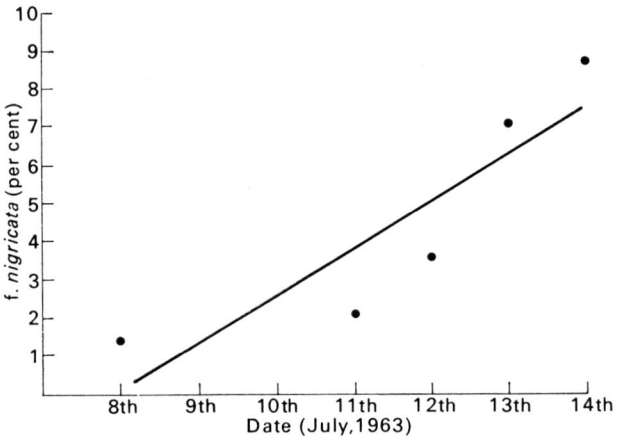

FIG. 10.1   Hatching of *Cleora repandata* and its black form *nigricata*. Black Wood of Rannoch, 1963

and by disturbing moths from pine trunks and by 'dusking', he took a sample of 651. During a period of seven days, the frequency of f. *nigricata* increased from 1·4 per cent to 8·7 per cent (Table 10.1 and Fig. 10·1). This suggests that the melanic form hatches later than f. *typica*, and could account for the high frequency of 12 per cent taken by me in 1969 in a small sample ($n = 25$), late in the season (August). Alternatively it could reflect the increased survival of this form. This is a reversal of the situation in *A. glareosa*, where the melanic f. *edda* appears to hatch earlier than f. *typica*.

The form *nigricata* is similar to the industrial melanic, f. *nigra* Tutt (Plate 5.6, right 5 and 6), which has spread today through built up areas in many parts of England. Though we have paired f. *nigricata* with the industrial f. *nigra* on numerous occasions in the last ten years, I am still unable

TABLE 10.1

The frequency of Cleora repandata f. typica and its black f. nigricata; wild samples from black wood or Rannoch, 1963 (Cadbury).

Date	f. typica ♂	f. typica ♀	Total	f. nigricata ♂	f. nigricata ♀	Total	Total	% melanic	% ♀♀
6 July	13		13				13 ⎫		
7 ,,	21	1	22				22 ⎬ 71	1·4	2·3
10 ,,	34	1	35	1		1	36 ⎭		
11 ,,	177	5	182	4		4	186	2·1	2·7
12 ,,	146	13	159	5	1	6	165	3·6	8·5
13 ,,	112	5	117	9		9	126	7·1	4·0
14 ,,	85	9	94	7	2	9	103	8·7	10·7
	588	34	622	26	3	29	651	4·5	5·6

There is significant regression of the melanic frequency in hatching date. $t_{(3)} = 3·300$, $0·05 > P > 0·02$.

to report whether they are allelic or not, owing to the inadequate number of broods reared. Both forms are dominant to f. *typica*.

Of the nine polymorphic species which I found in the Black Wood, *Cleora repandata* was obviously the one to choose for fuller field study, and, early in our investigations, I and three other observers noted that on taking flight f. *nigricata* was practically impossible to follow on the wing for more than a short distance, whereas f. *typica* could be seen flying up to a distance of one hundred yards or more. In order to learn more about the normal habits of this species, each morning many hundreds of pine trunks were searched, and each *C. repandata* was scored for camouflage efficiency, and a white indicator label was pinned a foot to its right. The pine trunks offered two backgrounds, a light lichen-covered one and a dark one; nevertheless, the majority of f. *nigricata* were scored as conspicuous and nearly all f. *typica* were assessed as inconspicuous, and the latter could frequently only be recognized on fanning the trunks with a net, thereby causing the wings to move momentarily. Later in the day the same trunks were examined and a record made of those which were missing. By this means it was found that on certain days as many as 50 per cent of the *C. repandata* had vacated the positions that they had occupied in the morning. On closer study it was apparent that these had taken flight owing to disturbance by a large black ant (*Formica rufa* group) or owing to direct sunlight which at this latitude strikes two-thirds of each trunk in the course of a day. We have, therefore, a situation in which the majority of melanics are at a slight cryptic disadvantage when at rest, but gain a considerable benefit whilst on the wing; also Ford (1940a, 1945) has shown the increased vigour of this form.

In this locality, large numbers of both forms have, in the course of a day, to fly because of disturbances from one cause or another. When this happens they may travel a considerable distance passing in and out of shade, but eventually taking up position again on a tree trunk. During these excursions we actually witnessed three specimens all f. *typica* captured on the wing by birds; two by Chaffinches, *Fringilla coelebs* L., and one by a Redstart, *Phoenicurus phoenicurus* L. This may well be part of the explanation of the presence of these melanic forms in these ancient woods, and it is my belief that the same species which today exhibit Industrial Melanism, have, in some previous era, found similar uses for their melanic forms under conditions very different to those of today.

This fact is again borne out by the Lepidoptera of North America where due to different reasons, melanic forms are common in the northern pine forests. Or again farther south in the genus *Zale*, occasional melanic forms turn up sporadically in many species and, in at least one Industrial Melanism has developed. In another, *Z. undularis*, the melanic form appears to have become fixed as the universal phenotype, the light patterned form *umbripennis* occurring in the population as a rarity. My work on Conifer Melanism confirms my belief that melanism is not a recent phenomenon, but a very old one.

(v) Melanism associated with fire-resistant trees

Fire has been an integral part of nature since the earliest times, and certainly long before Man discovered how to make it; fire from lightning or fire from volcanoes regularly set scrub and savannah aflame. In such regions fire-resistant trees have developed and here the blackened trunks of the living ones can be seen amongst the dead. I have had experience of several such areas both in the Northern and the Southern Hemispheres. Typical of each are the pine barrens of North America and certain regions in southern Africa. In both, melanic species of Lepidoptera or melanic polymorphisms occur.

Fire-resistant forests in North America—New Jersey Pine Barrens

In northern Europe one of the commonest trees is *Pinus sylvestris*, and prior to 8000 B.P. large areas of Britain were covered by it. In common with the majority of trees, it is vulnerable to fire and when cut down it never regenerates. Hence the disappearance of the indigenous pine which can only reappear by planting or, in the old Caledonian pine forests, from seed which requires special conditions for germination in nature (for example, Red Deer stamping-grounds).

Very different is the behaviour of some North American pine species, for example the Pitch Pine, *Pinus rigida*, the Longleaf Pine, *P. palustris* Miller, the Virginian Pine, *P. virginiana* Miller, and the Two-needle Pine, for here the trees withstand regular burning and regenerate from stumps after felling. Fire is indeed necessary for the seed of some species prior to successful germination. Such forests, interspersed as they are with a number of oak (*Quercus* spp.) which are also fire-resistant, are virtually indestructible except by gradual extinction due to climatic changes. This type of ecology has probably been shifting up and down the eastern side of the North American continent, from Canada to Florida, for thousands of years, depending on ice ages and warm periods. They have carried their insect populations with them, which themselves have had to adapt to the effects of regular burning.

Apart from blackened trunks such forests hold areas of shade, particularly in dense stands of trees like the Coastal Plain White Cedar, *Chamaecyparis thyoides*, whose bare trunks grow close together and whose overhead canopy casts semi-darkness beneath. Klots (personal communication) has suggested that melanism in such biotopes may be more likely to be due to shade and not to burning. He has pointed out also that there are conifer forests of similar density and shade in the north-west and also in the Olympic Peninsula in Washington, though he is unable to say whether they have served as reservoirs of melanism or, indeed, how prevalent it is there today. It would certainly be a worthwhile investigation to sample such localities, along with other indigenous pine forests: *P. murrayana*, the Lodgepone Pine in the Rocky Mountains, *P. taeda* L., the Lablolly Pine of the middle states coastal plain and elsewhere. I think that, until we have learnt more about such areas, shade melanism should be considered separately to that occurring in fire-resistant forests.

Apart from the New Jersey Pine Barrens, there are forests, or pockets, of fire-resistant trees throughout the North American continent (for example, Oregon State). I believe that here melanic polymorphisms have been maintained for millennia prior to Man's migration there and certainly before the advent of industrialization. The earliest archaeological records of the North American Indian are *c.* 30 000 B.P. and their use of fire from that time must have greatly enhanced the value of fire-resistance. Selection in favour of this trait in the flora must have had consequences in the fauna which depend on them.

In fire-resistant forests many of the backgrounds are blackened and charred, and here lichens are absent. Following a fire, however, in these specialized ecologies, pines resprout and oaks regenerate after only a short period. Many other plants and shrubs are fire-resistant so that a large number of food plants are available here for Lepidoptera larvae.

The New Jersey Pine Barrens extend over an area of nearly two million acres. In July 1967, I visited here for a few days, and sampled at three separate places, and took melanic forms of six species. The Pine Barrens are situated within 100 miles of both New York to the north and Philadelphia to the south. We cannot therefore altogether ignore the effects of pollution fall-out though this is likely to have been minimal. Nevertheless, the proximity of industrial areas has served to blur their importance as a reservoir of Non-industrial Melanism. For this reason experienced lepidopterists who have sampled the Barrens for years do not consider these regions as centres of melanism today, though they admit that such forms have occurred there from the earliest records. Their contribution to Industrial Melanism is discussed later in Part VII.

Fire-resistance in Cape Province, South Africa

Fire has been a constant hazard in Africa and indeed was recorded by Vasco da Gama there as early as the fifteenth century on his journeys south. Many plants, shrubs, and trees are fire-resistant here but it is unfortunate that the only record of a melanic polymorphism in the Lepidoptera found by me in South Africa is associated with an introduced tree, Rooikrans (*Acacia cyclops* Cunn.), which is indigenous to Australia but where, however, the fire risk is even greater. *A. cyclops* is now widespread throughout the maritime wastes in the Cape. In the sandy Fishoek-Kommetjie Valley which is 25 miles south of Cape Town I sampled Lepidoptera between 1949 and 1953 and I found one species in particular which had a melanic polymorphism, *Cleora tulbaghata* Felder (Selidosemidae) (Plate 10.3). Two per cent were black, and ten per cent were grey with a darker band. These were obtained by mercury-vapour trapping and by searching ($n = 350+$). I found that this species rests by day on the trunks and boughs of Rooikrans, but the larvae refused to feed on it. The single specimen of the black form I succeeded in finding was difficult to see at rest; f. *typica* (n = 85) less so. This valley has a high proportion of insectivorous birds (for example, the Cape White-Eye, *Zosterops c. capensis*). Other melanic polymorphisms are likely to occur, not only here, but also in similar fire-inflicted regions in other parts of the world, such as, perhaps, Australia.

PLATE 10.3 (p. 176). *Cleora tulbaghata* Felder (×1¼) and its more extreme melanic form (Cape Province, South Africa).

PLATE 10.5 (p. 169). *Euphyia bilineata* L. (×1½) and its melanic f. *atlantica* Kane.

PLATE 10.4 (p. 179). Alpine *Setina* spp. (Arctiidae) (×1) and their melanic forms from Central Europe.

PLATE 10.6 (p. 172). *Cleora repandata* L. (×1¼).
1. f. *nigra* Tutt. (industrial)
2. f. *nigricata* Fuchs.
3. f. *typica*.

PLATE 11.1 (p. 182). *Amathes glareosa* Esp. f. *typica* (×1½) and its melanic f. *edda* Stdg.

PLATE 11.2 (p. 182). *Amathes glareosa* f. *typica* (×1½) at rest in heather (Shetland).

PLATE 11.3 (p. 183). *Amathes glareosa* f. *edda* (×1½) at rest in heather (Shetland).

PLATE 11.4 (p. 183). Series of *Amathes glareosa* Esp. ($\times 1\frac{1}{2}$) taken in Orkney by I. Lorimer, 1971, showing continuous variation. Left, 1 and 2, are of f. *typica*, the rest are f. *edda*.

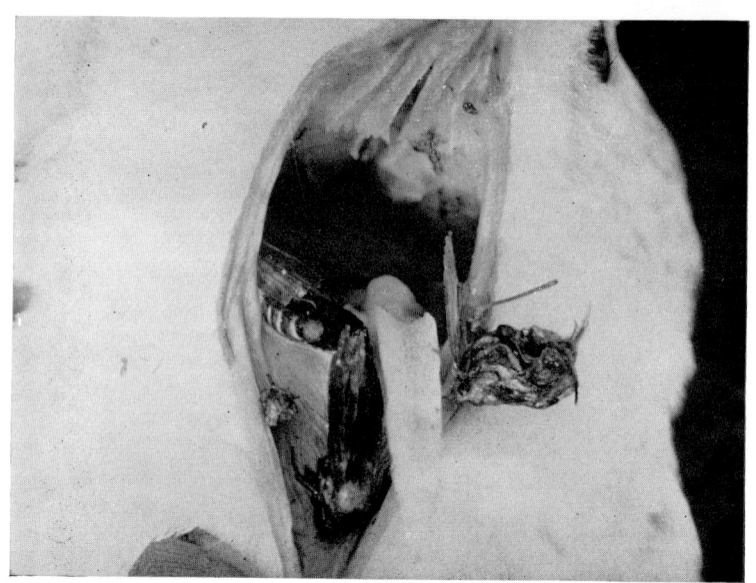

PLATE 12.1 (p. 201). Common Gull (*Larus canus*) regurgitating *Amathes glareosa* (Unst, 1960).

PLATE 12.2 (p. 201). Stomach contents taken from Common Gulls shot on Unst, 1960.

(vi) Pluvial Melanism

There are a few areas in the world today where the rainfall per annum is over 300 inches. These are to a large extent confined to high mountain forests such as those of South Island, New Zealand, or to parts of the Himalayas. Here the light intensity is low and the backgrounds dark because of the humidity.

Many regions of the earth must in the past have from time to time had pluvial periods maintaining similar conditions. For example, it is likely that northern Europe had such a phase about 1000 to 1200 B.C.

There is some evidence that many species are melanic today in such forests. Hudson (1928) after discussing mountain-top melanism in South Island, makes the following statement:

> A less pronounced melanism, or tendency to dark colouration, is often observed in Lepidoptera found in very damp localities, and many species which in their markings imitate moss and lichens, are more vividly coloured in wet than in dry localities. This is, I think, due to the darkening of the tree trunks, and the more luxuriant growth of lichens and mosses induced by almost constant and heavy rainfall, the protective colouring of the moths being correspondingly modified to meet the special environment. These conditions are, in fact, fulfilled on the west coast and in the extreme south of New Zealand. A similar darkening has been observed amongst Lepidoptera taken near the great manufacturing towns in England, where the tree trunks are darkened by deposits of smoke, the colouring of the moths having also darkened in order to afford them the needful protection. It is obvious that a light-coloured insect on a dark tree-trunk would be at once detected and destroyed, and that the darker forms alone would survive.

Though there are several conflicting statements here, the important point is that Hudson, a most reliable field entomologist, considered that melanism was associated with areas of high rainfall and there are many places in South Island where this amounts to over 300 inches per annum. That melanism occurs away from the west coast under such conditions has recently been confirmed by Dr. Salmon (personal communication). In the first place, it is clear that Hudson merges West Coast Melanism with that of the high rainfall forests, and in fact such a situation may exist. I have no evidence on the point. More confusing are his remarks about lichen-covered backgrounds which we know demand bold and light-coloured patterning for efficient crypsis. Yet he compares the dark-coloured insects occurring here with the industrial melanics of England.

I think it is likely that melanism is prevalent wherever there is constantly high humidity, be it western coast or mountain forest.

(vii) Thermal Melanism

Thermal Melanism is found in two differing ecological situations: first on high mountain summits and secondly at high latitudes. It is confined to day-flying Lepidoptera, which gain advantage by a more rapid absorption of heat than the light-coloured ones during fitful bursts of sunshine. This enables the darker individuals to take wing more rapidly for mating or ovipositing activities. Thus *Erebia* species (Rhopalocera), which frequent alpine or arctic biotopes, are always of dark colouration and they fly the moment the sun comes out. Probably for the same reason a very dark subspecies of the dark-green Fritillary butterfly, *Argynnis aglaia* ssp. *scotica* Watkins, is found off the north-west coast of Scotland on the Hebrides. Here, particularly in the female, the usual tawny-brown colour is replaced with blackish suffusion. Whilst this may enable the insect to become more active more rapidly during periods of sunshine than the light coloured f. *typica*, it is likely that at the same time the blackest females are at a considerable mating disadvantage. Magnus (1958) has shown experimentally that dark females in a related species, *Argynnis paphia* (f. *valesina*), are much less likely to obtain a mate than the normal yellow ones.

Here we can see the advantages and disadvantages conferred by blackness, and the intensive selection which must occur and which must vary in direction from season to season according to the amount of sunshine and the mean temperature during daylight.

Similar situations are found in a number of other Palaearctic butterflies such as *Pieris napi* L., of which many of the females in north Scotland have greatly increased black veining and the proximal portion of the forewings may even be smoky. A somewhat similar form occurs as ssp. *bryoniae* in the Swiss Alps. The genetics of these various forms and races have been most successfully worked out by Bowden (1962).

That Thermal Melanism is a true mechanism which promotes earlier activity has been shown by Watt (1968) under the heading of 'Variation of melanin pigment in relation to thermoregulation'. Using *Colias* species (Pieridae) inhabiting various altitudes (and latitudes) he showed that in colder zones the hindwing undersides have increased pigmentation; that insects orientate themselves presenting a maximum surface area to the sun's rays below 34–35°C, but above 37–38°C they rest parallel to the sun's rays thereby diminishing heat absorption to a minimum. In a series of experiments on *Colias meadii* and *C. nastes*, using small thermistor probes, he was able to verify the efficiency of increased pigmentation in producing heat both in living and in dead specimens. Watt has, I think, for the first time shown, by a series of finely controlled experiments on day-flying Lepidoptera, the selective advantages of pigmentation in regard to temperature control.

Amongst the moths, many mountain-top species in Britain are day-flying and have dark colouration. This is not surprising to those who have spent nights there, for the temperature at dawn, even during the summer months, may drop to around freezing point. In Scotland *Psodos coracina* Esper, the Black Mountain Moth, flies in sunshine and is found only between 2000 and 4000 feet. Another montane insect, *Isturgia carbonaria* Clerck, the Netted Mountain Moth, flies around noon in the sun only, but in this species there is a range of colour from grey to black.

These two species are confined to mountain tops only. They are relicts of a boreal or a sub-boreal fauna; more rarely, there are moths which have adapted themselves to living from sea level to great heights. They have a discontinuous distribution and only occur in local, though ecologically very different, habitats.

Setina irrorella Clerck (Lithosiinae), whose larvae feed on black and yellow lichens, is found in Britain in many diverse habitats: on the shingle beaches of Dungeness, Kent, just above the high tide mark on the edge of a large colony of breeding Terns (*Sterna hirundo*); it is common on chalk downs, for example at Dover, and Ventnor in the Isle of Wight; I have also found it on the rocky shores of western Ireland, and it occurs on Canna. Throughout its range, the males fly in sunshine, and in Britain the imago has a similar pale yellow colouration with small black dots on the forewings: I cannot distinguish individuals from one locality or another. Rarely specimens are found in which the proximal part of the nervures have black pigmentation and I bred a single female from the Dungeness shingles amongst 100 normal ones in which all four wings were dusky. On the Continent, *S. irrorella* has a montane and not a coastal distribution; it is found on the highest slopes of the Swiss and French Alps but here the population maintains a number of dark forms at a high frequency—f. *andereggii* with all nervures blackened, and f. *riffalensis*, a more extreme form (Plate 10.4). Several other nearly related species have similar forms (*S. roscida*, *S. aurita*, and *S. kuhlweini*).

In the Southern Hemisphere similar alpine melanisms exist. As early as 1928 G. V. Hudson wrote a most revealing paragraph in his book, *The butterflies and moths of New Zealand* (pp. 14 and 19):

On the tops of high mountains, and in certain other special localities, *Crambus* (which includes the well-known Grass moths) is replaced by the interesting and closely allied genus *Orocrambus*. The species included in *Orocrambus* are all day-fliers, and very darkly coloured, enabling them to absorb heat rapidly, and thus take advantage of the fitful periods of hot sunshine characteristic of alpine climates. . . . A few of the dark-coloured mountain species fly by day, but the majority rest with closed wings on rocks or tree trunks, immediately taking wing when approached and flying with considerable rapidity.

The black colouration, characteristic of most Lepidoptera inhabiting high alpine regions, has no doubt been acquired for the purpose of absorbing the heat of the sun, during the transient periods of intensely hot sunshine which occur in such localities. The same class of colouring also prevails amongst species found in high latitudes. Lord Walsingham, who first drew attention to these adaptations, made some very interesting experiments by placing a black and a white insect on snow, and observing the relative rate at which each sank when exposed to the sun's rays. It was found in every case that the black insect made a much more rapid impression on the snow than the white one, owing to its superior absorptive powers. It is clear that in a cold and stormy climate, like that existing in alpine or sub-arctic regions, the black insect, which matured rapidly, and was able to pair and deposit its eggs in the shortest possible time, would have a distinct advantage in the struggle for existence over the more slowly developing white insect, which, owing to its inability to rapidly absorb heat, could not fully avail itself of such short periods of hot sunshine.

Thermal Melanism can certainly not be described more aptly. It will be more convincing when further investigations (such as those of Watts) have been carried out, not only in the laboratory, but also in the field.

11. *Amathes glareosa* Esp. and its melanic form *edda* Stdgr. in Shetland: an example of Northern Melanism

(i) Choice of material

IN attempting to investigate and analyse the advantages and disadvantages of Non-industrial Melanism in the wild, under the difficulties imposed by climate and lack of facilities, and without guidance from previous work, it was necessary to choose a subject which offered as few complications as possible. Easy recognition of the melanic form was therefore essential, as was the choice of a species which was common, widespread and which could easily be collected. Mark–release–recapture experiments were envisaged in which the two contrasting forms would be released at the extremes of their frequencies (see Chapter 12). Furthermore, a clear-cut dimorphism would be desirable, and one which was comparable with those found in Industrial Melanism.

For these reasons, I decided to undertake a full investigation into the frequencies of *A. glareosa* Esp. and its dark form, *edda* Stdgr., throughout the Shetland Isles.

Nevertheless, *A. glareosa* cannot be considered as typical of all melanic moths in Shetland, for several reasons. Few species have such a uniform distribution there; many have discontinuous ones and are isolated from each other (see Chapter 10 (ii)). A clear-cut polymorphism (group 2) is not the general rule; nor are we able to report any other north–south clines in Shetland species, though amongst group 3 this could have been overlooked because of the difficulty of scoring intermediates, for example, in *Amathes xanthographa* Schiff. In spite of this, I think it likely that melanic clines will be found in many northern archipelagoes; for instance, on the Baltic Islands with such a species as *Simyra albovenosa* Goeze and its melanic form *murina*.

Northern species may, however, show local pockets of melanism and also of lighter forms. The large Noctuid moth *Xylophasia* (*Apamea*) *exulis* Lef. is widespread in Iceland where it is light and variegated. It is found also in many areas over 100 feet on the north Scottish mainland and there it has a blackish colouration and this form is referred to as ssp. *assimilis* Doubleday. Typical *X. exulis*, similar to those found in Iceland, were, until our recent expeditions to Shetland, believed to be the only form occurring there. Here it was found on the tops of peat hills at a height of about 1000 feet on the island of Unst and we obtained it in numbers. Approximately 2 per cent (sample 100+) were dark. We were surprised to find this species commonly at sea level near Hillswick, Shetland Mainland, but here the majority were dark and similar to the Scottish mainland ssp. *assimilis*. Between 2 and 3 per cent, however, were light and variegated as found on Unst and in Iceland. Such situations in which local populations are maintained independent of each other, contrast with that of *A. glareosa* where there is continuity throughout the length and breadth of the islands.

It was for these reasons that I chose *A. glareosa* in preference to others. I must here record my gratitude to R. P. Demuth who many years ago had attracted my attention to the potentialities of this insect in Shetland.

(ii) Description of morphs

Except in Shetland and the Orkneys (and a few coastal areas in Denmark), only the typical form of this species is found throughout its range (Plate 11.1) but see footnote, page 272. In this the ground colour is either light grey or light grey tinged with pink; however, the latter are commoner in the south-west of England. Occasional specimens with a dark ground-colour have in the past been recorded from parts of Yorkshire and also from Perthshire (Barrett, 1897). On each forewing f. *typica* has two conspicuous black wedge-shaped marks with their broad bases lying towards the costa. This pattern is without doubt disruptive, and its efficiency must therefore vary according to the depth of vegetation and the angle and brilliance of the sun. The hindwings are white; in common with most Noctuid moths the thorax and forewings alone are visible when at rest, and the black marks serve to break down the insect's outline in the shadows cast by overlying foliage (Plate 11.1, below and Plate 11.2).

The melanic form *edda* was first discovered on one island only, Unst, by McArthur in 1883 (Weir, 1884), but we now know that it is found throughout Shetland, including Fair Isle, and also in the Orkneys, where ffennell recorded three out of twenty of this form (15 per cent) and where more recently I. Lorimer has found it widespread at frequencies varying locally from 11 to 37 per cent. He sampled three localities, two on Orkney Mainland and one on Rousay. The f. *edda* frequencies show, not only rapid changes over short distances, but also fluctuations from year to year (Table 11.1); this must reflect corresponding selective pressures. None of these f. *edda* was as dark as those from the northernmost island in Shetland, Unst. *A. glareosa* does not occur farther north, in the Faeroes.

TABLE 11.1
Frequencies of f. edda *in Orkney and Fair Isle.*

Site	Year	Numbers caught over whole period of trapping			per cent frequency of f. *edda*
		f. *typica*	f. *edda*	Total	
Rousay (Orkney)	ffenell 1961	17	3	20	15·0
Orphit (Orkney Mainland)	Lorimer 1967	66	16	82	19·5
Orphit (Orkney Mainland)	Lorimer 1968	153	19	172	11·0
Binscarth (Orkney Mainland)	Lorimer 1968	37	22	59	37·3
Fair Isle	Hardy 1961	40	15	55	27·3

More recent sampling in Orkney by R. I. Lorimer (1971, personal communication) shows not only that f. *edda* there are less melanic than in Shetland, but that f. *typica* grades imperceptibly into f. *edda* (Plate 11.4). It would appear then that the blackest f. *edda* and disruptive selection have evolved in northernmost Shetland but that farther south in Orkney no such black individuals exist; nor indeed is there a clear-cut polymorphism. This interesting situation suggests that dominance has not yet been evolved in Orkney and that the heterozygote is more variable.

The form *edda* bears little resemblance to f. *typica*. It is of a dark brown-black colour with light orbicular and reniform outlines on the forewings. The thorax is, however, greyish in many specimens. It is important to note that the hindwings, which are only exposed in flight, are of a lighter colour. Further, not only is the dorsal surface of the abdomen black, but also the ventral. This is an unusual occurrence in melanic Lepidoptera that have light-coloured hindwings, and must have a survival value. In its normal resting position in fine weather on a peaty background, f. *edda* is most difficult to see (Plate 11.3). I have observed that, on the commencement of rain, the moths leave their exposed resting sites and run beneath debris (such as small stones on Muckle Heog, Baltasound) and then sit vertically with only the grey thorax visible from above. The form *edda* therefore reflects the substitution of a disruptive pattern by a highly efficient cryptic one when resting on peat and peat-stained backgrounds.

(iii) Life History

There is a one-year life-cycle, but it is possible that occasionally, in Shetland, in common with other northern species, *A. glareosa* may pass a second winter in the pupal state. The moth lays its eggs singly, probably on heather in the wild, but in captivity on heather, muslin, or cardboard. The ova are spherical and ribbed and of a pale yellow colour. After a few days, the fertile ones can be recognized by flecks of red appearing around the upper pole. After two or three weeks the eggs darken and by the beginning of October all are blackish-grey. The following eight weeks certainly constitute a most vulnerable period in the life history of the species; in 1959, 100 per cent of our stock died at this stage, and on dissection the eggs were found to contain fully formed larvae. In 1960, by keeping ova cool and in a damp atmosphere, we succeeded in hatching a large proportion of them, but the period over which this took place varied from 30 to 80 days, and within one brood (g/24/60) from 47 to 80 days. It seems in fact as though the larva goes into hibernation for a time within its eggshell. This is a well-known phenomenon in certain species such as *Lasiocampa trifolii* L.

In contrast with my Shetland stock, ova from an Orkney female f. *edda* obtained by I. Lorimer in 1969 hatched within 48 hours of each other in early October when kept under identical conditions to the earlier Shetland ones.

A large proportion of young larvae of *A. glareosa*, no matter under what conditions they are kept, die 24 hours after hatching. Although our broods were distributed to many of the best breeders in Britain, all had the same results. It is likely that a heavy mortality takes place in nature at this same period in the life history of the species. Those larvae which survive the first forty-eight hours attain full growth with a comparatively low mortality rate. In warmth in the laboratory we accomplished this in under two months. In nature it is probable that the small larvae feed slowly on warm nights throughout the winter. This much we know for certain: that when full-grown they have been found but rarely in the spring, and then only at night. They were feeding on plants such as Hyacinth (Barrett, 1897). They pupate in April or May, and they do this in cocoons below the surface of the ground. Both larvae and pupae are therefore largely protected from visual predation.

(iv) The genetics of the two forms (Kettlewell, 1961c)

Table 11.2 gives a list of 46 wild-caught females, with their local gene-frequencies and the numbers of each form which were bred from them. From these data, it appears that f. *edda* is dominant to f. *typica*. The form *edda* varies from a light brown to black, but there is no clear-cut segregation within the f. *edda* of our broods, because, except for two, the numbers

TABLE 11.2
Breeding results: Amathes glareosa

Brood no.	Phenotype of ♀ parent	Locality	f. *typica* local gene-frequency per cent	f. *typica*	f. *edda*
g/37/60	f. *typica*	Unst	16·4	0	4
g/40/60	f. *edda* (light)	,,	16·4	0	3(2 dark)
g/29/60	f. *typica*	Mainland Reawick	63·7	0	2
g/2/60	,,	Aith	66·3	5	7(light)
g/31/60	f. *edda*	Skellister (S. Nesting)	61·6	3	1
g/13/60	,,	N. border of Tingwall Valley	86·9	1	1
g/19/60	,,	,, ,,	86·9	0	6
g/20/60	,,	,, ,,	86·9	1	2
g/21/60	,,	,, ,,	86·9	1	0
g/22/60	,,	,, ,,	86·9	0	12(3 dark)
g/14/60	,,	S. border of Tingwall Valley	88·7	3	16(2 dark)
g/15/60	,,	,, ,,	88·7	2	0
g/15/60	,,	,, ,,	88·7	1	1
g/8/60	,,	Boddam	99·0	0	2
g/9/60	,,	,,	99·0	0	1
g/11/60	,,	,,	99·0	1	0
			Total bred	18	58

were too small. Brood g/14/60 (from south Tingwall) showed a reasonable 1:2:1 ratio, and brood g/22/60 (from north Tingwall) not such a good one; it is likely that in the latter brood there was a deficiency of homozygous f. *edda*. Females are usually darker than the males and this applies particularly to the hindwings. In the north of Shetland (Unst), the very black form is common and the pale f. *edda* less so. In the south, however, where f. *edda* forms only 2 per cent of the population, the black form is absent, the dark one rare, and nearly all the f. *edda* are less extreme. We have further proof that these two forms are controlled genetically in that we took a halved mosaic in south Shetland in which one side was pale f. *edda* and the other f. *typica*. I discuss the question of dominance modification in the following paragraph.

(v) Dominance modification

As early as our first expedition in 1959 we noted that the f. *edda* samples from Unst were blacker than those taken farther south. Furthermore, the samples from central Mainland, where the frequency was between 30 per cent and 70 per cent, showed f. *edda* varying from light to dark brown individuals. The southernmost population (from Dunrossness) seemed to contain morphs other than light brown f. *edda* only rarely. Subsequently we attempted to score this morph into either a light or a dark class, ignoring worn specimens. Table 11.3 shows the result of thus classifying some 6000 f. *edda* from four localities from north to south Shetland. On the assumption that the darker individuals were the homozygotes, one was able to assess the expected number of heterozygotes from the Hardy–Weinberg ratio. There was a significant degree of agreement in all but the Unst sample. It is therefore fair to assume that the lighter specimens reflect heterozygosity.

TABLE 11.3
Numbers of light and dark f. edda *in wild-caught samples.*

Site	Per cent frequency of *typica*	Numbers of			total	Observed* and expected† percentage frequencies of			
		light f. *edda*	dark f. *edda*	f. *typica*		f. *edda* heterozygotes		f. *edda* homozygotes	
						observed	expected	observed	expected
Baltasound	3·0	300	2162	77	2539	11·8	28·8	85·2	68·2
Hillswick	23·6	602	355	295	1252	48·1	50·0	28·3	26·5
Tingwall	73·6	287	33	893	1213	23·7	24·4	2·7	2·0
Dunrossness	97·5	19	5	933	957	2·0	2·5	0·5	0

* Assuming that light f. *edda* are heterozygous for the gene.
† Assuming that the Hardy–Weinberg distribution is followed.

On Unst, however, there was a deficiency of light insects. Of the 12 818 f. *edda* caught, 2539 were scored, of which just under 12 per cent were classified as being 'light f. *edda*'. The Hardy–Weinberg expectation is that 29 per cent of all f. *edda* would be heterozygotes (Kettlewell and Berry, 1969).

Furthermore, the blackest f. *edda* to be found in Shetland were on the island of Unst. It is likely under the intensive visual predation which occurs there that selection has favoured darker f. *edda* both as heterozygotes and homozygotes. Such a situation is likely to evolve when an allelomorph is at a high frequency in a population (Fisher, 1928, Haldane, 1956, Clarke and Sheppard, 1960a). I have had to assume that f. *edda* is controlled by a single identical allele throughout Shetland and it is likely that this is so. This is in contrast with such an assumption in *Biston betularia* and its melanic f. *carbonaria*, and still more so in its so-called 'f. *insularia*' which we know is not so controlled (Chapter 7). Unless the homozygotes are at an overall advantage to the heterozygotes, which is unlikely, an increase in dominance must have evolved in the f. *edda* on Unst.

Conversely, further south in Orkney, it would appear that both heterozygous and homozygous f. *edda* are less extreme and here in fact a few individuals are difficult to differentiate from f. *typica,* a situation not found in Shetland (see p. 183).

(vi) The distribution of the two forms

It is a remarkable fact that a gradual change of gene frequency as great as 1 per cent to 98 per cent should occur over a distance of 54 miles (from north to south) covering the three major islands in the Shetland archipelago, and that in spite of the two sea barriers between them a constant frequency-cline should be maintained (Table 11.4, Fig. 11.1) (Kettlewell

TABLE 11.4

Constitution of population samples of Amathes glareosa
(from north Shetland to south) 1959–62.

Area	Numbers caught			Overall per cent frequency of f. *edda*
	f. *typica*	f. *edda*	Total	
Unst				
1. Muckle Heog (Baltasound)	379	12 818	13 197	97·1
Fetlar				
2. Tresta	34	162	196	82·6
Yell				
3. Gutcher	38	15	54	27·7
4. West Sandwick	139	224	364	61·6
5. Burravoe	23	29	54	53·7
North Mainland				
6. North Roe	29	47	77	61·0
7. Hillswick	442	1402	1844	75·9
8. Mossbank	58	61	119	51·3
9. Voe	63	90	153	58·8
10. Sandness	119	95	214	44·4
11. Aith	61	74	139	53·2
12. Reawick	39	57	96	59·4
13. Kergord	14	9	23	39·1
14. South Nesting	22	34	58	58·6
Tingwall Valley: north border				
15. Catwalls (Gott)	467	218	685	31·8
16. Scalloway	38	9	47	19·1
Tingwall Valley: south border				
17. Braewick (Dales Voe)	27	4	31	12·9
18. Burn of Dale (Dale of Voe)	121	33	154	21·4
Dunrossness				
19. Cunningsburgh	96	3	99	3·0
20. Boddam	4364	85	4449	1·9
21. Levenwick	173	2	175	1·1
22. Scousburgh	375	14	389	3·6

and Berry, 1961). Still more remarkable is it that a greater change of frequency occurs near to, but not actually at, a geographical fault, the Tingwall Valley, which runs across the Shetland Mainland (the central and largest island of the three (Kettlewell and Berry, 1969). Here, only one half to

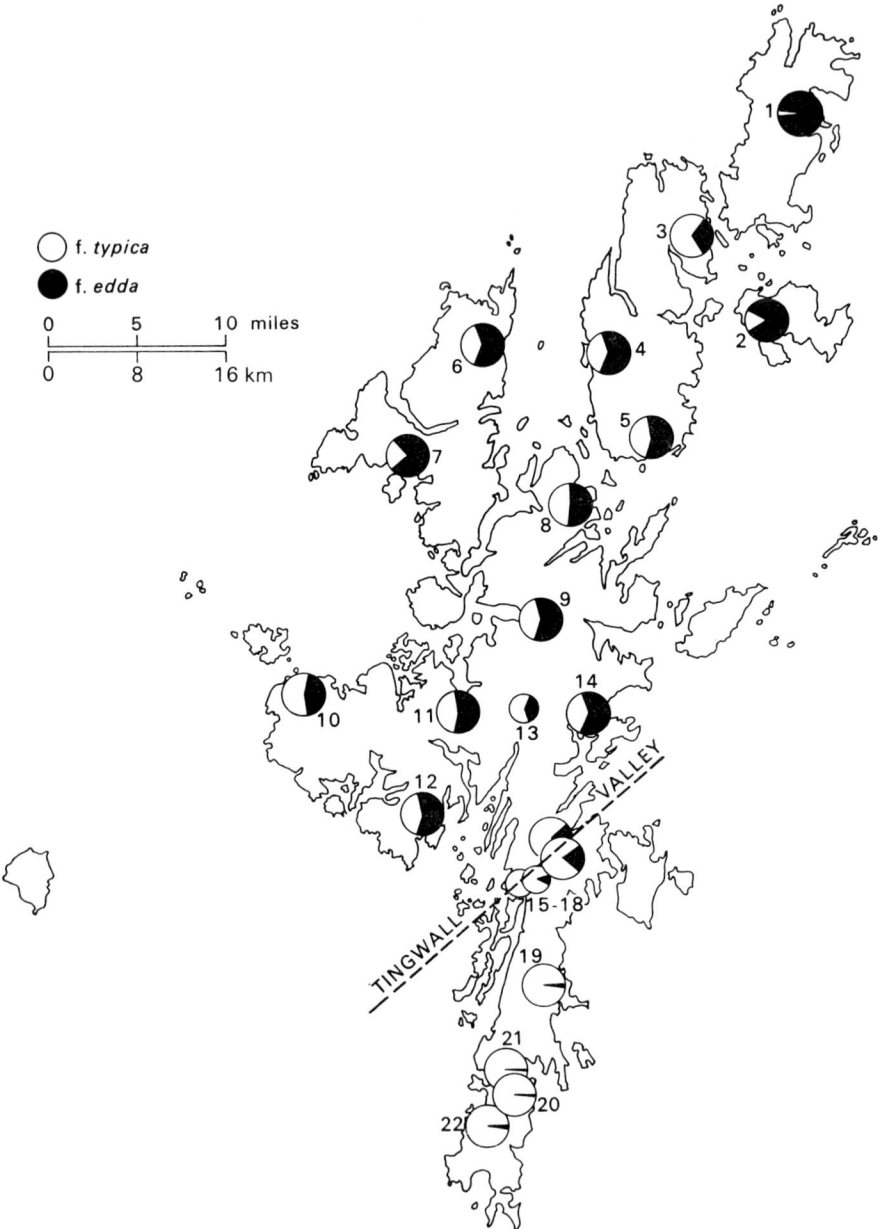

FIG. 11.1 A frequency map of *Amathes glareosa* and its melanic form *edda*, comprising 20 000 records from 22 localities in Shetland

two miles of arable land (and not water) has to be crossed from the northern to the southern population. Yet here the maximum disruption of the cline takes place.

By making a series of approximate measurements of the distances of the sampling places apart, and ignoring minor effects, such as the number of miles from the sea, it was possible to plot a curve of gene-frequency along the length of Shetland (Fig. 11.1). Measurements were made of the shortest distances overland from one sampling centre to another. This graph shows both the gene frequency and the phenotype frequency in different areas. It can be resolved under three headings:

(a) The incidence of f. *edda* remains high over the whole of the northern half of Shetland, falling but gradually (40 per cent in 45 miles) in a southerly direction.

(b) In the south Mainland where its frequency is low, the rate of change is even less (2 per cent in 12 miles), being only 3 per cent as far north as Cunningsburgh.

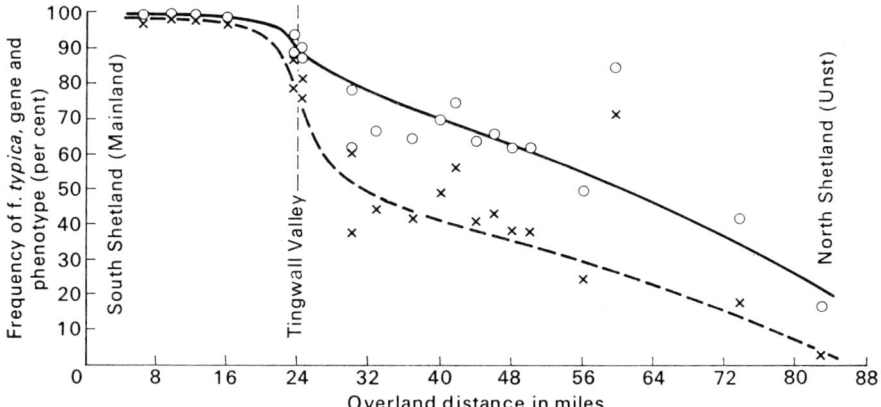

FIG. 11.2 Decline in frequency of wild type *Amathes glareosa* and replacement by f *edda* with distance northwards in Shetland. Phenotype frequencies are represented by X, gene frequencies by O. The dotted line shows the cline in phenotype frequency, the continuous line that in gene frequency

(c) It will be seen from Fig. 11.2, however, that a different state of affairs exists in the intervening region. Here the phenotype frequency of f. *edda* drops 35 per cent in about 8 miles, although the rate of change of gene frequency in the same area is proportionately less. This change takes place at the Tingwall Valley.

The analysis of this cline can therefore be fairly discussed under three headings: (*a*) northern Shetland, (*b*) southern Shetland, and (*c*) the area of the Tingwall Valley.

(a) The distribution of f. edda *in northern Shetland* (Kettlewell, 1961*d*)

The islands of Unst and Fetlar and the northern third of Mainland present rugged headlands with many windswept areas of heather-type vegetation. There is no place, in fact, at a greater distance than three miles from the sea. The hill of Muckle Heog on Unst, and also Fetlar have, however, a different geological basis from the rest of Shetland in that such heather lies on serpentine rock (Spence, 1957). It is here that so many other plant species grow (for example *Scilla*) and that the maximum populations of *A. glareosa* are found. The species is certainly polyphagous in its tastes but at the same time selective in its choice of a number of food plants.

Northern Shetland is also a temporary resting ground for thousands of birds migrating southwards from Scandinavia (and Iceland) in the second half of August and early September. Most of these are insectivorous; in particular the Common Gull, *Larus canus* L., the Peewit, *Vanellus vanellus* L., the Golden Plover, *Charadrius apricarius*, the Hooded Crow, *Corvus cornix* L., the Wheatear, *Oenanthe oenanthe* L., and the Starling, *Sturnus vulgaris* L. There are two points which are worth mentioning here. First, the size of the first four species must demand a large intake of a moth as small as *A. glareosa*. Secondly, the majority of migrating birds on their journey southwards make landfall on the northern coast when they are exhausted and they tend to remain in that vicinity for a period of time depending on the strength and direction of the wind. With the exception of the Wheatear, I have never seen evidence of concentrations of migrating species in the middle third of Shetland.

I believe that, after resting in the north, the birds move rapidly south on a favourable wind but that some rest again at the southern extremity of Shetland Mainland where there are extensive coastland areas of sand hills. On such a background f. *edda* could offer no cryptic advantages.

I am certain that heavy selective predation takes place all along the northern periphery and I shall give evidence of this in the next chapter. I think that visual predation is relaxed throughout the middle third of central Mainland. We have little evidence of predation amongst the southern sand dunes; this is because our mark–release–recapture experiments here were conducted on inland areas of heather-peat only, before we had realized that a high population of the species (98 per cent f. *typica*) existed amongst the sand hills, a very different type of ecology from that of the north.

(b) The distribution of f. edda in southern Shetland

In south Mainland all the populations sampled contained less than 5 per cent of f. *edda* and, in fact, dropped only 2 per cent over 12 miles. This is remarkable in view of the 98 per cent frequency of this same melanic form only a few miles farther north.

The ecology to the south of the Tingwall Valley area differs from that to the north in many ways. Geographically the coastline has fewer headlands, ecologically there are larger areas of arable land, this in particular around the periphery of the coast—the heather-type of terrain is, in fact, confined largely to small inland districts at somewhat higher altitudes. Also, as already stated, there are considerable stretches of sand dunes.

Until our recent expeditions, f. *edda* had not been recorded from the southern third of Shetland, nor indeed from anywhere other than the northernmost island, Unst. This was, no doubt, due to the less efficient methods of sampling in the past.

We have, therefore, an interesting situation in which a northern population consists largely of the melanic form *edda* and a southern one of f. *typica*. The intervening area of the Tingwall Valley shows a rapid change of gene-frequency: 50 per cent over about 15 miles, in fact. Nevertheless, the cline is maintained throughout.

That the f. *edda* is sensitive to the degree of latitude need not be considered as a direct cause of the frequency changes because on Orkney Mainland, 100 miles to the south, the frequency can be as high as 37 per cent and at its lowest 11 per cent (see Table 11.1). Nowhere is it as low as in southern Shetland.

(c) The area of the Tingwall Valley

The Tingwall Valley runs in a south-west to north-east direction across the centre of Shetland Mainland. It is the result of a limestone fault and varies in width from one-half to two miles. It consists of fertile farmland, and is quite unsuitable to the requirements of *A. glareosa*.

Here, in 1961, we concentrated on sampling at 16 separate sites, both to the north and to the south of the Valley (Fig. 11.3). As always, the method used was to attract the moths to light traps. These had 125W mercury-vapour bulbs which rested over special large capacity cylinders, each containing 20–30 egg trays to give increased surface areas for resting, and with these no anaesthetic was ever necessary. Traps were turned on from dusk to dawn. Both sexes of *A. glareosa* came freely to light.

The following somewhat surprising facts emerged: the average gene-frequencies on the two sides of the valley were effectively the same, a similar situation to what we had found farther north on the opposing sides of the two sea-barriers between the islands. Secondly, that immediately to the north of the Tingwall Valley the frequency of f. *edda* increased rapidly (10–15 per cent in the first two or three miles); this change in frequency did not coincide with the actual physical barrier.

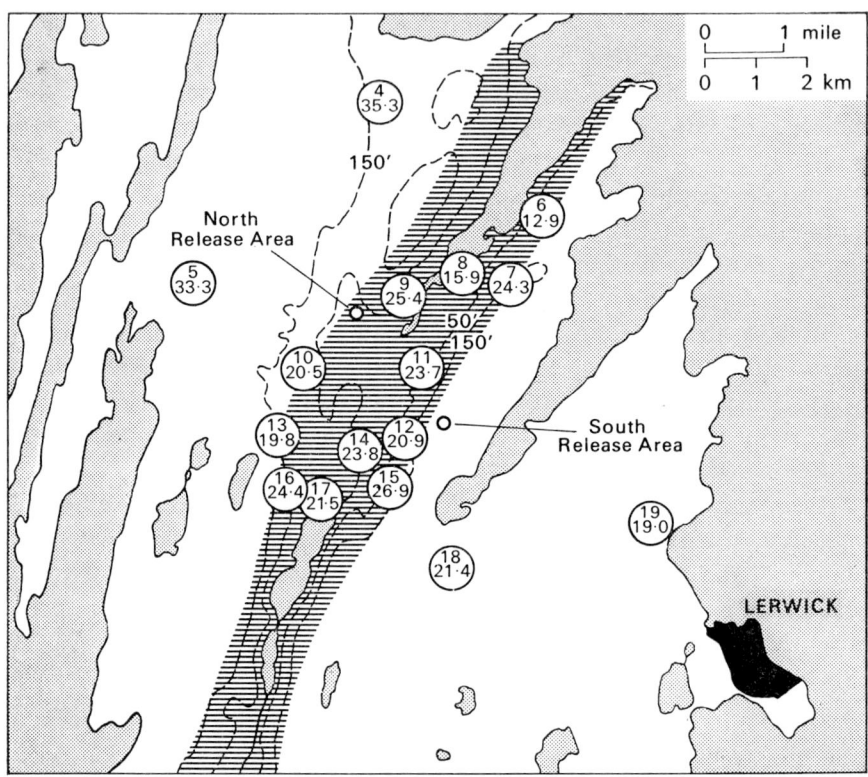

FIG. 11.3 Map of the Tingwall area (the valley proper is shaded) showing the percentage frequencies of f. *edda* in 1961 (except for sites 6 and 18 which were only sampled in 1960)

This at first appeared to us a somewhat surprising finding. Yet similar clines to that of *A. glareosa* in Shetland have been recorded elsewhere. Sumner (1926–30) in a series of papers described a cline he found in the coat colour of the mouse *Peromyscus polionatus* in Florida. His well-documented data showed that on the off-shore island of Santa Rosa, which consists of a bank of extremely light sand, the mice were almost white. On the mainland one quarter of a mile opposite, where the ecology was quite similar, a somewhat paler form occurred. For the next 40 miles into the mainland interior, regardless of dark backgrounds, this form continued to appear. Only after this distance inland was the typical dark-coloured *Peromyscus* found. However, this change took place in two stages. One 20 miles from the coast and another at 40. At both, over a short distance, darkening factors increased and at both the colour-phase changed over a distance as short as three miles. More recent sampling has shown that this situation is no longer maintained however (Selander, R. K., 1969, in press). Sumner's breeding experiments demonstrate that coat colour was determined multifactorially, so presumably one or more alleles for 'light' were lost at each of the two points; the important observation being that neither of these coincided with an obvious change in background colouration, in rather the same way as for *A. glareosa* on Shetland Mainland.

For this reason we (Kettlewell and Berry, 1961) attempted to analyse our *A. glareosa* cline data on the basis of the model put forward by Haldane to account for the *Peromyscus* cline (Haldane, 1948) and one in which he envisaged, incorrectly, that mouse colour was controlled by a single autosomal 'dominant gene' and its allelomorph, though such an interpretation is correct for the *edda* form of *A. glareosa*. Haldane (personal communication) agreed that his model was valid for our *A. glareosa* data, except that he had conceived a border where extensive predation was taking place and not a barrier. He went on to suggest that such a barrier as the Tingwall Valley might be as effective as ten miles of heather. Accepting his limitations and our reservations it seemed to us that his model fitted our data.

The use of models in general has been criticized by Ford (1964) on several grounds, not least that a mathematical model can suggest only the average advantage along a cline and demands gross over-simplification. Nevertheless, if one meets situations where the data appear to fit, comparisons with such a model can be of use though some may be misleading. In the present instance it was certainly helpful, in spite of the fact that the cline of *Peromyscus* reflects a colour-phase and that of *A. glareosa* a morph-ratio.

At the same time we suspected that the Tingwall Valley must act as a partial barrier to gene flow but we had no proof of this. Alternatively, the average conditions in the north favoured f. *edda* and in the south f. *typica*. It seemed likely that both factors must contribute to the distribution we had found. It was clear that without experimental releases and a full investigation into the survival, behaviour, and habits of the two forms, both to the north and to the south of the valley, we could get little farther.

12. Mark–release–recapture experiments on *Amathes glareosa* Esp.

(i) Gene flow in the Tingwall Valley district

BEFORE attempting to analyse any cline it is of paramount importance to know whether the cline is maintained each year and whether the relative frequencies at the various points sampled do not vary significantly from one season to another. Such variation is, of course, likely to occur if some climatic factor, affecting the whole area, favours one form more than another. Such a situation has been shown to occur during the different seasons of a year: for example, in the multivoltine species, *Drosophila pseudoobscura*, in the south-western states of the U.S.A. (Dobzhansky et al., 1964) or again in the pre-winter and post-winter samples of the melanic forms of the Two-spot Ladybird, *Adalia bipunctata* L., in Germany (Timoféeff-Réssovsky, 1940).

We have provided evidence, however, from five places sampled for more than one year that both forms of *A. glareosa*, when collected throughout the season, remain at a fairly constant level from one year to another (Table 12.1). More recently this has again been corroborated in a sample numbering 1901 collected by R. J. Berry in 1968 from eleven localities previously

TABLE 12.1

Frequencies of A. glareosa *f.* edda *at sites sampled on more than one year.*

Site	Year	f. *typica*	f. *edda*	Total	Overall per cent frequency of f. *edda*
Baltasound	1959	77	2 462	2539	97·0
	1960	302	10 356	10 658	97·2
Hillswick	1959	4	19	23	82·6
	1960	21	66	78	75·9
	1961	222	683	905	75·5
	1962	195	634	829	76·5
Vatster	1961	116	69	185	37·3
	1962	88	48	136	35·3
Catwalls	1960	33	102	135	24·4
	1961	337	87	424	20·5
	1962	97	29	126	23·0
Scousburgh	1959	95	4	99	4·0
	1960	280	10	290	3·4

TABLE 12.2

Constitution of population samples of Amathes glareosa (from north Shetland to south) 1968 (Nine year later), (figures provided by R. J. Berry).

Area	Numbers caught			Overall per cent frequency of f. *edda*
	f. *typica*	f. *edda*	Total	
1. Baltasound	16	513	529	96·9
2. Uyeasound	7	104	111	93·7
3. South Yell	29	32	61	52·5
4. North Roe	56	117	173	67·6
5. Hillswick	9	33	42	78·6
6. Voe	53	95	148	57·4
7. Vatster	79	65	144	45·1
8. Catwalls	155	46	201	22·9
9. Grimista	149	32	181	17·7
10. Quaff	187	41	218	14·2
11. Boddam	87	6	93	6·4

worked by us (Table 12.2). This showed, with one possible exception, the same frequencies as those obtained six to eight years previously (shown in Table 11.4).

In 1960 we carried out small-scale preliminary sampling on either side of the Tingwall Valley. At that time the two forms appeared to fluctuate wildly in numbers from night to night, and I came to the conclusion that the frequencies of the morphs on either side of the Valley varied in inverse proportion to each other (Kettlewell, 1961d). In view of our subsequent discovery of a flight-habit difference between the two forms (or maybe between the northern and the southern populations) such a thing may occasionally take place. However, the following year when sampling on a large scale continued throughout the hatching period there was no evidence of this.

In 1961 four of us undertook large-scale mark–release–recapture experiments in order to find out whether individuals crossed the Valley and also to get some idea of the distances traversed by the two forms (Kettlewell and Berry, 1969).

The design of experiments

Each undamaged moth was marked with quick-drying cellulose paint applied to the underside, both colour and positioning varying each day. Thus five colours on four wings gave 20 distinctive markings. There were two releasing points, one on the north and one on the south side of the

Valley. Each of them was one quarter to one half a mile distant from the nearest light trap. Marked moths were shaken into the heather at dusk (about 9.00 p.m.) so as to avoid visual predation, but before their normal time of flight. Our recaptures were collected at sixteen mercury-vapour traps distributed throughout the district as described in the previous chapter (Fig. 11.3).

By these means we were able to build up a pattern of flight-behaviour and to discover whether the wind funnel of the Tingwall Valley acted as a barrier to gene flow.

Results

A total of 1682 marked moths (533 f. *edda* 1149 f. *typica*) which had been caught locally were released at the two points; 847 on the north side and 835 on the south (Table 12.3). Only 65 (3.9 per cent) were recaptured

TABLE 12.3
Summary of mark–release–recapture experiment in the Tingwall Valley.

	Released on north side of Valley		Released on south side of Valley		
	f. *typica*	f. *edda*	f. *typica*	f. *edda*	Total
Number of moths marked and released	578	269	571	264	1682
Number of moths recaptured	25*	13*	21	6	65
Moths recaptured (per cent)	4.3	4.8	3.7	2.3	3.9
Direction and distance of movement between release and recapture — SW 550 yards	—	—	2	1	3
SW 700 yards	1	1	1	1	14
SW 880 yards	16	8	—	—	24
SW 1050 yards	—	—	—	1	1
SW 2290 yards	4	—	—	—	4
N 700 yards	—	—	8	3	11
E 1230 yards	1	3	—	—	4
E 1410 yards	1	—	—	—	1
E 2110 yards	1 (crossed valley)	—	—	—	1
Mean distance SW	1140 yds	860 yds	680 yds	770 yds	
N or E	1320† yds	1230 yds	700 yds	700 yds	

* Site of recapture of one moth not recorded.
† Omitting moth which crossed the valley.

(19 f. *edda* and 46 f. *typica*). Of these only one crossed the Valley ($1\frac{1}{4}$ miles) and this was an f. *typica* which had been released 13 days earlier at the northern release site. Yet the mean distance travelled by each along either

side of the Valley was in the order of half a mile. This means that though considerable gene flow was taking place on each side little occurred across the Valley (about 1½ per cent). From this it can be stated with some confidence, that the Tingwall Valley is an effective barrier to gene flow and virtually divides the Shetland *A. glareosa* into a northern and a southern population. Intensive selective pressures must be taking place on either side in order to maintain similar gene-frequencies on each. Yet the problem remained as to why in the next 2½ miles northwards the frequency of f. *edda* should increase by 15 per cent. It certainly suggested that factors other than crypsis must be contributing to the cline in the areas of the Tingwall Valley. Some of these factors became apparent incidentally in our extensive mark–release–recapture experiments, which were designed to test the degree of selective predation. These are described on the following pages.

(ii) Selective predation on *Amathes glareosa* in Shetland

Mark–release–recapture experiments at the opposing ends of the Shetland islands

All balanced polymorphisms must, in order to be maintained, be subjected to selective pressures on a number of characters controlled by the same alleles that are responsible for visible distinctions. It is my belief that the colour difference in the two forms of *A. glareosa* is the most important of these, particularly in northern Shetland.

Whenever the frequency of two morphs constitutes a cline, this situation offers opportunities for discovering the different advantages and disadvantages of each. I hold the view that in order to analyse these the optimum points for investigation are two, namely where the highest concentration of each of the two forms occurs; for here the disadvantages of the rarer one are likely to become more obvious. This I had successfully undertaken in my work on Industrial Melanism, choosing the extremes in frequency of f. *typica* and f. *carbonaria* of *B. betularia* (chapter 8). It appeared to me that the *A. glareosa* situation in Shetland should be tackled in the same way. Here was an indisputable non-industrial melanic with a cline comparable with those so commonly found in industrial melanic species. How wrong I was! In 1960, after an extensive survey throughout Shetland I decided to undertake mark–release–recapture experiments at the opposing ends of the islands, namely on Unst in the north where f. *edda* forms 98 per cent of the population, and at Dunrossness in the south where this form is under 2 per cent. I have previously referred to the geographical, ecological, and agricultural differences between these two poles, yet they are only 54 miles apart.

First, I want to recount the story of the failure of myself and five co-workers to establish a significant differential predation using the same technique of mark–release–recapture which I had previously found so satisfactory (Kettlewell, 1961d). Secondly, I would like to boost the morale of research workers who fail to find immediate answers to their problems in field experiments. For whenever the results of these conflict with those anticipated, the reason is likely to be of the utmost interest. One of the fallacies today is to expect those research biologists who are attempting higher degrees to obtain 'positive results'. Negative answers are of equal importance. If one method of investigation fails to reveal a particular line of inquiry another must be substituted. Hence when we found that the recapture figures reflected a behavioural difference rather than selective predation we had to resort to shooting predators in order to examine their stomach contents. When we suspected that crepuscular predation was taking place on *A. glareosa* when in flight, we impaled equal numbers of the two forms on small hooks attached to a length of nylon which permitted flight. This was stretched across an area where the birds had been working. It is essential to change one's techniques in the course of any new investigation.

The main difficulty we encountered was that both in the northern and the southern experiments we failed to catch a sufficiency of the rarer of the two morphs, for here they formed only two to three per cent of their respective populations. We resolved this by running a shuttle service of f. *typica* from the south to Unst and of f. *edda* from the north in the opposite direction. Each of these journeys involved the usage, each day, of three buses and two ferry boats; to the credit of the Shetlanders on no single occasion over the weeks did they fail to deliver the containers. The commoner morph which did not have to travel thus, was given comparable treatment and accompanied us in cars on each occasion in our daily trips before being marked and released. In all, 10,645 of the 30,000 *A. glareosa* we caught were marked and released; of these 819 were recaptured.

Selection experiments on Unst, north Shetland
We had noted that by far and away the largest capture of this species was to be obtained on serpentine outcrops. Furthermore, these same areas were the centres of bird activity. For example, parties of a dozen or more Common Gulls, *L. canus*, could be seen daily spread out and feeding on many areas of Muckle Hoeg. For this reason we eventually chose such an area, which was approximately one-quarter square mile in extent, for our

releasing centre. On the periphery of this we placed three mercury vapour light traps (125W) at distances of 200 to 500 yards from the releasing area. Here 4344 marked *A. glareosa* (f. *edda* 2260, f. *typica* 2084) were subjected to the normal hazards of predation, wind, and rain. Of these we recaptured 132 f. *edda* but only 95 f. *typica*, when, on a corresponding basis, 121 of the latter were expected. Because equal numbers of each form had not been released each day, a simple χ^2 test could not be used but it could be applied to the two expected values which are 125 and 104 ($\chi^2_{(1)} = 1\cdot40$). Though there was no significance in the number of f. *edda* recaptured nor in the number of days survived (440 days f. *edda* to 256 f. *typica*) the figures did suggest a 7 $\pm 6\cdot5$ per cent advantage for f. *edda* (Kettlewell, 1961*d*).

This was indeed disappointing in view of the size of the release and the extent of the predation we witnessed daily. The local population was too vast for a selection predation difference to be shown by the mark–release technique.

Selection experiments at Dunrossness, south Shetland
Using similar techniques to those we had developed in the north, two of my co-workers marked and released 2144 *A. glareosa* (f. *edda* 961, f. *typica* 1183) at Dunrossness. Unfortunately, the release site I chose here was not characteristic of the southern ecology, nor at that time did I realize that the main reservoir of the species here was among the sand-hill areas. No predation was witnessed near the releasing site which was a small heather-moor, and no sign of selective predation was suggested from the recapture totals (observed f. *edda* recaptures 50, expected 49·034).

However, a most important and unusual observation was made in both the northern and the southern experiments. As we daily charted the recapture results on two histograms which hung on a wall of the fisherman's house where we lived, it became apparent in the first week that more marked f. *edda* were recaptured on the second day after release than on the first. This was entirely contrary to any mark–release–recapture experiments I had previously conducted on any other species. My interpretation of this is given later in this chapter (Fig. 12.1).

In the meantime, in 1960 we had to admit that we would not advance our knowledge on differential predation by such release–recapture experiments, and it seemed to me that a more direct approach had to be made and that an examination of the stomach contents of birds seen feeding in the areas was indicated.

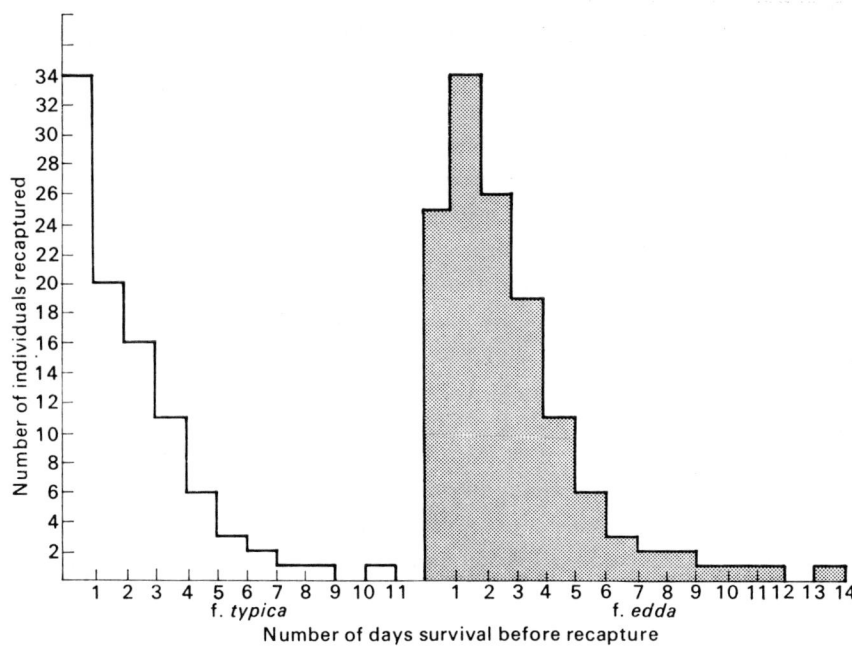

FIG. 12.1 Histograms of marked recaptures of both forms of *Amathes glareosa* showing the number of days' survival in the wild population (from 19 separate releases), Unst 1960

Bird stomach contents

Accordingly in 1960, having obtained the necessary permission, I shot five Common Gulls, *L. canus*, near Baltasound, which we had observed feeding on insects in an area where *A. glareosa* was common; two of these had empty stomachs. Three out of 18 identifiable *A. glareosa* in the stomachs of the other three were f. *typica*. In 1962 I shot two Common Gulls and six Hooded Crows, *Corvus cornix*, in the same place. *A. glareosa* was found in the stomachs of both species, and three out of eleven scorable were f. *typica*. The overall frequency of f. *typica* in the stomach contents was therefore 21 per cent in an area where the frequency of the morph was 3 per cent (Plates 12.1 and 12.2). (The predation frequency of f. *typica* was unfortunately transcribed as 11 per cent in our original papers, 1969.) The stomach contents of the 13 birds shot showed therefore a highly significant excess of f. *typica* and certainly indicated that the feeding habits of two species of bird were selective and that f. *typica* was at a high cryptic disadvantage on Unst.

(iii) Habit differences

Differential flight-behaviour between the northern and the southern populations

In the majority of Lepidoptera species, and certainly all that we had previously worked on, there is a clear-cut diurnal or nocturnal flight-behaviour, which takes place at fixed times in 24 hours. All our *A. glareosa* samples were caught at light traps; this signifies therefore that they were taken in flight. We naturally anticipated that this species would fly each night and that as more of our releases must be available on the first night after release than on the second that this would be reflected in our recapture histograms. This, in fact, happened in all three releases of *A. glareosa* f. *typica* from Dunrossness and was confirmed by P. Harper in a small pilot experiment he undertook in 1961 in south-western England (N. Devon) where, of course, this form occurs alone.

The recapture pattern of f. *edda* of Unst origin was, to our surprise, entirely different. There was a significant deficiency ($P > 1$ per cent) of this form on the first day as compared with the second (Fig. 12.1 and Table 12.4). Furthermore, this same phenomenon took place regardless of

TABLE 12.4

A comparison of day 1 and 2 recaptures from three mark–release–recapture experiments, Unst, Shetland, 1960.

	Day 1	Day 2
Observed f. *edda*	30	52
Expected f. *edda*	43·984	45·699
Variance	18·170	18·930
	$P = 0.001$	$P = 0.15$

whether Unst f. *edda* were released on the home ground or at Dunrossness in the south (Fig. 12.2). This therefore excluded any effects due to local meteorological conditions.

At the same time I decided to test whether there was any direct evidence of flight-behaviour differences by a small pilot experiment. Twenty-five f. *typica* from southern Shetland and 25 f. *edda* from Unst were marked and released in to an area of approximately 10 square yards at 8.00 p.m. All the moths took up resting positions after having previously fed on the heather flowers. From then until dark I was in a close position to observe them and see that no bird predation took place. The first sign of activity took place at late dusk (9.55 p.m. G.M.T., 18th August). Fifteen f. *typica* were seen to fly before 11 p.m. but only 5 f. *edda*. This could be accounted

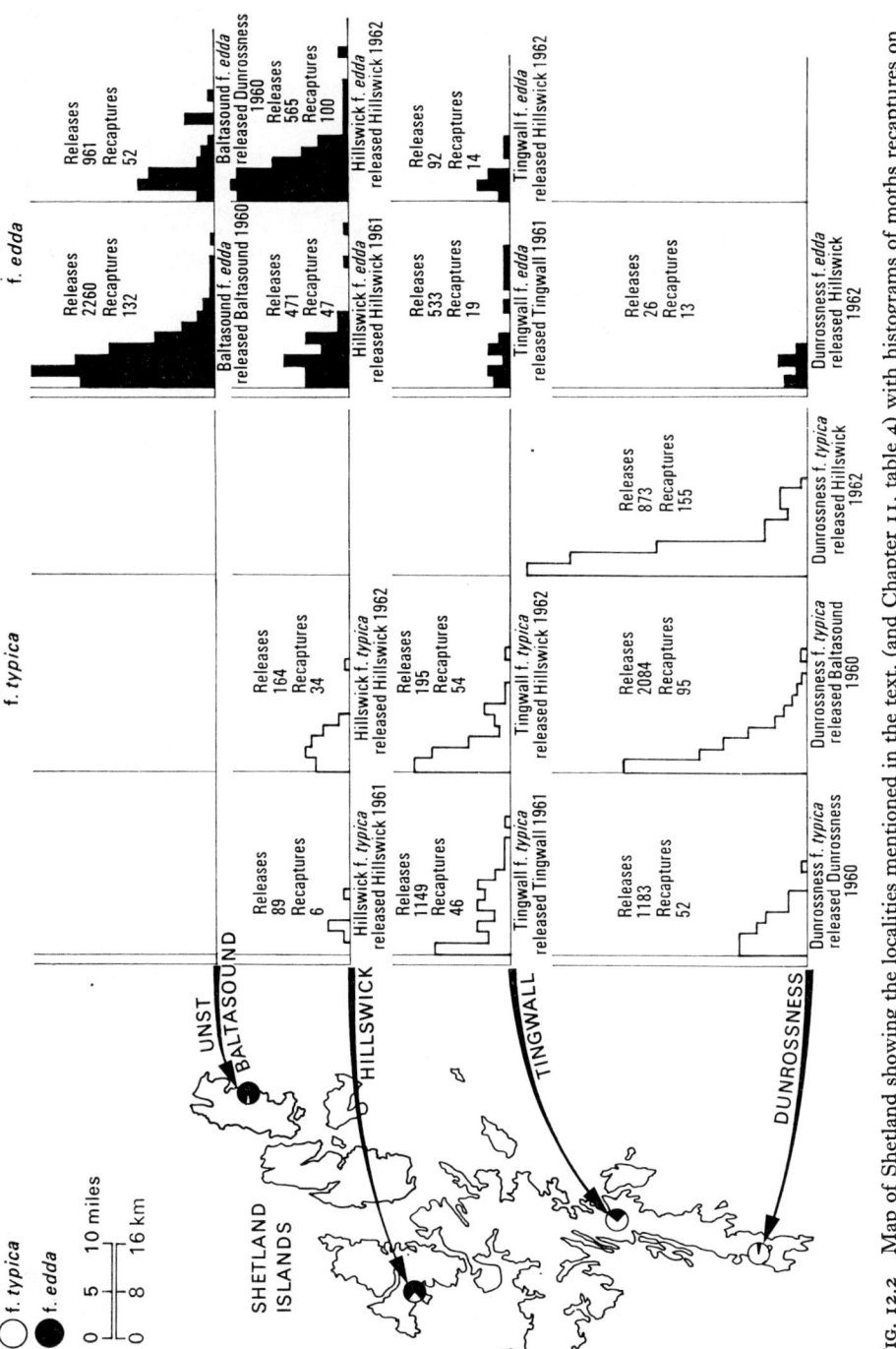

FIG. 12.2 Map of Shetland showing the localities mentioned in the text, (and Chapter 11, table 4) with histograms of moths recaptures on each day after release

for because of the greater ease of seeing the light coloured f. *typica* on the wing. However, using a torch a count of 12 *edda* to 3 *typica* ($\chi^2_{(1)} = 4\cdot 3$) was made at midnight as they fed or rested on the heather. It was apparent that most of the f. *typica* had flown. Furthermore, twenty-four hours later on the following evening 5 marked f. *edda* were found but no f. *typica* in spite of this form being so much more conspicuous. It is hard to construe these observations in any way other than that the Unst f. *edda* took flight less frequently than south Shetland f. *typica* (Fig. 12.3). As early as 1960

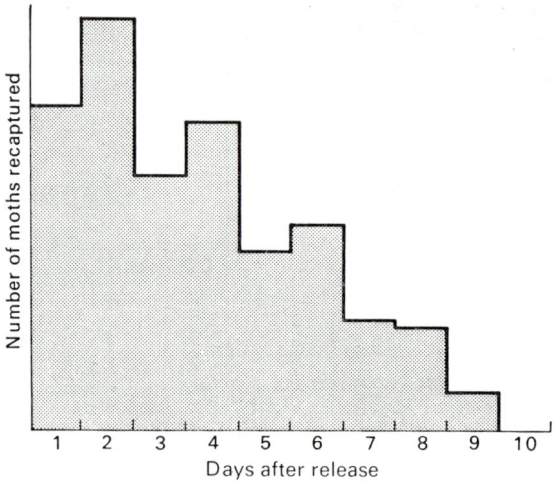

FIG. 12.3 Model histogram of recaptures on assumption of 70 per cent of moths flying each night and 30 per cent every other night

I was convinced that there must be an innate habit-difference between Unst f. *edda* and Dunrossness f. *typica* (Kettlewell 1961d). This called for further mark–release of samples collected from other places along the length of Shetland: no longer with the idea of discovering differential survival but rather of examining the flight-behaviour from each.

Accordingly in 1961 and 1962 we attempted to determine the recapture pattern of both forms obtained from three separate places, Hillswick, Tingwall, and Dunrossness. Insects from there were released at Hillswick each with recognizable markings.

Figure 12.2 shows comparative histograms from all sample releases at four sites. From these it is clear that there is a habit difference between the northern population where the frequency of f. *edda* is high and the southern population where f. *typica* occupies from 70 to 98 per cent of loci.

Once again it is the feature of the Tingwall Valley which attracts attention for it is here that two co-adapted gene-complexes meet. To the north it appears that flight activity is progressively diminished; to the south normal nightly flight takes place. Yet we are still unable to state categorically whether the diminished-flight habit is associated with the *edda* form or is a property of the northern population, though the latter situation appears more likely from the Hillswick releases.

Survival differences in areas of little predation

In the course of the flight-behaviour experiments we had at the same time an opportunity to test the mean survival of the local populations along with that of the individuals we introduced from more distant localities. With one exception *A. glareosa* released on their home ground had longer survival than when released elsewhere. That trauma in transit could play a part in this is discounted by the one exception: this is that f. *edda* from Unst survived longer when released in the south at Dunrossness (4·0 days) than when released at their site of origin (3·3 days).

This does suggest that all *A. glareosa*, regardless of origin, are not homogeneous. More likely is it that local adaptations of the gene-complex to local conditions are the cause of a differential survival and that such adjustments are more important for f. *edda* than for f. *typica*.

Evidence of a different time of emergence

Crypsis in Lepidoptera must always imply a genetic response on the part of the prey in order to minimize predation. When a polymorphism occurs in such a species and when the various morphs are all cryptic, yet quite distinct, as in *A. glareosa*, it is strange that we have had so little evidence of different hatching times between the morphs. For various predators must react differently to each. Such a happening is more likely to be found in areas of temporary but heavy predation. Foremost amongst these must be the landfalls along the routes of migrating birds.

Williamson (1956) recorded that in Fair Isle, the most southerly Shetland Isle and where *A. glareosa* is common, f. *edda* (phenotype frequency 30 per cent) emerged before f. *typica*. In 1955 three f. *edda* were caught on August 4th, yet no f. *typica* were taken until one week later. On Unst in 1960 we found a similar hatching difference (Fig. 12.4). During the period of emergence the frequency of f. *typica* rose from 1·5 per cent to 7 per cent at the end of the month ($n = 292$ in 10 651); thereafter it fluctuated widely, no doubt because the population was collapsing. During August, Fair Isle is a recognized centre for migrating birds and many of these depend on insects for food in order to continue their journey south. A similar situation

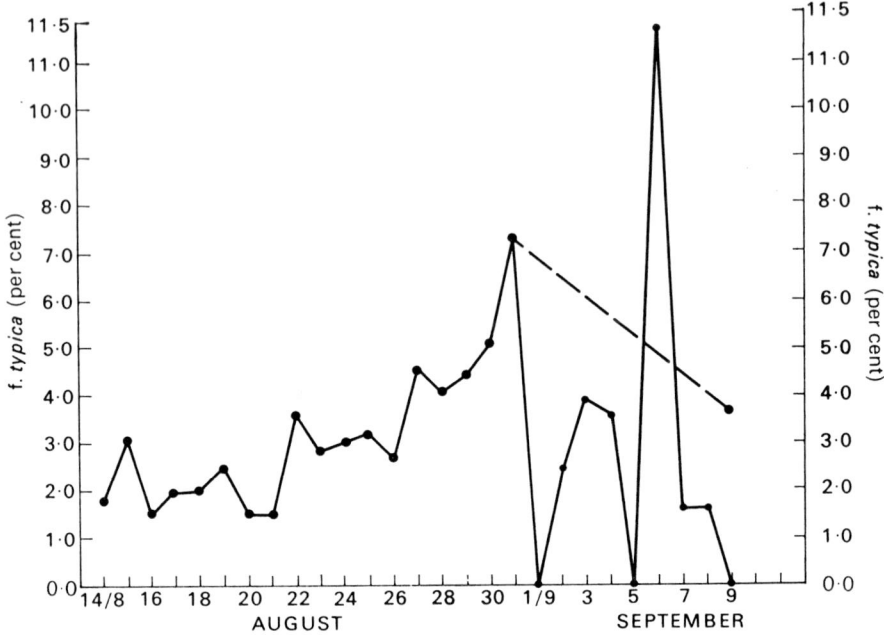

FIG. 12.4 Graph showing daily percentage of f. *typica* in the wild population in Unst 1960

occurs on Unst but here we have proved the point and shown that 60 per cent by volume of the stomach contents of some birds consists of Insecta. After the last week of August bird migration is not so intense and in particular Common Gulls (in my opinion the main predators) have departed south. Throughout central Shetland where we have observed little sign of intensive predation we found no evidence of a difference in time of emergence.

I think that a likely explanation is that in areas of heavy predation there is a premium on camouflage and in particular the cryptic pattern of f. *edda* is favoured. In the north, when predation is relaxed later in the season, the physiological advantages of f. *typica* again become more important, as they are throughout the range of the species elsewhere. This could have led to selection in favour of f. *typica* hatching later in the season in such specialized circumstances. Alternatively it could reflect the direct effects of diminished predation.

A résumé of results of field observations on A. glareosa

Apart from the colour-morph cline favouring melanism in the north there are several other less obvious, but equally interesting changes in the

Shetland population: a behavioural difference between the northern and the southern *A. glareosa* representing a decreased flight activity in the northern moths. I have suggested that the hazards of flight on these windswept headlands may be the reason for this. A greater degree of dominance modification has been assimilated in the Unst population. Here also there may be a difference in the time of the emergence of the two morphs. Both these last adaptations must reflect genetical responses to intense predation. The point of overriding importance is that we have shown that by such changes *A. glareosa* has adapted to local conditions and that individuals survive better on their home-ground than elsewhere.

The melanism in *A. glareosa* is most certainly non-industrial and of ancient origin. It has served to demonstrate some of the uses of melanism in the past and the behavioural and physiological changes which may be associated with it.

Part V. The origins of melanism

13. Recessive Melanism and rare melanic mutants

Definition. Melanism which in general occurs at a low frequency and is found in many genera of aposematic Lepidoptera; it is less frequent in cryptic species, and in Britain these melanic forms have spread only in one or two instances, under special circumstances. The degree of pigmentation is frequently extreme. The homozygous recessive melanics are, usually, subviable or semi-lethal.

The rarity of Recessive Melanism
There are few industrial melanics whose inheritance is not fully dominant; it is one of their chief characteristics. By this we mean that the heterozygotes are to the eye indistinguishable from homozygous black individuals; only rarely can we recognize the one from the other. This no doubt is, by analogy, because of the fact that whether 10 or 20 minims of a black dye are added to one millilitre of water, no difference is discernible between the two: visual saturation is established in both. In only a few industrial melanic species are the heterozygotes recognizable as intermediates, or they may vary from such intermediates to individuals which are indistinguishable from the homozygous melanic.

By contrast, the heterozygotes of recessive melanic forms are not recognizable from the wild type, indeed the forms would not be recessive if they were, and the homozygotes usually occur in the laboratory unexpectedly and at a low frequency from inbred stock.

It is worth considering an important point here. Because of the rare appearance of the homozyygote and the fact that this is frequently sub-lethal, we can have no idea of the true frequency of the gene in the population. We can certainly anticipate that it will often be in excess of the expected figure when calculated from the Hardy–Weinberg ratio. If we admit equal viability amongst the three genotypes, the Hardy–Weinberg table demonstrates that though a gene-frequency is as high as 2 per cent in the population, the homozygous recessive character will occur in only 1 in 1000 individuals. It is possible, in fact, that in some instances genes controlling recessive characters in Lepidoptera are present at a level which constitutes a true polymorphism—the gene may in fact be common and contribute materially to the benefit of the species because of the heterozygous advantage conferred.

It is likely that in future more heterozygotes will be identifiable by one means or another, other than visual (e.g. electrophoresis, chromatography).

Viability

In most instances homozygous recessive melanic forms of aposematic species have lowered viability, for instance the Syntomid-like form of *Panaxia dominula* f. *nigra* (Plate 13.1) is a rare aberration in which all the wings have green-black colouration but with the normal white patterning of the forewing stencilled in brown. The warning crimson of the hindwings is also replaced by black. It has only been taken in one isolated colony near Deal, Kent; eleven in all. For 25 years (1922–1947) I bred and inbred several thousand individuals from this colony but never have I hatched this aberration. Yet on three occasions I have recognized its presence after opening partly emerged pupae; furthermore, some of the existing eleven specimens were crippled. Scoring phenotypes from imagines extracted from pupal shells may certainly be misleading because such individuals may have died at a moment when the process of patterning was incomplete, different pigments being laid down at different times. But two of the three f. *nigra* were still alive but had failed at eclosion. Similar situations have been found in many other aposematic species whose melanic forms are excessively rare.

Recessive melanics in *cryptic* species are uncommon. For example, in Britain the only known melanic of *Ectropis punctulata* Schiff, has occurred as a halved mosaic. Presumably the gene conferring melanism is present in the species but is lethal except in the physiological requirements provided by f. *typica* (Murray, 1928). This species (Selidosemidae) is cryptic and spends the day at rest on tree trunks and fences.

Later (Chapter 14), I shall show, however, that the gene frequency of certain recessive melanics can, under special ecological pressures (which are probably of recent origin) occupy over 80 per cent of an available locus in a population. Any physiological disadvantage which may exist is outweighed by the cryptic advantage of the melanic phenotype.

So far we have discussed two species only as an introduction to Recessive Melanism, one aposematic and the other cryptic; in both such melanism is rare. Yet Recessive Melanism is widespread throughout the Lepidoptera. Clearly before we attempt to interpret its significance, it is essential to know more about the habits and behaviour of each individual species and in particular the way each passes the daylight hours. Such an analysis leads to a classification which cannot be considered clear-cut nor am I satisfied that we can accept that our interpretation of the method of inheritance is correct in each instance. Because of lowered viability, pairings are usually impossible to obtain.

CLASSIFICATION OF RECESSIVE MELANISM

(i) **In aposematic species**

Panaxia dominula L. ab. *nigra* Spuler-Hofman
Zygaena filipendulae L. ab. *chrysanthemi* Borkh.

(ii) **In cryptic species**

(*a*) Associated with flight only. Substitution of 'flash' colouration of hindwing by black.
(*b*) Associated with resting position.
   (1) Resting on trees (trunk, boughs, and twigs)
   (2) Resting on the ground (including debris, earth, stones, and rocks)
      Meristis trigrammica Hufn. f. *obscura* Tutt
      Antitype chi L. f. *olivacea* Stephens
   (3) Resting on foliage (green and brown)

(iii) **A first step to polymorphism**

Lycia hirtaria Clerck f. *nigra* Cockayne

(iv) **Recessive melanic polymorphism**

Lasiocampa quercus L. f. *olivacea*. Tutt f. *olivaceo-fasciata* Cockerell
(This is fully discussed in the next chapter.)

DISCUSSION

(i) **Melanism in aposematic species**

In Europe two families which are largely comprised of aposematic species are the Zygaenidae and Arctiidae.

The former are always gregarious though colonies of them may be local. All fly by day and many of them exhibit scarlet spots on the forewings at rest and in flight crimson hindwings. They are highly toxic (Turner, 1971).

In *Zygaena filipendulae* a rare mutant has been found in several localities where it recurs, in which all the red is replaced by brown-black colouration (ab. *chrysanthemi* Borkh.). Similar forms have been taken in the wild in *Z. trifolii* (ab. *nigricans* Oberth.). In each, the melanic is recessive to f. *typica*. All Zygaenidae have black bodies, some with a red abdominal belt, and many alpine species which have to fly in every gleam of sunshine, have semi-transparent wings covered with black hair-scales. It would be interesting to test the thermodynamics in these.

In the Arctiidae some species fly by day, some at night, in others one sex flies by day and the other at night (see *Cycnia mendica*, Chapter 17). The majority have black and white disruptive patterns on their forewings and spend the day at rest beneath foliage. The hindwings in many of the Arctiidae are brightly coloured (red or yellow); these are only exposed on attack by predators, and frequently in conjunction with the secretion of droplets of distasteful fluid from the cervical glands. This then is a second line of defence. If the attack is driven home, *A. caja* (and other species) may stridulate (Lane and Rothschild, 1965, Rothschild and Haskell, 1966). Having developed so complicated a defence system it is not surprising that the majority of black mutants of day-resting Arctiidae are rare recessives. For example completely black specimens of *A. caja* and *A. villica* are occasionally bred from wild larvae or from inbred stock, and also *P. dominula* (see p. 212).

Recessive Melanism, I believe, implies that in the long past history of a particular allele, it has not contributed to survival as a homozygote. The gene-complex is, therefore, not adjusted to enhance such a mutant when it occurs, but is adapted to make its effects recessive.

(ii) Melanism in cryptic species

(a) Associated with flight only. Substitution of black for 'flash colouration' of hindwings.

A large number of a palatable species in widely separated genera of moths, whilst having highly cryptic forewings, have brightly coloured hindwings which are conspicuous only on taking flight. The colours are red, yellow, blue, or, more rarely, white. A comparable but even more frequent usage of *flash colouration* is found in the *Locusta*, and it is always a second line of defence to crypsis.

In the Lepidoptera, some have the colour of the hindwings replaced by all-black colouration, either as a rare mutant, or as a species characteristic when, perhaps, the gene responsible for black hindwings has become homozygous; the flash mechanism has then been eliminated.

Flash colouration is always associated with a definite behavioural pattern which is the opposite to that exhibited by those insects which display warning colouration. These, even on attack, never take flight, but feign death. By contrast, flash colouration demands rapid and sudden flight when the over-powering recognition is the coloured hindwing. Equally suddenly this disappears on landing when the cryptic forewings take over protection. Such species, depending on this for their survival, are always easily disturbed when at rest. Many of them are large and heavy Agrotidae in which it normally would be difficult to take wing instantly. However, the majority of species which depend on this mechanism, rest in the open either on the ground or in exposed situations, and hence are subjected to the sun's heat; thus they can take flight immediately. Such, for example, is *Triphaena pronuba*, the Common Yellow Underwing.

Catocala is a genus of moths, many of great size, which are found widespread throughout the Palaearctic, especially North America where they have undergone a burst of speciation, represented by over one hundred species. All have highly cryptic forewings, usually sculptured to match the bark of a particular tree on which the larva feeds. The hindwings in the majority are brightly coloured. In Britain we have three indigenous species and one, *C. nupta*, is common. It rests on dead wood such as telegraph poles, which have offered it a modern niche. I have on many occasions found more than two on a single post (5 in Huntingdonshire, 1930) and the species may gain advantage from being gregarious in the same way as may the Noctuid moths *Amphipyra pyramidea* L. and *A. berbera*, the Copper Underwings, from their communal resting habits, beneath loose bark.

In hot weather *C. nupta* flies erratically at the slightest touch and then the bright red of the hindwing is most conspicuous. On alighting it immediately takes up a resting position and the red hindwings are hidden behind the highly cryptic forewings.

In southern England, there is a rare mutant in which a chocolate-brown colour is substituted for the red of the hindwing (ab. *brunnescens* Warren). It has to date been taken in the wild on less than a dozen occasions, mainly in the vicinity of London. Yet in several North American *Catocala* a phenotypically similar form constitutes 100 per cent of their populations. Amongst the score of species which have forfeited their flash colouration in this way are: *C. agrippina* Strecker, which occurs in the region of New York and the Great Lakes, *C. vidua* Abbott and Smith and *C. dejecta* Strecker, both of which species occur in Canada, as well as *C. retecta* Gröte, *C. obscura* Strecker, and others. Yet in none of these can I find a record of a behavioural difference in the field such as one would expect from a profound colour change which necessitates the complete abandonment of a major mechanism of defence.

It may be that flash colouration has conferred advantages in the past, and still does to the majority of species of *Catocala*, but that in a slightly changed environment (for example, in colder conditions or one in which there is more shade) sudden flight became impossible because of a metabolic impasse. This would lead to a breakdown in the advantages offered by conspicuous coloured hindwings and rapid flight. If rapid flight, in fact, became physiologically impossible, brightly coloured hindwings could provide attraction to predators of a palatable insect.

That dark-hindwinged individuals should occur as rare mutants in a species in which the flash mechanism normally gives protection (as in *C. nupta* ab. *brunnescens*), is indicative of the substitution in the past of similar or identical mutants to those which today are a species character in North America. Unfortunately, we have no knowledge, as yet, of the inheritance of these black-hindwinged mutants.

(b) In cryptic species associated with resting positions

Introduction. A basic tenet in our knowledge of the behaviour of cryptic insects is that they must distribute themselves at low densities on the particular background, frequently highly specialized, on which they rest in daylight; with rare exception, gregariousness is avoided in cryptic edible species. Available surface areas must, therefore, play an important part in this concept. Lepidoptera in particular have spread their risks over most of the backgrounds in nature and these fall somewhat uneasily into three categories:

(1) Tree resters (trunks, boughs, and twigs).
(2) Ground resters (earth, stones, rocks, dead leaves, and detritus).
(3) Foliage resters (green and brown).

If we attempt to analyse cryptic Lepidoptera in this way we can state that those falling into the first two categories have throughout periods in the past had to adapt to *varying* backgrounds brought about by a multitude of influences, such as the level of humidity, the degree of light or darkness, the presence or absence of lichen coverage, and, more important in group (1), the fluctuations in tree species (for example, conifer versus deciduous). Foliage sitters (group (3)) however, have had a different past history to those insects in the first two categories. Living leaves have always been green and dead foliage brown or yellow but rarely are the colours of these backgrounds affected by extrinsic factors. Melanism is usually recessive or absent altogether in group (3). Recessive Melanism is also rare in group (1) but for a different reason: here melanic forms are common but nearly all have *full dominance*. Melanism is common in group (2) where it has in many instances become a race characteristic and is controlled polygenically: only rarely is it unifactorial and recessive.

PLATE 13.1 (p. 212). *Panaxia dominula* L. f. *typica* (×1½) and the rare recessive f. *nigra*.

PLATE 13.2 (p. 221). *Selenia bilunaria* Esp. f. *typica* (×2) and ab. *harrisoni*.

PLATE 13.3 (p. 223). *Lycia hirtaria* Clerck f. *typica* (×1½) at rest on Lime trunk.

PLATE 13.4 (page 223). *Lycia hirtaria* Clerck f. *nigra* (×1½) at rest on London Lime trunk.

PLATE 14.1 (p. 227).

Left

♀ and ♂ *Lasiocampa quercus* ssp. *callunae* f. *typica* (×1) (Caithness).

Right

♀ and ♂ *Lasiocampa quercus* ssp. *callunae* f. *olivacea* (×1) (Caithness).

PLATE 14.2 (p. 237). *Lasiocampa quercus* ssp. *callunae* f. *typica* wild ♀ (×1¼). Caithness 1971, at rest (Photograph by D. Lees).

PLATE 14.3 (p. 237). *Lasiocampa quercus* ssp. *callunae* f. *olivacea* ♀ (×1¼) at rest on heather (Photograph by J. Cadbury).

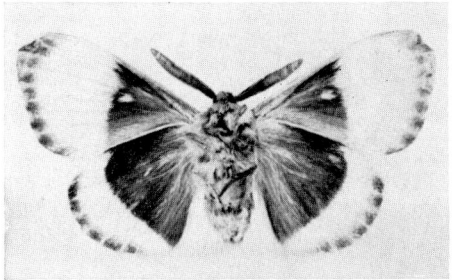 

PLATE 14.4 (p. 230).

Left
The underside of ♂ *Lasiocampa quercus* ssp. *callunae* f. *typica* (×1) (Yorkshire).

Right
The underside of ♂ *Lasiocampa quercus* ssp. *callunae* f. *olivacea* (×1) (Yorkshire).

PLATE 14.5 (p. 251). Two forms of the larva, 'chocolate' and 'normal' (×1) (Yorkshire).

Plate 15.1 (p. 265).
Melanism in *Papilio* (× 1).

1. *Battus philenor* L.

2. *Papilio machaon* ssp. *britannicus* Seitz f. *nigra* Reutti.

3. *Papilio machaon* ssp. *britannicus* Seitz f. *typica*.

4. *Papilio asterius.*

5. *Papilio glaucus* ♀ melanic form.

6. *Papilio glaucus* ♀ yellow form.

PLATE 16.1 (p. 272). *Polia nebulosa* Hufn. (× 1½).
1. f. *typica*.
2. f. *robsoni*.
3. f. *thompsoni*.

PLATE 16.2 (p. 273). *Spilosoma lutea* f. *typica* (× 1½) and incomplete dominant f. *zatima*.

PLATE 17.1 (p. 293). *Cycnia mendica* Clerck (× 1½).

Left
1. Typical English ♂.
2. Typical English ♀.

Right
1. Normal Irish ♂, race *rustica*.
2. F$_2$ ♂ from British × Irish cross.

1. f. *melaleuca* View.

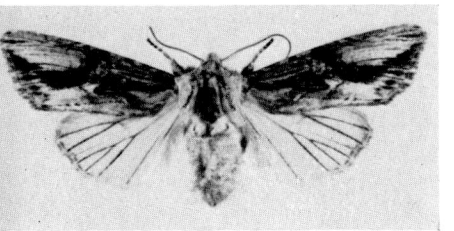

2. f. *intermedia* Tutt.

3. f. *typica*.

PLATE 17.3 (p. 298). Three forms of *Xylomyges conspicillaris* L. (× 1¼).

PLATE 14.6 (p. 248). Black-headed Gulls (*Larus ridibundus*) in the act of taking female *Lasiocampa quercus* f. *typica* from heather.

PLATE 14.7 (p. 228). (× 1). *Lasiocampa quercus* ssp. *callunae* female f. *typica*, and male f. *olivacea* copulating on heather stems.

(1) *In cryptic species which rest on tree trunks.* A large number of cryptic Lepidoptera throughout the world choose tree trunks as a resting place. This, without doubt, is on account of the advantages such a site offers: a multitude of backgrounds are readily available because the trunk of each tree species is different; there is freedom from drowning in storms, and shade protection from a hot vertical sun; also the predation and disturbance risk from mammals is lower here than on the ground. Trunk-sitters thereby minimize predation by scent, but in so doing they become immensely vulnerable to visual predators: in particulars, birds. For this reason perfect crypsis has been demanded, but, excluding melanism, polymorphism among tree-trunk sitters is uncommon. This is in contrast with ground-resters where polymorphism, except in species which have adapted to specialized geological situations, is frequent.

This then is the overall pattern of survival by crypsis; but to it must be added the fact that over two-thirds of cryptic trunk-sitters in Britain have today melanic polymorphisms and that over 90 per cent of these are inherited as Mendelian *dominants*. Recessive Melanism in this class is, in fact, extremely rare, but a few exceptions stand out.

First *Lycia hirtaria* Clerck and its melanic f. *nigra* and secondly the Sphingid moth, *Hyloicus pinastri* L. As to why these two species have failed to conform with the majority of tree-resting species I can offer no theories other than that both may be recent mutations which have not as yet had time to adapt their gene-complexes. Melanics of *L. hirtaria* have been bred many times from wild-caught British specimens and those of *H. pinastri* at least twice. In every brood in both species they have been shown to be unifactorial and recessive. Yet we have demonstrated that the homozygous black *L. hirtaria* are at a cryptic advantage in the London squares where this form has been taken for over thirty years and there is evidence that recently it has been getting more common (de Worms, 1961): the gene-frequency, in fact, must be in the order of 2–10 per cent in some London populations. More recently five f. *nigra* have been recorded in small samples (under 100 per annum) taken over four years at Stanford-le-Hope, Essex (R. Tomlinson, personal communication), so that here the frequency of the heterozygotes must be approximately 20 per cent of the population. For this reason *L. hirtaria* f. *nigra* can be considered as a polymorphism in the making, occurring at a low but increasing frequency in London and its adjacent counties. I give details of our investigation on this species later in this chapter. The recessive polymorphisms of *Lasiocampa quercus* and its black forms are discussed separately in Chapter 14.

Melanics of *H. pinastri* referrable to ab. *unicolor* Tutt have been found in Britain on two occasions only: the first in Dorset and the second in a wood in Suffolk, where, by what must be one of the most extraordinary coincidences on record, a wild caught specimen was at the same time heterozygous for albinism (Cockayne, 1926c). Melanic *H. pinastri* are, however, common in industrial areas in Czechoslovakia where it is stated to be dominant (personal communication); this form may be f. *brunnea* Spuler—I have received specimens of these and they are phenotypically identical with our British melanics. Although *H. pinastri* has in the last 40 years spread throughout southern Britain where it abuts on to several industrial areas (Southampton, Reading, and even London), no Industrial Melanism has so far been recorded there.

A third example of a trunk-sitter with a recessive melanic form is *Catocala fraxini* f. *mürens* Fuchs. It is common on the continent of Europe where it can be considered as maintaining a polymorphism. In the National Collection of British Lepidoptera (R.-C.-K.) at the British Museum of Natural History, London, there are a few examples only, but the mode of inheritance is clear. Denham bred an F_1 generation which were all f. *typica*, but in the F_2 f. *mürens* appeared. More recently Goater, who was sent a female f. *typica* from Shetland, bred approximately one quarter of this form. The original insect was without doubt a primary migrant, probably from Scandinavia as there are no trees in Shetland. I have never found f. *mürens*, though I have searched for it in Europe where it is associated with Ash and Poplar as well as Aspen. In Britain *C. fraxini* seems to be confined to stands of Aspen and here in Kent I found f. *typica* on three occasions when it was most difficult to see. Disregarding other pleiotropic advantages the melanic form must presumably give the species an opportunity of extending its range into woodlands other than Aspen. I have no knowledge of any association between f. *mürens* and industrial areas.

The overall situation appears to be that nearly all tree-trunk resters have melanic forms which are *dominant*. By contrast, we shall see that melanics occurring among foliage-sitters are either unknown or recessive, but never dominant.

(2) *In cryptic species which rest on the ground*. A minority of ground-sitters have developed highly specialized shapes and patterns, such as *Phalera bucephala* L. which resembles a broken twig to perfection. Though dark-coloured *P. bucephala* have been recorded, from Sweden for example (G. Howard, personal communication), recurring melanism is unknown amongst such species.

As has been pointed out earlier, those Lepidoptera which have been adapted to different geological backgrounds (for example, chalk and peat) have attained their local crypsis polygenically and melanism is rarely controlled unifactorially.

Many ground-resting species, however, are polymorphic. In Britain the range of forms in *Celeana secalis* L. and the various species in *Procus* and *Triphaena* are good examples, and one or more of the morphs is usually of a black form. Their genetics for the most part are unknown as many of them are difficult to breed in the laboratory continuously, but their morphs are distinct.

The Caradrinid moth *Meristis trigrammica* Hufn. spends the day on or near the ground amongst dead fallen leaves or around tufts of grass. It has one or more constantly recurring melanic forms. One, f. *obscura* Tutt (=*bilinea* Haw.) occurs throughout its range but varies in frequency from place to place. It is found commonly in Ireland (Howth) and Wales. Cockayne (1925) believed its inheritance was recessive, on the evidence of one brood: the thirteen offspring from a female f. *obscura* were all classified as f. *typica*; but one had a slightly darker ground colour. It is possible that there are several different melanic forms controlled by different genes in this species. The one under discussion came from the chalk downs near Polegate, Sussex—a very different geology to that of Ireland or Wales, where the melanic form may constitute a polymorphism and may have another origin and method of inheritance.

It is a remarkable fact that even though Heslop Harrison attracted attention to the melanic polymorphism in *Antitype chi* L. (Lithophaninae) in 1919, no one has studied the natural history of this species. *A. chi* occurs throughout central, western, and northern Britain, where it sits on rocks, boulders, walls, and, less frequently, on isolated tree trunks, but it is rare or absent in southern Britain. In Oxfordshire, the four specimens I have taken in 18 years of trapping at Steeple Barton have been uniformly white, similar to those found throughout Europe. Farther north, there are at least three separate melanic forms (usually collectively referred to as '*olivacea*'), but we know little of their methods of inheritance. Fifty years ago I used to find both f. *typica* and the melanic forms in large numbers by searching the rocks on the Knaresborough moors in Yorkshire. In the yearly samples ($n = 50+$) taken in this way on three successive years, 40–50 per cent were melanic. From the distribution of the melanic forms in Wales and central and northern England, I think they must be considered as primarily of non-industrial origin. Nevertheless, the darkest forms (including f. *suffusa* Robson) are to be found around the industrial centres where today it must be considered an industrial melanic.

I must here quote from the writings of Barrett in 1897.

It is a very curious circumstance that the dark varieties of the moth by no means select dark walls, but make themselves obvious where the masonry is light coloured; nor, on the other hand, do the usual white forms avoid walls blackened by the smoke of the great manufacturing towns, but even seem to prefer to sit conspicuously upon them. The same may be said of places in which trees are more available than walls. The late Mr. J. Sang, one of our most keen and accurate observers, pointed out long ago the habit of the whitest moths of this species of sitting on the blackest tree trunks and the darker varieties upon the whitest walls. Similarly, in the north of Ireland, white forms may be found quite at home on the black basaltic rocks of County Derry, where strange to say, darker varieties are never known to occur.

It is hard to understand the quite exceptional behaviour recorded here, but surely the most likely explanation is that these observations referred to a minority of individuals only, possibly ones which had just hatched from pupae close to the resting site. Be that as it may, I believe that Barrett's statements written in 1897 greatly influenced Heslop Harrison in 1919 and were in part responsible for his strictures on the lack of selective predation on *Antitype* (*Polia*) *chi* on walls around Newcastle (Harrison 1920*a*). It led him to believe in fact that this species demonstrated 'that the effect of natural selection is quite negligible as a factor in progressive melanism'.

The genetic inheritance of the various melanic forms (*nigrescens, suffusa*, and *nigra*) is not clear. Maddison (1893) states that in two broods from *A. chi* f. *typica*, each produced 75 per cent f. *typica* and 25 cent '*olivacea*'. South (1904) records a brood in which the parents were melanic × melanic, in which 43 offspring were '*olivacea*'. The evidence suggests that at least some melanic forms (probably all) are recessive and that these forms constitute a high proportion of some populations. If this is so, *Antitype chi*, together with *Meristis trigrammica*, should be considered under the heading of recessive polymorphism discussed in the next chapter.

The time is indeed ripe for us to work on these two interesting species; in the laboratory to elucidate their genetics, and in the field for mark–release–recapture experiments as well as those on background choice.

(3) *In cryptic species which rest on foliage (green and brown)*. Throughout large areas of the world in temperate climates foliage both green and brown offers surface areas greatly in excess of any other background in nature; but in the Palaearctic this is so only during the summer and autumn months. In England this extends over a period of eight months from April

to November; green is the prevailing background in deciduous woodlands for the first six but according to the season, earlier or later, foliage turns brown or yellow and, in many trees and shrubs, such leaves remains attached to the twigs and branches until November or maybe, as in young Beech (*Fagus*,) throughout the winter.

Green-coloured species occur throughout the summer and most of these sit exposed on foliage: for example the 'Emeralds' (*Geometra, Euchloris,* and *Hemithea* in the Geometridae) and, in the Cymbidae, species in *Earias* and *Pseudoips*. In none of these have melanic forms been recorded.

Many genera of moths which occur in the autumn consist entirely of species with brown or yellow colouration which match dead leaves; but rarely in the summer do we find this (for example, *Gastropacha quercifolia* L., but here an individual dead leaf is copied to perfection). Foremost amongst these autumn genera are *Ennomos, Selenia,* and *Xanthia* (the 'Thorns' and the 'Sallows'). These may occur commonly in some localities which support mixed deciduous trees where, in Britain in the month of October, 50 per cent of the local species may fall into this category.

Melanism is found in this group but only in a few genera. Here the melanic forms are always *recessive*. In other genera such as *Xanthia* melanism is unrecorded though many of the species are highly variable. I have on a number of occasions found *X. fulvago* and *Citria lutea* in the wild. They sit on the upper surface of yellow leaves of *Salix* and Aspen (*Populus*) which are still attached to the trees. Similarly in other genera, for example, *Hydraecia hucherardi* Mab. rests on the dead yellow leaves of *Althaea officionale* on the roots of which plant the larvae feed. In all these species melanism is unknown.

The same habit is found throughout the genus *Ennomos*, the Thorns, (Selidosemidae) but here they rest with their wings dorsiflexed. All hatch in the autumn and all spend the daylight hours in and amongst dead yellow foliage. Surprisingly, all have rare dark or melanic forms but their inheritance is recessive (for example, *E. autumnaria* ab. *brunneata* Cockayne and ab. *schultzi* Siebert) (Cockayne, 1952b, Bretschneider, 1936, Minnion, 1957). A melanic form ab. *perfuscata* Prout of *E. quercinaria* was at one time taken regularly in Regents Park, London, so that it must then have had a high gene-frequency in that population.

A nearby related genus *Selenia* maintains three species in Britain and two of these have similar rare recessive melanic forms, *S. bilunaria* Esp. ab. *harrisoni* (Plate 13.2) (Harrison and Garrett, 1926; Harrison, 1927c, 1928b) and ab. *glaucescens* Smith and *S. tetralunaria* Hufn. ab. *nigrescens* Cockayne and Kettlewell, (1949) and ab. *notabilis* Th.-Mieg. There is some evidence that these are found more frequently today than previously but this may be due to larger sampling by using M. V. traps.

The questions posed are, therefore, why is melanism unknown amongst green leaf-sitting Lepidoptera? Why is melanism also unknown in many species in dead-foliage-resting genera? Yet Recessive Melanism constantly recurs in *Ennomos* species.

An indication may be provided in other related, but quite distinct genera, *Gonodontis* and *Crocallis*, each represented by one species in Britain. In both, melanic forms occur, and both rest by day with their wings *flattened* on to the surfaces on which they sit. *G. bidentata* Clerck is normally a trunk-sitter, and a melanic f. *nigra* Prout (Plate 5.6, left 4 and 5) is widespread throughout industrial Britain, where it occurs in up to 75 per cent of the population near Leeds and 50 per cent at Cannock near Birmingham, and where it rests on pine trunks. The melanic form of *G. bidentata* is controlled by a single gene which has dominance and on pine trunks or on privet in the London Squares, this form has a high cryptic advantage. Many years ago the advantages of studying this species were pointed out by me for, by contrast with *Biston betularia*, gene-flow appears to be minimal (see also Chapter 5(**vii**)).

Throughout southern Britain *Crocallis elinguaria* L. is a pale-yellow moth with brownish bands on the forewings which are exposed when at rest, usually amongst vegetation. In Scotland these bands are frequently absent and occasionally smoky-dark forms (f. *fusca*) occur; I have taken these both in Perthshire and Aberdeenshire and it has been recorded from Skye, and western Ireland. Recently I have seen two similar examples taken by Bretherton in Surrey in a district where Industrial Melanism occurs at a high frequency (Ottershaw, where *Biston betularia* has approximately 87 per cent of melanic form ($n = 2354$)) The genetics of the melanic forms of *C. elinguaria* are unknown and we are at present breeding them.

Suggestions on the origin of melanism in the cryptic foliage-resters. Can it be that in the highly specialized green-coloured Lepidoptera there are no chemical paths leading to black pigmentation and never have been? Can it be that in those genera whose species regularly sit exposed on the upper surfaces of dead yellow foliage melanism has but rarely conferred an advantage? By contrast may it not be that throughout the distant past the environment has imposed melanism on ancestors; those of *Ennomos* for example? May it not be that in the last few thousand years ecological conditions have not presented opportunities to such forms; yet genes controlling the production of an enzyme leading to the formation of black pigment are still open to rare mutation? Is it possible that in *Gonodontis* melanism has, throughout the ages, been maintained during different periods in the past—because of the habit of sitting with all four wings flattened, and on pine trunks, which so frequently offer advantages to melanic Lepidoptera? Prior to 10 000 B.P., before the succession by deciduous trees, *G. bidentata* must have largely depended on pine species for food plant; it is likely that during this period melanic polymorphisms were widespread among this and many other species. Of the fourteen British species listed by Scorer (1913) as feeding on *Pinus sylvestris*, plus two he omitted, twelve have melanic forms, all except one having dominance.

Today, under prevailing conditions in and around industrial regions, melanic forms offer cryptic advantages, not only to such species as *G. bidentata*, but occasionally for the first time in the recent past to the recessive melanics in *Ennomos* and *Selenia*. Because of the rarity of such recessive homozygotes they are unlikely to spread, though this is possible in the future even without heterozygote advantage which, however, will then evolve if the selection pressures for crypsis outweigh the physiological disadvantages (see *Lasiocampa quercus*, Chapter 14).

(iii) A first step to polymorphism: *Lycia hirtaria* Clerck f. *nigra* Cockayne (Plates 13.3 and 13.4)

The form *nigra*, the jet-black recessive form of *L. hirtaria* has been taken only in and around London and Essex though another melanic, f. *fumaria* Haworth, much less extreme and never to be confused with f. *nigra*, is found to be more widespread (for example north of Oxford). The form *fumaria* was recognized as early as 1803 though f. *nigra*, the more extreme one, was not named until 1948. It is likely therefore that it has appeared more recently. Another melanic f. *hannoviensis* occurring in Europe is apparently an incomplete dominant (Malan, 1918).

Elsewhere in Britain the species is found in two other types of ecology, very different from that of London. First, in the Scottish Highlands and, secondly, locally in the countryside of southern England, Wales, and Ireland.

Because of its distinctive appearance, its recent increase, and its strange ecology in isolated London squares, f. *nigra* appeared to offer special opportunities for field research into recessive melanism and this we commenced in 1961. In this, Cadbury has been responsible for most of the work latterly and I am grateful to him for relieving me of the many complicated situations that must necessarily arise when one undertakes research both by day and by night in the London squares!

Haldane (1956) has postulated that however great the advantages may be of a rare recessive mutant in a changed environment it must be difficult for such a form to spread through a population as, because of its scarcity, it will be impossible for natural selection to modify the gene-complex. This must usually be brought about by way of heterozygote modification.

It appeared, therefore, that the artificial 'flooding' by f. *nigra* (the homozgyote) in circumscribed and isolated areas such as the London squares might, *a priori*, give selection an opportunity to adjust the mutant under natural conditions. It was anticipated that this would be a long-term project, extending over a number of years. We have, in fact, used five London squares as outdoor breeding boxes. We did this because the female *L. hirtaria*, though fully winged, rarely, if ever, flies. The males alone must be entirely responsible for gene flow. One must immediately ask why the female has kept its large wings whilst other members of the family are apterous. As the advantages of flight are excluded it must surely be that the wings are retained for the purpose of providing crypsis.

In 1961 I had substantiated that, when placed on London tree trunks, f. *nigra* had an approximately 50 per cent cryptic advantage to the human eye. By this I mean that f. *typica* could on average be seen at a distance of half as far again as f. *nigra*. I had also found out that extensive bird predation took place in the squares largely by three species, the Great Tit, *Parus major* L., the Robin, *Erithacus rubecula*, the House Sparrow *Passer domesticus*, and two others (but less frequently), the Dunnock, *Prunella modularis* L., and the Nuthatch, *Sitta europaea* L. Predation also occurred, but to a lesser extent, on trees which bordered the main thoroughfares, for example Cromwell Road. Cadbury, by extensive mark–release experiments determined the following points:

(1) After seven hours the differential mortality between the two forms released in Holland Park (which numbered 198 in approximate equality) was in favour of f. *nigra* ($P > 0.001$). Elsewhere in the smaller squares it favoured f. *nigra*, but not significantly.

(2) The smaller Kensington squares (such as Thurloe, Hereford, and Cadogan Place) each maintained at least one pair of Great Tits, Dunnocks, and Robins as well as innumerable sparrows and these were constantly working the limited number of tree trunks in each square. In fact, over 90 per cent of all releases were eaten by the end of a day. Cadbury was able to verify this by direct observation.

(3) There is consequently a low population of the species in such squares and the main reservoirs are centred on the small gardens bordering the main streets and also in the larger parks (such as Holland Park). This is in contrast to my own earlier observations just 50 years previously when each year I collected many hundreds of specimens which were resting on the privet and lilac stems in all the squares to the north of Hyde Park; but I never saw f. *nigra* amongst them. During this period the species occurred in huge numbers and I believe that the House Sparrows obtained ample food from the nose-bags of the hundreds of cab-horses which I can so well remember. Also I think that other bird species were rare in the squares at that time, probably due to extensive egg-collecting which was so popular then amongst boys.

(4) Cadbury showed that there was a deficiency ($P = 0.05$) for f. *nigra* in his eight back-cross broods ($n = 805$). The homozygote f. *nigra* may therefore have disadvantageous physiological effects, and this may also be reflected in the larger number of cripples of both sexes which fail to dry their wings.

(5) On subsequent sampling we have found no evidence as yet of an increase in frequency of the melanic form in the centres where we have released many thousands of individuals both heterozygous and homozygous.

Summarizing our work we can see that a rare recessive melanic mutant provides considerable cryptic advantage. The physiological disadvantages of the homozygote, however, still outweigh other considerations, and insufficient time has elapsed for a readjusted gene-complex to have been selected.

14. Recessive melanic polymorphisms

(i) *Lasiocampa quercus* L. f. *olivacea*, f. *olivaceo-fasciata*, and f. *lurida*, and its separate polymorphisms in Britain

Introduction

In the classification of melanism in the Lepidoptera (Chapter 4) I discussed two divisions which, however, are not completely clear-cut—Ancient and Industrial Melanism. Though normally distinct, Ancient Melanism finds itself at an advantage when such relict forms come into juxtaposition with present-day industrial environments, when the two divisions may merge. Both, in fact, have a number of common denominators: both maintain melanic polymorphisms, both have dominant inheritance and both have similar melanic phenotypes. However, laboratory proof of an identical allele being responsible for both is so far lacking.

The previous chapter attempted to classify the third division, Recessive Melanism. So far I have recorded no balanced polymorphism in this group; nor can this be expected to occur frequently if my theories of the origin and past history of melanism in Lepidoptera are correct. *Lycia hirtaria* f. *nigra*, I suggested, is in the process of such a change as are possibly *Meristis trigrammica* and *Antitype chi*.

It is of interest, therefore, to consider a definite instance, or rather several instances in one species, of this rare phenomenon, where a species maintains polymorphisms, largely I believe, by selection acting on the recessive melanic mutants, due to differential predation.

The melanic recessive forms of *Lasiocampa quercus*, the Oak Eggar, fall under the names of f. *olivacea* Tutt (1902), f. *olivaceo-fasciata* Cockerell (1889), and f. *lurida* Cockayne (1951b, 1952a)†. In Britain there are three areas where such polymorphism occurs: we have shown in Caithness, North Scotland, that a melanic form is in a state of balanced polymorphism with a gene-frequency as high as 80 per cent (Plate 14.1). On the industrial moors of Yorkshire, however, the frequency of the melanic forms appear to fluctuate. Here they are again recessive, phenotypically identical but genetically different from the north Scottish forms, and here they have increased in frequency during the last 60 years, but only to a maximum gene-frequency of 20 per cent. The polymorphism in Yorkshire must therefore be considered as transient. A point of additional interest is that

† These do not refer to the three separate polymorphisms. All were named from Yorkshire specimens.

a proportion of black Oak Eggars in Yorkshire come from black larvae which are recessive to wild-type larvae (Bell, 1909).

A third and quite distinct polymorphism occurs on the Lancashire/ Cheshire coastline of western Britain and this is in the one-year life-history race *L. quercus* ssp. *quercus*. Here again there are black larvae.

These three recessive polymorphisms of *L. quercus* and ssp. *callunae* are of exceptional importance because they may reflect the earlier stages which melanism went through before developing dominance, as is the case in the majority of cryptic species. For this reason I discuss the black Oak Eggar polymorphisms at some considerable length.

Lasiocampa quercus L. ssp. *quercus* and ssp. *callunae*

Life History. In Britain there are two races separated, more or less completely, by different food plants, lengths of life-cycle, and, to a lesser extent, by geography. The 'Southern Eggar' is *L. quercus* ssp. *quercus* (Edleston, 1860); the 'Northern Eggar' is referred to as ssp. *callunae*.

In spite of the efforts of the taxonomists, no clear phenotype differences between the two forms have been found (Newman 1865a, 1865b, Thompson, 1896, Tutt, 1896, 1902), though it is possible to relegate certain individuals of either to their correct positions.

L. quercus ssp. *quercus* has a one-year life-history, but ssp. *callunae* a two-year life-history, the imagines appearing only in odd years and their full-grown larvae in even. These coincide with a similar life-history of this subspecies in Scandinavia and in North Germany (heather moors). (In montane Europe there are dark races with a three-year life-history—ssp. *alpina* (Pictet, 1931).)

There is a clear-cut behavioural difference between the two sexes. The males of all subspecies fly rapidly in sunlight when they assemble to virgin females, which rest motionless throughout the day on the foliage above the cocoons from which they hatched (Plate 14.7). Hence the females of *L. quercus* ssp. *quercus* find themselves among dead yellow leaves near the ground beneath deciduous shrubs on which the larvae feed. Subspecies *callunae* must usually rest by day on its foodplant, *Calluna vulgaris* (heather).

Throughout most of mountainous Scotland, however, heather is intermixed with two very different types of foliage, Birch bushes (*Betulus* spp.) and the Bracken Fern, *Pteris aquilina*. The dead, yellow leaves of Birch and the brown bracken fronds from previous years provide an alternative background throughout these areas. I had not appreciated the importance of these to female Oak Eggars until I heard a casual remark by a most observant Scottish naturalist, Le Measurier, of Aviemore, Central Highlands, who stated that he could regularly find the females of this species by searching around birch trees whose dead and yellow leaves are caught up in the heather. In this situation they are highly cryptic at rest. On pure heather moor on black peat, they are, on the other hand, conspicuous.

In Britain the females, after mating, fly at night, when they scatter their eggs at random over heather; only rarely, in hot weather, do old females fly short distances by day. In Denmark, however, females are regularly seen on the wing in the afternoon, when they oviposit (Hoffmeyer, personal communication).

L. quercus ssp. *quercus*, the nomino-typica Oak Eggar, feeds on a wide range of deciduous plants and shrubs: on chalk downs bramble, *Rubus* sp., on sand dunes *Salix*. All ssp. *callunae*, however, feed on heather, *Calluna vulgaris* L. usually on open moorland, and also on *Vaccinium* growing in such ecologies.

The imagines of *Lasiocampa quercus* ssp. *quercus* hatch in July; the small larvae commence feeding in August and go into hibernation in the third instar. They are full grown the following year in May or early June.

The imagines of ssp. *callunae* fly in late May and June and the offspring hibernate the first winter as larvae and the second as pupae.

Geographic distribution. In Britain ssp. *quercus* (the Southern Eggar) is found in a number of niches from chalk Downland, to sand dune, hedgerows, or woodland. All are associated with deciduous trees and shrubs; hence ssp. *quercus* occurs locally throughout southern England and as far north as the coast-lines of Durham and Lancashire.

Subspecies *callunae* in Britain is limited to those areas where heathermoor and pine-heath ecologies occur. It is therefore confined to Scotland, the Pennines, Wales, and the Dartmoor–Exmoor regions of south-west England. Only where heather hills abut on to sand-dune ecology do the two subspecies overlap.

The imaginal phenotypes. Lasiocampa quercus (and its subspecies) is sexually dimorphic. There is little resemblance between the two sexes. This is in common with a general rule that where each sex follows a different behavioural pattern—such as day or night flight—this must inevitably lead to a difference in colour or markings.

Yet we know that in nature the two sexes must be in approximate equality. Such sexual dimorphism must throw a considerable strain on one or the other sex when an environmental change takes place which effects one more than the other: for example, night-flying males versus day-flying females as in *Cycnia mendica*, discussed in Chapter 17, (or those species in which the female is apterous). In both these instances a different set of predators may be at work on each sex. It appears to me that little work has been attempted on this aspect of dimorphic species which have, today, to adapt to a rapidly changing world. *L. quercus* certainly falls under this heading.

Description. MALE. Antennae highly pectinate; all four wings dark chocolate-brown with bright yellow bands which occasionally in the southern race spread across the forewings as far as the outer border. The yellow band in ssp. *callunae* males never spreads in this way.

FEMALE. In the southern race this large moth is a clear yellow with an inner area on both forewings and hindwings faintly darker insider a paler yellow band. In the northern race, ssp. *callunae*, this inner area is more pronounced and may be a shade of brown, whilst the whole insect may have a more variegated appearance. The overall impression is that of a *yellow* and not of a dark-coloured insect in the females of both subspecies.

The melanic mutants

So-called 'black eggars' are controlled respectively by a number of genes which differ in the various localities. In no instance do these forms have dominance. There is considerable confusion over their nomenclature, which has arisen for several reasons, not least a fading of wild-caught specimens.

Three melanic forms can be recognized as phenotypically separate, but only when they are in bred condition: f. *olivacea* Tutt (1902)—in both sexes the whole insect is greenish-black so that the male appears black from beneath instead of yellow (Plate 14.4): f. *lurida* Cockayne (1951)—colour is rufus-black instead of green-black; f. *olivaceo-fasciata* Cockerell (1889)—the yellow bands of the male are replaced by greenish colouration. This is particularly evident on the underside. There is but little evidence that f. *olivaceo-fasciata* is the heterozygote of f. *olivacea*.

Distribution in Europe

Apart from Britain, 'black eggars' have been recorded from the coast of Holland (Lempke, 1960) and Denmark, where they occur regularly on coastal sand hills, more rarely inland (Hoffmeyer, 1948). In East Germany 'black eggars' were taken regularly at the beginning of the century along the River Havel near Brandenburg (Frings, 1905, Niepelt, 1911). The males are described as being violet-brown and the females as olive-green, both with an olive-green band across the forewings. Frings named these f.

paradoxa but retracted, probably wrongly, because he believed that this form was synonymous with f. *olivaceo-fasciata*.

Distribution in Britain

'Black eggars' have been recorded casually as rare specimens from many places over a wide range—from Poole Harbour, Dorset, in southern England to the Dawlish Warren to the west, and in north Kent and the Norfolk Broads to the east. We must take note here that all these localities in the southern half of England are coastal and therefore subject to gull predation.

Melanic polymorphisms, however, only occur in three completely separated areas. The first of these is in Caithness, north-east Scotland, over an area of approximately 800 square miles. At Keiss the melanic phenotype frequency has in the past often been as high as 70 per cent (=gene frequency 83·7 per cent). From this point there are definite north/south and also westerly clines (Fig. 14.1), the latter being steeper than the former. The second area is in Yorkshire, central England, on the east side of the Pennines, where only in the last 40 years similar phenotypes have contributed up to 5 per cent of the population. Finally, the third area is on the west coast area of Cheshire and Lancashire (the Wirrall Peninsula and Lancashire sand hills). The first two polymorphisms occur in the two-year life-history ssp. *callunae*. The third is in ssp. *quercus* and is found yearly.

(ii) *L. quercus* ssp. *callunae*: the Caithness polymorphism

Caithness is situated in the extreme north-east of Scotland between latitudes 58° and 59° N. It differs from the rest of Scotland in that the whole of the northern two-thirds is non-mountainous, geologically consisting of shelly boulder-clay. This area is largely covered by a sponge of wet peat-moss overspread with heather, which is split up into moors by strips of arable land. Most of Caithness is today practically treeless though prior to 4000 B.P. extensive Birch and Pine forests existed. The ecology during this period must, therefore, have changed considerably.

Though 'black eggars' have been taken or seen rarely in other parts of Scotland (two from Aviemore, Inverness-shire, and one from Aberdeen-shire) only in Caithness do they form a polymorphism (and in the Orkney Island of Hoy to the north (Table 14.1).) It is of interest to record that of 222 larvae collected by the Oxford Scientific Society in 1970, from more than 8 areas in the neighbouring county of Sutherland, which is mountainous, we did not breed a single f. *olivacea* (Table 14.5).

TABLE 14.1

L. quercus ssp. callunae—north-east Scotland: phenotype frequencies of imagines (wild-caught).

Locality (with ordnance survey no.)	Year	Sample	f. typica Number observed	Intermediate Number observed	Per cent observed	Per cent expected heterozygotes	f. olivacea Number observed	Per cent observed
(1) Dunnet Head ND 2075	1967	61	49	5	8·2	44·3	7	11·5
	1971	2	2	0	—	—	0	—
(2) Mey Mill ND 3173	1967	50	32	3	6·0	49·5	15	30·0
	1971	8	4	1	—	—	3	—
(3) Warth Hill ND 3769	1965	21	8	1	—	—	12	57·1
	1967	86	38	8	9·3	43·1	40	46·5
	1969	178	98	20	11·2	48·7	60	33·7
	1971	73	41	6	8·2	48·0	26	35·6
(4) Keiss ND 3362	1965	126	12	24	19·0	26·5	90	71·4
	1967	140	48	11	7·9	36·3	81	57·9
	1969	111	49	4	3·6	20·6	58	52·3
	1971	75	29	5	6·7	38·3	41	54·7
(5) Black Hill ND 3058	1967	37	6	7	—	—	24	68·6
(6) North Watten Moss. ND 2459	1971	129	37	23	17·8	39·0	69	53·5
(7) Loch Olginey ND 0959	1967	54	31	1	1·9	46·1	22	40·7
	1965	10	—	3	—	—	7	—
	1967	33	11	1	—	—	21	63·6
	1971	13	4	0	—	—	9	—
(8) Mybster ND 1750	1967	52	14	8	5·3	36·3	30	57·7
	1969	57	19	3	3·6	40·2	35	61·4
	1971	284	66	36	12·7	32·0	182	64·1
(9) Loch More ND 0846	1967	50	34	4	8·0	50·0	12	24·0
(10) Badlibister ND 2549	1967	63	24	4	6·4	37·7	35	55·6

Recessive melanic polymorphisms

TABLE 14.1 (continued)

Location	Year							
(11) Tannach ND 3046	1965	327	111	86	35.8	46.6	130	39.8
	1967	145	61	23	7.5	45.6	61	42.1
	1969	198	106	15	7.6	46.9	77	38.9
	1971	129	47	19	14.7	42.0	63	48.8
(12) Camster ND 2642	1965	37	12	11	29.7	47.3	14	37.9
	1967	41	15	10	24.4	46.9	16	39.0
(13) Rumster ND 1941	1967	51	16	10	19.6	42.0	25	49.0
	1971	136	24	29	21.3	34.2	83	61.0
(14) Dunbeath ND 1231	1965	53	22	14	26.4	49.1	17	32.0
	1971	120	48	18	15.0	44.2	54	45.0
(15) Berriedale ND 0931	1965	33	24	4	—	—	5	15.1
	1969	22	19	0	—	—	3	13.6
(16) Golspie, Suth. NC 8203	1969	56	54	2	—	—	0	—
	1971	8	7	1	—	—	0	—
(17) Reay NC 9264	1967	26	22	3	—	—	1	—
	1969	45	33	1	—	—	11	24.4
	1971	7	1	1	—	—	5	—
(18) Trantlemore Suth. NC 8950	1967/69	32	28	1	—	—	3	(9.4)
(19) Forsinard Suth. NC 8837	1967/69	19	17	0	—	—	2	—
(20) Kinbrace Suth. NC 8631	1969	31	31	0	—	—	0	—
(21) Kildonan Suth. NC 9121	1969	49	48	0	—	—	1	2.0
(22) Hoy, Orkney ND 2099	1969	24	24	0	—	—	0	—
	1971	112	102	4	3.6	18.0	1	1.3

Since 1960 we have carried out work each year over large areas of Caithness sampling larvae in even years and imagines in odd years. This has proved to be both exciting and rewarding. The following facts have been established. There are at least two different melanic phenotypes in the Caithness population: a dark green-black form, phenotypically similar to f. *olivacea*, and a variable 'intermediate' similar to f. *olivaceo-fasciata*.

There is little evidence that this 'intermediate' form is the heterozygote of f. *olivacea*. Of 377 bred and wild-caught specimens from Keiss in 1965 and 1967, 221 were f. *olivacea* (59 per cent). The *expected* heterozygotes and typicals from the Hardy–Weinberg ratio are, therefore, 36 per cent and 5 per cent respectively. In fact, of the *observed* heterozygotes, 15 per cent were classified as intermediates and 26 per cent as f. *typica*. The majority of those individuals scored as f. *typica* must, therefore, have been heterozygous for f. *olivacea* (assuming that the intermediates are not less viable than either of the homozygotes). On the other hand, we have had one brood in which f. *olivaceo-fasciata* segregated in 1:2:1 ratio, though the same results could be accounted for by postulating another and separate mutant. In many broods known heterozygotes are as yellow as those found in districts where f. *olivacea* does not exist whilst others are intermediates. The heterozygotes of f. *olivacea* may therefore be variable in expression in Caithness.

The highest frequency of f. *olivacea* coincides with the highest density of the species (Fig. 14.1, Table 14.1). This, until 1971, has been at Keiss in the lowland country to the north of the county near Wick where in 1965 wild-caught melanic males amounted to 71·4 per cent ($n = 126$) and in 1967 to 57·9 per cent (n 140), giving a gene-frequency of 84 per cent and 76 per cent respectively. From Keiss there was a cline which ran in a southerly direction for a distance of 30 miles or more (Fig. 14.1). Thus, at Berriedale, which lies close to the Sutherland border where the country becomes more hilly, the frequency of f. *olivacea* was 15 per cent in 1965 ($n = 33$). Going progressively northwards from here to Keiss, the intervening samples showed frequencies of 32·0 per cent ($n = 53$) on the Dunbeath Moors, 37·9 per cent ($n = 37$) at Camster, and 38 per cent ($n = 327$) at Tannach which is nine miles south of Keiss and 21 miles north of Berriedale.

Proceeding west from Keiss, the peat-moss areas which are well isolated here by intervening strips of arable land, maintain a high frequency of f. *olivacea* as far as Loch Olginey (65 per cent), 20 miles to the west.

To the north-west of Keiss, however, towards the northern coastline of

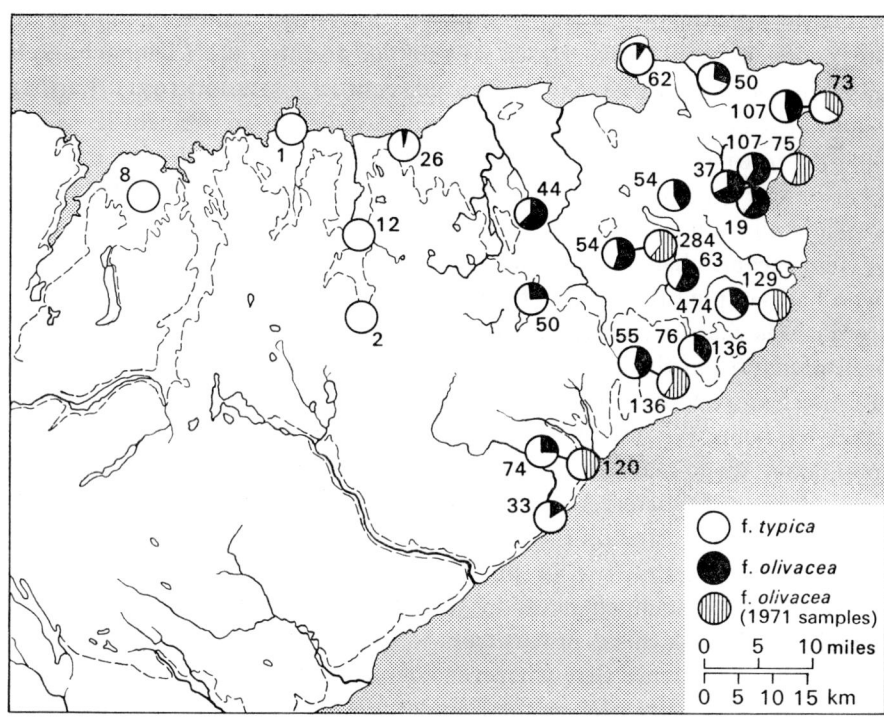

FIG. 14.1 Distribution map of the morphs of *Lasiocampa quercus* in Caithness 1965–71 (cross-hatching = 1971 samples)

Caithness, there is another cline which drops rapidly so that at Mey, five miles distant, the frequency is 30 per cent ($n = 50$) and at Dunnet Head, only 9 miles from Keiss, it has fallen to 11·5 per cent ($n = 62$). There must, therefore, be strong selective pressures affecting the two forms between the windswept headland of Dunnet and the peaty mosses of Keiss; and also between them and the mountains to the south. A major effort in our investigations has been to attempt to identify the forces involved in them.

That strong selective pressures are indeed at work has been demonstrated by our more recent work in 1971, for here we found that several changes had taken place. Firstly, the centre of melanism had shifted south. On the one hand, the percentage of f. *olivacea* has dropped in the northeast, from 71 per cent to 51 per cent at Keiss, at nearby Black Hill from 68·6 per cent ($n = 37$, 1967) to 53·5 per cent ($n = 129$), and from 46 per cent to 35 per cent at Warth Hill. Mybster, which is situated in the central north, varied but little (60 per cent, 1967/69 ($n = 109$), 64 per cent, 1971 ($n = 284$)).

On the other hand, the large samples we took in 1971 showed an increase in f. *olivacea* in all the localities we sampled to the south (Tannach, 39·8 per cent ($n = 327$) in 1965, to 48·8 per cent ($n = 129$) in 1971). Farther south at Rumster, the frequency had increased from 49 per cent ($n = 51$) in 1967 to 61 per cent ($n = 136$) in 1971, and at Dunbeath, from 32 per cent ($n = 53$) in 1965 to 45 per cent ($n = 120$) in 1971. The centre of the melanic clines has shifted.

The second change, more difficult to define, but obvious to a person who had sampled these areas for four consecutive hatching years (1965 to 1971), was that in contrast to previous years, there was no evidence of gull predation in the north. Gulls, however, were constantly working the Rumster and Dunbeath moors to the south where on earlier occasions I had seen none. We saw no predation taking place at Keiss (where the total population was much reduced) nor in other northern areas, but it was much in evidence at Mybster. Also the large colony of Black-headed Gulls which had previously existed on the island in Loch Olginey had vanished. It is a salutary thought that two or three boys with a rubber dinghy could be responsible for such changes, in their efforts to satisfy an egg-hungry public, yet I think this is in part true.

It is a general rule that wherever colour polymorphisms are found in living things of a size such as the Lepidoptera (as distinct from *Drosophila*) the colour difference is itself likely to be of fundamental importance, though pleiotropic effects such as viability and behavioural differences must also contribute to the balanced polymorphism.

In both sexes the *olivacea* form, when newly hatched, bears little resemblance to f. *typica*. A dark form, which is monomorphic, is to some degree substituted for the distinct sexual dimorphism of f. *typica*. This is a remarkable fact when we consider the differences in habits between the two sexes. Foremost among these is the flight habit of the males which only takes place in daylight and is always directed to finding the virgin females which rest among the heather (neither sex visit flowers for nourishment, neither the male by day, nor the female by night). This sexually monomorphic melanic, however, becomes rather less remarkable when we take into account the local meteorological conditions. In Caithness in June it may frequently rain for days in succession (or more rarely snow). The males as well as the females are therefore compelled to spend this time at rest on heather, when both are subjected to visual predation.

We have carried out many series of experiments on the cryptic advantages of the two forms, in both sexes, first to human recognition and secondly to known predators.

Placing them on heather stems in their normal resting position (Plates 14.2 and 14.3), we compared the females of both forms and the males separately. The mean 'disappearance distance' was then scored by a number of individuals. This was assessed by walking slowly backwards but, because of variations in the amount of light, the humidity which affects the background colour and other variants, only those observations taken at the same time on the same day could be compared. Thus in one series of comparisons by two observers, the females of f. *olivacea* were not visible beyond 14 m whilst yellow females could be seen up to 40 m. The overall estimate was that f. *typica* females were recognizable at just under three times the distance to f. *olivacea*. In males, the distance was less because f. *typica* of this sex have, except for the yellow bands, dark chocolate colouration; also the males are considerably smaller than the females. The respective disappearance distance was f. *olivacea* 5 m, f. *typica* 9–11 m (Table 14.2).

TABLE 14.2
The cryptic efficiency of L. quercus
f. typica *and f.* olivacea *at rest on heather.*

Rombalds Moor, Yorkshire—on polluted heather. Clear sky

Sex	f. *typica*	Intermediate (from Caithness)	f. *olivacea*	Observer
Male	49 m		15 m	H.B.D.K.
			11 m	,,
			10 m	,,
	30 m		11 m	C.J.C.
	26 m		11 m	,,
Female	57 m		15 m	H.B.D.K.
	51 m			
	52 m	31 m	21 m	C.J.C.
	50 m	26 m	18 m	,,

Caithness—on unpolluted heather on damp peat. Overcast sky

Male	11 m		5 m	H.B.D.K.
	9		5 m	C.J.C.
Female	40 m	21 m	14 m	,,

'Disappearance distance' in metres

If visual recognition by predators was anything of the same order, black eggars would have a colossal cryptic advantage in Caithness.

Visual predation

In work I had done on the black eggars earlier in Yorkshire, I had been greatly surprised to discover that heavy predation by gulls (*Larus* spp.)

took place. At this early date (1956), I had not anticipated that such seabirds could have any effect whatsoever on Lepidoptera. Spärck (1950) had already pointed out, however, that the Common Gull (*Larus canus*) was largely insectivorous.

When Cadbury and I commenced our work in Caithness we had, therefore, an indication of this and even a list of other possible predators. Neither of us was prepared, however, for the degree of predation we were to witness on the Oak Eggar in the alternate (odd) years.

In present-day Caithness, two species of gulls have taken over large areas where in May and June they breed in vast numbers: the Black-headed Gull, *Larus ridibundus*, and the Common Gull, *L. canus*. Each has entirely different habits—feeding, resting, and behavioural—and together, during the brief period of the appearance of the moth, they destroy large numbers of the species.

The Black-headed Gull in Caithness nests in densely populated colonies on islands in lochs. From such centres it forages afield during nesting time up to seven or eight miles distant. It feeds by searching whilst in flight, with the head in a forward position and the eyes looking ahead. By contrast, the Common Gull nests at low density in colonies in which each pair is separated by a distance from others over an area of moss. It hunts in an entirely different manner from the Black-headed Gull. The head is flexed ventrally during feeding flight and the birds look vertically downwards beneath themselves. The same areas of moorland are frequently covered by both species. The maturing of the young gulls coincides with the hatching period of the moths.

In Caithness, where eggars may be locally at a high density, gulls and other birds search for females at rest by day and I have observed predation by the Short-eared Owl, *Asio accipitrinus*, at dusk. The male moths also are taken on the wing by Black-headed Gulls which we have frequently seen working up wind and over-taking them when they are assembling to the newly hatched females. Some predators even follow the males into the assembling traps and we have extracted the Meadow Pipit, *Anthus pratensis*, on several occasions and, once, a weasel. Wheatears (*Oënanthae oënanthae*) also regularly catch and eat the males and may take up station on top of the assembling traps awaiting arrivals. In 1971 Lees found a pair of Merlins (*Falco columborius*) which were catching large numbers of flying eggars whose wings we were able to collect from their 'feathering sites' around the nest, which contained three young. This was at Mybster, when the f. *olivacea* frequency in that year was 64 per cent ($n = 284$). Discounting doubtful wings, Lees was able to score with certainty that they represented a minimum of 32 moths (24 ♂ and 8 ♀). The f. *olivacea* frequency of the Merlin sample was only 46·9 per cent. This form would seem to have a distinct visual advantage on the wing against Merlins. (For ♂♂, $\chi^2_{(1)}$ = 3·830, 0·1 > P > 0·05.) The main predation is, however, by the two species of gulls on the females; in many localities near to their breeding colonies but few *typica* of that sex can survive.

Proof of selective predation
Under such conditions colour differences must play a major role in survival. A method of carrying out release experiments on resting females had to be devised in which, ideally, equal numbers of both forms could be put to test. This, as can be imagined, created considerable problems on featureless heather-moor. We developed an appropriate technique by using 100 yard lengths of cord and chicken-rings of two colours. The cord was pulled taut over a suitable area of moorland and the moths released onto the heather stems along its length. A coloured ring was fixed to the string at the nearest point to the release, always under two feet distance, and the choice of colour decided which of the two forms it indicated. The cord and rings were then picked up, leaving a marker-stick in position at either end. Not more than 30 moths were released in the same experiment, 15 of each form, and the moths were on average 10 feet apart. Using binoculars we were able to count the number of moths taken; we did so by observing gulls dropping into the heather. When approximately half had been eaten we returned to the releasing site and replaced the cord in its old position—absent individuals were recorded. The bias therefore was against that phenotype which survived the better, because f. *olivacea* could frequently have been overlooked by us even when marked by the rings.

What I had not anticipated was the degree of selectiveness I was to find in the predation of the two forms by gulls. Of the 50 female f. *olivacea* released onto heather moors, not one was eaten in six separate tests. The gulls seemed unable to see them. Of 49 female f. *typica*, 33 were predated. f. *olivacea* had in fact an 80 to 90 per cent cryptic advantage over f. *typica* (Table 14.3).

The release of marked males was more difficult because such experiments could only be carried out on cold, wet, or dull days, otherwise the moths would take wing. We therefore used freshly killed males in these experiments. Under these conditions the selective advantage for f. *olivacea* males was found to be similar, but not so intense, as that of the females. Release experiments were carried out in three separate localities and it

TABLE 14.3
Lasiocampa quercus ssp. callunae: *release experiments in Caithness and Sutherland (Selective predation by Black-headed and Common Gulls).*

Female		f. typica		f. olivacea	
Place	Date	Released	Predated	Released	Predated
Loch Olginey, Caithness	24.6.65	10	8	10	0
	29.6.65	7	7	8	0
	2.7.65	2	2	2	0
		19	17	20	0
Mybster	22.6.71	15	11	15	0
Dunrobin	24.6.71	15	5	15	0
Golspie		30	16	30	0
Total female predation		49	33	50	0

Males					
Place	Date				
Loch Olginey, Caithness	26.6.65	6	5	6	2
	29.6.65	2	2	2	1
	2.7.65	6	4	6	0
	3.7.65	4	4	4	1
	6.7.65	4	4	4	0
Total male predation		22	19	22	4

The difference in the proportion of each form taken is highly significant. $P < 0.001$.

appeared to us that few f. *typica* of either sex could survive for long in any of them, certainly not the females, which are eliminated at rest under all weather conditions.

Physiological differences

The cryptic advantage of f. *olivacea* in Caithness implies comparable disadvantages throughout the greater part of the range of the species where only f. *typica* occurs. This disadvantage could be because the melanic mutants fail to give cryptic advantages elsewhere. If this was so, it would be comparable to the case of the industrial melanic forms of *Biston betularia*, which I have shown to be at a visual disadvantage in a non-polluted countryside. Alternatively, there might be a relaxation of predation elsewhere, but this would provide no answer to the difference in colour patterning.

Accordingly in 1971 we undertook release experiments at Golspie, Sutherland, where a sample of 64 males contained no f. *olivacea*. The site was some 300 yards from a large colony of Black-headed Gulls. Whilst all 15 f. *olivacea* females survived, one third of the f. *typica* were predated (Table 14.3).

The genes which are responsible for the black forms and which confer such cryptic benefit in Caithness must at the same time control other important characters, and some or all of these characters are likely to be disadvantageous.

A comparison of the physiological and behavioural characters of the two forms
Though we know that pleiotropic effects exist, as yet we have little knowledge of their selective intensities. In the laboratory broods there is no evidence that one form hatches earlier than the other. In wild sampling, however, in Yorkshire in 1959, 1961, and 1963, the frequency of f. *olivacea* was slightly, but consistently higher in the later part of the sampling periods, though this is not true for 1957. The number of f. *olivacea* caught was small and the difference has no significance ($\chi^2_{(1)} = 1.849$, $0.4 > P = 0.3$).

Such a situation is theoretically likely to be found, as a direct consequence of predation, particularly in years of abundance.

In laboratory broods we have not observed a deficit in the homozygous melanic forms, though this may well occur in the wild, unrecognized, under the more stringent conditions imposed.

The fact that in the majority of samples we have found fewer f. *olivacea* amongst wild-caught males than occur amongst the progeny of larvae collected the previous year from the same localities suggests some physiological or behavioural differential (Tables 14.4 and 14.5). Thus from the 1964 sample of larvae collected at Tannach, of the 53 which hatched 49 per cent were f. *olivacea*; of the 327 males we assembled on the same ground, the following year only 38.8 per cent were of this form. From similar samples taken at Camster, 60 per cent of the 25 moths which hatched were f. *olivacea* whilst only 38 per cent of the 37 which assembled in 1965 were melanic. These figures, however, are, separately, not significant (Tannach $\chi^2_{(1)} = 12.69$, $P = 0.2$, Camster $\chi^2_{(1)} = 2.12$, $P > 0.1$). There is no heterogeneity, however, between the Tannach and the Camster figures. Combining the two, the difference is only just outside the 5 per cent significance level. On comparing similar sets of figures for 1970/71 (Tables 14.1 and 14.5), two of the three go in the same direction (Keiss:

TABLE 14.4

L. quercus ssp. callunae—*Caithness, Scotland. Phenotype frequencies of imagines reared in the laboratory from wild larvae collected in 1964, compared with those of 'assembled' males in 1965 in two localities.*

Locality	Sample	f. *typica* intermediate	f. *olivacea* Number	Per cent
Tannach:				
Bred	53	27	26 {13♂♂ / 13♀♀}	49·0
'Assembled'	327	197	130	38·8
Camster:				
Bred	25	10	15 {7♂♂ / 8♀♀}	60·0
'Assembled'	37	23	14	37·9

There is no heterogeneity between the two sets of figures. The difference between the bred and caught samples when combined is significant (P > 0·05).

(f. *typica* and the 'intermediate' are combined because of the difficulty of separating the two in these localities.)

TABLE 14.5

L. quercus ssp. callunae—*Northern Scotland. Phenotype frequencies of imagines bred from wild samples of larvae collected in 1970*

Locality		Sample total	f. *typica*	'Intermediate' Number	% obs.	% exp. (het.)	f. *olivacea* Number	%
(1) Keiss	♂	36	11	1	2·8	29·7	24	66·7
	♀	46	19	2	4·3	39·0	25	54·3
Total	♂♀	82	30	3	3·7	34·9	49	59·8
(2) Mybster	♂	40	8	5	12·5	29·7	27	67·5
	♀	58	12	3	5·2	24·0	43	74·1
Total	♂♀	98	20	8	8·2	26·5	70	71·4
(3) Tannach	♂	35	18	2	5·7	45·1	15	42·9
	♀	29	15	1	3·4	44·2	13	44·8
Total	♂♀	64	33	3	4·7	44·7	28	43·8
(4) Auckingill	♂	3	2	0			1	
	♀	1	0	1			0	
Total	♂♀	4	2	1			1	
(5) Sutherland; mixed localities	♂	97	96	1			0	
	♀	125	123	2			0	
Total	♂♀	222	219	3			0	

bred 59·8 per cent, caught 54·7 per cent; Mybster: 71·4 per cent to 64·1 per cent). But the Tannach sample went in the opposite direction (bred 43·8 per cent, caught 48·8 per cent). None of these are individually significant. The winter of 1970/71 was an exceptionally mild one however. Further sampling will, no doubt, give the answer, though comparable differences may not occur in each generation. Such a frequency change could be accounted for in at least two separate ways. Either the full-grown larvae and pupae of f. *olivacea* have a higher mortality rate; or f. *olivacea* flies less freely to the assembling traps on which we depend for sampling. The indications we have so far are that the second of these alternatives is almost certainly true. We have released equal numbers of marked males of both forms 300 yards down-wind to virgin females. In competition of this kind, it appears that f. *typica* frequently arrived first and in nature there must be considerable selective pressure for speed of flight. The practical difficulties in carrying out such experiments are great. Though equal numbers are released at the same moment, a number fail to take flight, others may be blown down-wind and may appear half-an-hour later.

Other influences are brought to bear during flight. The two forms of the male appear very different on the wing and to my eyes the black eggar from above is less conspicuous against the peaty background. Many birds, in particular, the Swift, *Apus apus*, overcome this by skimming under the insects and attacking them from beneath, when silhouetted against the sky. Under these conditions colour and pattern can play little part in survival.

I, and my co-workers, in particular J. Cadbury, have attempted to find other physiological and behavioural differences. In order to test mating preferences, Cadbury devised a method in which he set up two batches of virgin females at a distance of 20 yards apart and placed in a cross-wind direction. One contained f. *typica* and the other f. *olivacea*. The males which assembled showed no assortative nor disassortative choice and random attraction appears to be a rule. He did, however, record that fewer males came to females of the *olivacea* form (35 (of all phenotypes) to f. *typica* ♀♀, 15 to f. *olivacea*, Caithness 1965).

There is one further possibility in regard to flight and colour differences. It could be postulated that the black males would heat up more rapidly and take flight earlier during periods of fitful sunlight. Cadbury also tested this, using f. *typica*, f. *olivaceo-fasciata*, and f. *olivacea*. He found no significant difference in 'take-off time' between the three phenotypes. It seems, therefore, that the black colouration of f. *olivacea* and f. *olivaceo-fasciata* does not lead to earlier flight as in Thermal Melanism to which I have previously referred. I would think that *L. quercus* has too much bulk for this to be effective. As yet we have not tested this by using thermocoupling.

Summary of, and problems posed by the Caithness data

The melanic polymorphisms in ssp. *callunae* which exist throughout Caithness are maintained by a balance of cryptic advantages and physiological disadvantages. In the mountainous country farther south and west and throughout the rest of Scotland, only f. *typica* is found, but f. *olivacea* and f. *olivaceo-fasciata* may occur as rare mutants.

The exceptional interest in this situation is that neither of the black forms has dominance. Such recessive polymorphisms maintaining melanism at a high frequency in a population are of rare occurrence.

(iii) *L. quercus* ssp. *callunae:* the Yorkshire polymorphism

The Yorkshire heather-moors extend intermittently in an easterly direction from the Pennines as far as the North Sea coastline. They are isolated from each other by areas of industrialization and all of them are heavily polluted by smoke which drifts over the mountains from Lancashire and by smog from such centres as Leeds, Huddersfield, and Halifax. Areas of arable land and woodlands, which are the home of Industrial Melanism in many species, intervene in the valleys and usually separate the towns from the moorlands. The ecology is, therefore, very different from that of Caithness. The common denominator is heather growing on a peat soil. This, for the most part, is much higher and better drained than in Caithness which lies nearly 500 miles to the north.

Subspecies *callunae* occurs in Yorkshire on all such moors and from time to time builds up populations which assume plague proportions. On such rare occasions the heather may be devastated over large areas, as it was in 1954 and 1956 on Rombalds Moor which has been our centre of work since 1956.

Form *olivaceo-fasciata*, the original 'black eggar' was first taken on this moor in 1893 by the well-known lepidopterist, S. T. Porritt (1906), whose collection can still be viewed in the Huddersfield Museum, and during the next 50 years many black eggars were bred or assembled on this moor by local collectors. Black eggars were also found to occur regularly during the last quarter of the nineteenth century on other local moors, Crosland, Norland, and Roydes Edge, situated on the outskirts of Huddersfield and Halifax, but since then destroyed by man. Form *olivacea* and f. *olivaceo-fasciata* may, in fact, have been present as a polymorphism in Yorkshire, but at a low density, on many moors, but only in the last 90 years or so.

When we commenced investigations in 1956, we were unaware of the comparable, but different situation, which existed in Caithness; we only knew that phenotypically similar forms occurred there and these were being sold to collectors by dealers. The earlier evidence we accumulated at this time convinced me that, even though inherited as a recessive, black eggars in Yorkshire came under the heading of Industrial Melanism; and indeed they may. The heather is black from pollution fall-out; there is little bracken and most moorland trees such as Birch and Aspen have long since vanished except in steep valleys.

One of the first points we had to establish was whether the black eggars of Yorkshire were controlled by the same gene as those of Caithness. When the two melanic phenotypes were paired, however, the offspring were 100 per cent f. *typica*. Different genes, at separate loci, therefore, control the two forms.

A further and important difference between the Caithness and Yorkshire black eggars is that the majority of Yorkshire f. *olivacea* arise from dark-coloured larvae. These were absent among the many hundreds we collected in Caithness. I discuss these larval characters later in this chapter.

Frequency of f. olivacea *in Yorkshire*

Because of the confusion in differentiating between f. *olivacea*, f. *olivaceofasciata* and f. *lurida*, particularly in the earlier records, I have referred to these probably distinct forms as 'f. *olivacea*' or 'black eggars' in the present context.

Prior to 1954 we have little knowledge of the structure of the Rombalds Moor populations. Black eggars were, however, occasionally taken there, more particularly by breeding from larval samples, but the number of f. *olivacea* from these was frequently not recorded. In 1956, however, a sample of 519 larvae was collected there (Gordon Smith) and these produced 16 f. *olivacea* giving a frequency of just over 3 per cent (Table 14.8). I assembled a sample of 493 males in 1957 in a mark–release experiment and of these 4·7 per cent were of this form, making the expected frequency of the heterozygotes 34 per cent if the three genotypes be equally hardy, the highest recorded here (Table 14.6).

Mark–release–recapture experiments with the object of assessing the population size of ssp. *callunae* on Rombalds must be more inaccurate than in other species with which we have worked (for example *Biston betularia*, *Amathes glareosa*, and *Panaxia dominula*) because of the extent of the moorlands and the violent flight of the males leading to widespread dispersal. Nevertheless, Cadbury and I continued these experiments for four consecutive hatching-years and as the same errors must largely be common to all the experiments they must reflect the relative densities.

Table 14.6 lists these experiments, along with the percentage of f. *olivacea* and heterozygote frequencies estimated on the assumption of

TABLE 14.6
L. quercus ssp. callunae—*Rombalds Moor, Yorkshire.*
Phenotype frequencies of imagines.

| | | | | f. olivaceo-fasciata | | |
| | | f. olivacea | | observed | expected heterozygote per centage | Mean daily population ♂♂ |
Year	Sample	Number	Per cent			
1955	519	16	3·1 ± 0·8	?	(29·0)	unknown
1957	493	23	4·7 ± 1·0	?	(34·0)	661 ± 194
1959	463	6	1·3 ± 0·5	0	(20·2)	392 ± 259
1961	142	2	1·4 ± 1·0	3	(20·9)	151 ± 90
1963	313	3	1·0 ± 0·6	1	(17·7)	247 ± 91
1967	214	1	0·5 ± 0·5	0	(13·2)	unknown

1955 ex wild larvae (Gordon Smith, 1956)
1957 'assembled' male imagines (Kettlewell, 1959)
1959–67 'assembled' male imagines, (Kettlewell, Cadbury, and Lees, 1971, Cadbury, 1969).

equal viability. The table shows that after 1957 the numbers fell suddenly and this is born out by the size of each larval sample taken on each subsequent even year (Table 14.7). In 1956, 14,000 larvae had been collected by Collinson and 2000 by my assistant in only a few hours. At the same locality in 1958 after four days searching by two of us, only 134 were collected.

Table 14.6 also shows that during such a fall it is the f. *olivacea* component of the population that suffers most. Its frequency on Rombalds Moor dropped from 4·7 per cent in 1957 to less than 1·5 per cent in 1959. The frequency change of this form seems to be directly proportional to the size of the population.

It can also be seen that the heterozygote proportion of the population is less affected and that therefore in optimum seasons the homozygote recessive melanics can increase rapidly during a period of expansion. Though Rombalds is the best documented moor, many others in the district have been investigated, foremost among which are those around Penistone and the adjoining Thurlstone and Longsatt Moors which are situated at the

TABLE 14.7
L. quercus ssp. callunae—*Rombalds Moor, Yorkshire.*
Phenotype frequencies of wild-collected larvae.

Year	Observer	Sample	Number of melanic larvae ('black silky' in brackets)	Percentage of melanic larvae
1932–4	Collinson	280	0	
1936	Collinson	2000	0	
1938	Collinson	3000	13(1)	0·4
1940–44		No Samples		
1946	Collinson	1000	19 (2)	1·9
1948	Collinson	700	16 (1)	2·2
1950	Collinson	700	0	} 0·3
	Hewson	50	2	
1952	Collinson	700	16 (3)	2·3
1954	Collinson	6000	34 (4)	0·6
	S. G. Smith	681	48	7·1
1956	Collinson	14 000	189 (32)	1·3
	Kettlewell	2000	82	4·2
1958	Collinson	619	46	7·5
1960	Collinson	27	2	} 8·6
	Cadbury	8	1	
1962	Collinson	30	2 (1)	(6·3)
1964	Collinson	106	7 (2)	6·9
1966	Collinson	118	3	2·5

References: Hewson (1953), Gordon Smith (1956), Kettlewell (1959b), Cadbury (1960), (unpubl.), Cadbury and Lees (1966, unpubl.), Collinson (personal communication).

centre of the industrial triangle of Huddersfield, Sheffield, and Manchester (Hewson, 1953).

Here in the 1920s B. Morley reared a number of black eggars from 'blackish' larvae. These actual specimens are extant today in the Tolson Museum, Huddersfield. I have examined them and both f. *olivacea* and f. *olivaceo-fasciata* are represented.

More recently, in 1957 and 1959, in combined samples of 191 males assembled on Penistone by Cadbury and myself, none was scored as a black eggar though three were slightly darker than f. *typica*. Nor did we see any gulls during the hatching period in these two years, except half-a-dozen Herring Gulls, **Larus argentatus**, which appeared to be searching the heather. There has been a deterioration of these heather moors in the last 50 years due to pollution and uncontrolled burning and maybe the eggar population is not sufficient to attract other than minimal predation.

This has not occurred on other moors near Rombalds, however, as, for example, on an isolated heather tract near Keighley which lies six miles to the south-west. Collinson's records again give us information since 1956. Of the 145 males he has assembled there, between this date and 1964, 10·3 per cent were f. *olivacea* and a further 13·8 per cent f. *olivaceo-fasciata*. One quarter of the population here is, therefore, represented by the black forms of the eggar. Gull predation was not noted, nor at this period anticipated by him.

Selective predation

I began working on Yorkshire eggars in 1956 at a peak period for this species, in its larval phase.

In June 1957 the males were flying in large numbers and in this year I first witnessed gull predation. Flocks of Black-headed Gulls, *Larus ridibundus*, came up from the valleys, where previously they had been feeding on the refuse tips or following the plough, and ravaged the moors. Though they frequently attempted to catch flying males, they dropped on the resting females on which they gorged and much time was spent on filming these observations. Also, because of the difficulty in hatching laboratory-bred stock at the correct time, I was unable to carry out the extensive experiments I had p'anned for testing the degree of selectivity. On one occasion only did I attempt a mark–release experiment with females, and this was undertaken with the only two f. *olivacea* available. These along with 44 f. *typica* were distributed over 1000 square yards of heather where the gulls were feeding. After two hours, only one of the 46 releases remained and it was a f. *olivacea*. To us, this form appeared most inconspicuous when resting on the polluted heather. The 'disappearance distance' was 15 metres compared with 57 metres for f. *typica*, almost the same figure that we obtained in Caithness (Table 14.2). Here, also, Swifts (*Apus apus*) came up from the towns in numbers and attacked the males, and this also we were able to record on film. Wheatears, *Oënanthae oënanthae*, frequently caught the males in flight and on Penistone Moor I observed a Meadow Pipit, *Anthus pratensis*, take one: all these were the same species of birds which we were later to record as predators in Scotland; but there were others also (Plate 14.6).

In spite of the number of ramblers at the weekends on Rombalds Moor, there were occasional pairs of nesting Merlin, *Falco columborius*, which constantly hawked male eggars. From a 'feathering ground' I examined it seems likely that ssp. *callunae* constituted a main article of diet, when available. No f. *olivacea* were among the five dozen wings picked up, nor was this to be expected if the sample was random, as it was too small. The moor had a large population of Red Grouse, *Lagopus scotica*, Golden Plover, *Pluvialis ceplicarius*, and Curlew, *Numenius arquata*, and it is likely that all of these large birds contributed to the massacre of the eggars, but we have no evidence of this. Gulls are the main predators.

The hatching of the Oak Eggar, which may be limited to two to three weeks, reminded me of the swarming of termites in Africa. At such a time many predatory species moved in from afar and concentrated for a feast which appears so rarely and for so brief a period. In swarm years any cryptic advantage of a morph must, under such intense predation, increase its frequency in the population, maybe only temporarily, in spite of other pleiotropic disadvantages which may be present during some part of the life history. The predation pressure in swarm years is, however, entirely different from that which takes place on the more frequent occasions when the population is at a lower density. We have observed no predation whatever then; nor have flocks of Black-headed Gulls been present on the moors.

The control of population density. Larval and pupal mortality

Though selective visual predation is largely responsible for the changes in the frequency of the various morphs, the density of the species, according to our own observations, is decided by other factors.

In 1956 larvae were in plague proportions and this was followed by a sudden fall in their numbers. In fact, the following season, 1958, as already stated, only 134 were found in a week of searching. A number of dying larvae were observed on the moor during the 1956 season and they had the appearance of being infected by acute nuclear polyhedrosis. This was confirmed in the laboratory from our own samples from Rombalds Moor.

The following spring I visited that area and found 311 cocoons in the peat beneath dead heather which had been killed by the ravages of the larvae, the previous year. Only 43 (14 per cent) hatched into moths. Of the remaining 86 per cent, the cause of death was established in the majority of instances at post-mortem. 81 had been attacked by Magpies, *Pica pica* L. (predation by these birds had been noted during the previous winter months (Fletcher, personal communication), so that 26 per cent had been eliminated by visual predation.

It has been stated that melanic larvae produce black or dark cocoons and this is substantially true; nor is it surprising when one considers that their hairs are incorporated in the silk. Black cocoons are certainly more difficult to see on the surface of peat, but at the same time the pupae within must experience the disadvantage of heat absorption when in exposed positions.

What is not realized is that many brown larvae make dark cocoons also. Apart from the interwoven hairs, there must also be considerable variability in the colour of the silk itself. This is similar to cocoon colour in many other species, for example *Bombyx mori* where the colour is genetically determined (Tazima, 1964, Kikkawa, 1953, Yokoyama, 1959).

Of the remainder of the 1957 cocoon sample, approximately 60 per cent had died from other causes, and 17 from Ichneumonid parasites of which 4 contained living solitary Ichneumonidae undergoing a second year of diapause. (13 had already hatched). The remainder (170), over 55 per cent, came under the heading of 'disease', viral, fungal, or bacterial. It is my belief that these, rather than attacks by birds or parasites, controlled the density of ssp. *callunae* in 1956-7, the year of a population collapse. Disease however, is unlikely to affect all genotypes equally and recessive homozytgotes may be more at risk due to physiological disadvantages.

The complexity of the population control is emphasized by considering the part played by the insect parasites of this species. From the 1956-7 samples only 5 per cent of deaths fell into this category, yet five Ichneumonid and six Tachinid species were involved. Of the former, *Peltocarus dentatus* was the commonest. Yet within this one species a clear-cut polymorphism may have developed: one with a two-year life-history depending entirely on the larvae of ssp. *callunae* for host, and another, a one year race, with an alternative host, in particular *Saturnia pavonia* L., which occurs yearly. I understand, indeed, that there is some taxonomic evidence which suggests that subspeciation may already exist in *P. dentatus* following such a clear-cut isolation in time and host preference. Be it a polymorphic or a subspecies difference it does provide additional evidence that ssp. *callunae* may have existed in Britain for a long period of time, though not necessarily in its usual present-day habitat of open moorland.

P. dentatus, though having little effect on the population density of ssp. *callunae* may nevertheless, in my opinion, be responsible for its clear-cut two-year life-history. Most specimens of *P. dentatus* have been bred from cocoons of *Lasiocampa quercus*, in particular ssp. *callunae*, which, as has already been pointed out, rarely overlaps with the one-year life-cycle race. If ssp. *callunae* gets out of phase, as occasionally happens in a small number of individuals in nature (for example, Rombalds Moor 1946, Collinson, Caithness 1958, Rosie) and *P. dentatus* has a preference for parasitizing this species to others, such individuals would rapidly be eliminated.

Polyhedrosis is likely to be the major contributing factor in the control of numbers, particularly at peak periods when the larval density ensures that a highly virulent virus is spread by contagion: birds (together with physiological differences) are largely responsible for the polymorphism, and the parasites for the two-year life-cycle.

This is the hypothesis which 15 years' work on Rombalds Moor suggests. As to the interplay between one component and another, we have as yet little knowledge.

Larval melanism in Yorkshire. The frequencies

Unlike ssp. *callunae* in Caithness, where melanic larvae are not found,† in Yorkshire we have records of them from Rombalds Moor since 1932 (Table 14.7): these are the results of regular biennial sampling by A. Collinson, a naturalist who lives within a few miles of the moor. He has shown that there are at least two separate forms of melanic larvae here; a rarer and a darker one which he called 'silky black' and a 'chocolate form' which is variable (Plate 14.5). In the first three seasons he sampled (1932–6) none of either melanic form appeared in a sample of 2280. In 1938 from 3000 larvae he obtained 12 'chocolate' and one 'black silky'; the combined frequency was then 0·4 per cent. Collinson has also shown from his records that this has increased over 20 years, but irregularly, to 7·5 per cent ($n = 619$) in 1958, a year of abundance; nor has it dropped appreciably since then in the recent period of comparative scarcity of f. *olivacea*, being 6·9 per cent ± 2·5 in 1964. This is surprising because it contrasts with the deficiency of black eggar moths occurring in the population since 1958. It is even more remarkable because Collinson states that 'black silky' always gives rise to f. *olivacea*, that 18 per cent are bred from 'chocolate', but that they are also bred from typical light-brown larvae—about 2 per cent (Table 14.8).

In my opinion these apparently conflicting pieces of evidence can be reconciled by one hypothesis only. The gene-frequency of black eggars has had to be estimated from the male *olivacea/typica* ratio in the population. From the experimental evidence we have produced (including that of Caithness) it is likely that in swarm years, (particularly wet ones) when intense predation takes place, f. *olivacea* males survive in greater numbers than f. *typica*—which include the heterozygotes; and that the Hardy–Weinberg constant is distorted by this. More likely is it that the apparent dearth of black eggars since 1958 reflects the true gene-frequency in the population.

In a polymorphism which largely depends on crypsis, this is the price

†Though one extreme and quite different phenotype of a melanic larva was found there in 1972 (H.M.K.) (Plate 17.4).

TABLE 14.8
L. quercus ssp. callunae—*Rombalds Moor, Yorkshire.*
The relationship between melanic larvae and imagines.

Observer	Larval Colouration	Imaginal Colouration			Per cent f. olivacea
		f. typica	intermediate	f. olivacea	
Gordon Smith	(a) 'melanic'	26	—	7	21·2
	f. typica	477	—	9	1·9
Kettlewell	(b) 'melanic'	7	—	15	68·2
	f. typica	359	—	7	1·9
Collinson	(c) 'black silky'	0	0	16	100
	'chocolate'	194	0	42	17·6
	f. typica	1767	0	34	1·9
Cadbury and Lees	(d) 'chocolate'	1	2	3	—
	f. typica	41	1	0	—
Cadbury	(e) 'chocolate'	13	10	13	31·7
	f. typica	35	3	2	4·7

a, b, c, and d, are wild samples, e, from four broods. 'Melanic' in a and b, may have included 'black silky'. There is no significant difference in the proportion of f. olivacea from f. typica larvae in four sets of data. (Gordon Smith, 1956, Kettlewell, 1959b, Cadbury, 1969, others unpublished.)

paid by a recessive character. The melanic larval frequency may, in fact, bear a more realistic assessment of the imaginal gene-frequency than the moth samples themselves.

The genetic control of the larval colouration

Throughout the range of *Lasiocampa quercus*, and its ssp. *callunae*, various larval polymorphisms may occur. In southern Europe larvae of f. *meridionalis* Tutt are said to have the usual rufus hairs replaced by white ones. Along with these, there are normal larvae with reddish-brown, short hairs, which are 'partially dominant' to f. *meridionalis*. The imagines reared from both are similar (Bacot, 1901, Warburg, 1901, Tutt, 1902). Cadbury (unpubl.) cites Lempke and states that over the greater part of Holland *L. quercus* larvae are grey-brown, but a reddish phase which resembles the normal British form occurs at a high frequency in the three northern provinces. In a sample of 49 larvae from this region, only three were grey. Grey-brown is dominant to the reddish colour (Lempke, 1951).

I, myself, am worried about this statement because we have found that the degree of pigmentation in the larvae varies within individuals according to the instar and, in particular, the final one. Thus, for a period immediately following ecdysis, a larva may indeed appear to have light hairs. Such individuals may change to grey and finally rufus.

Melanic larvae behave somewhat similarly. In some broods it is possible to segregate the melanic individuals in the third instar: in others it is impossible even at the commencement of the last. For this reason it is essential to score all broods at full growth. We have found that all melanic larvae are always recessive to the normal. 'Black silky' on Rombalds Moor may be controlled by a gene or a super-gene which is responsible for an enzyme which confers melanism on both the larva and the imago and may give similar advantages to both if predators are present and the background favours the melanic forms of each. It is possible that this is the same mutant gene which at Brandenburg, East Germany, controls black larvae which are reputed always to produce black moths. Similarly all melanic larvae from the Wirral Peninsula have always metamorphosed into black eggars, and no crossing-over has been demonstrated.

The position of the 'chocolate' larva in Yorkshire is entirely different as reference to Table 14.8 shows. Collinson's 1958 sample of 2053 larvae contained 16 'black silky' all of which produced black eggars; of the 236 'chocolate', only 17·6 per cent ($n = 42$) of the imagines were of this form. The fact that my own sample from Rombalds Moor (Kettlewell, 1959) (22 'dark larvae' in 388) produced 68 per cent ($n = 15$) f. *olivacea* could be accounted for by a locally high frequency of 'black silky' which at that date I did not recognize as a separate entity. Also at this time Smith and Collinson included a deep reddish-brown form of the larva in the 'dark' category. These are not uncommon and have no connection with f. *olivacea*, and should be listed as non-melanic larvae. In this group, in spite of the variability of the normal larvae ($n = 2653$), each of the three wild samples produced 1·9 per cent of black eggars (=50 f. *olivacea*) (Table 14·8).

We can say that for 'chocolate', the gene controlling larval melanism is situated on the same chromosome as the one for f. *olivacea* and that crossing-over takes place frequently. Only recently have we been able to estimate the correct cross-over value. Lees (Kettlewell, Cadbury, and Lees, 1971) has found that in one brood it was 8·9 per cent, and in another, 25·6 per cent, when f. *olivaceo-fasciata* was classified with f. *olivacea* as I think it should be, but 10·1 per cent and 27·1 per cent, respectively, if included with f. *typica*.

The larval melanism is not as simple as this, however, and in wild samples and in some broods there are, occasionally, individuals of intermediate appearance and frequently these produce f. *olivacea*.

Larval crypsis and predation

The eggar larvae, quite apart from colour, are protected by having hairs which are highly irritant. These I believe to be limited to the shorter ones, the longer being largely of tactile use. Prior to pupation these shorter hairs are bitten off and incorporated in the silk of the cocoons which are also highly poisonous to touch and produce large urticarial blotches on contact. In spite of this the cocoons are opened and the pupae eaten in numbers by Magpies, *Pica pica*, during the winter months. I am unaware of the chemical nature of the toxin, but it is likely to be a substance allied to acetylcholine producing a local release of histamine.

Why, therefore, have *L. quercus* larvae not made use of aposematic colouration as indeed other related species have, particularly in their earlier instars when they are conspicuous, with black and yellow rings? If such chemical warfare had succeeded in the past, the world would have been filled with such insects which had found a final answer to survival. Highly irritant as these larval hairs are, beneath them lies a palatable insect with considerable food value. For this reason certain species of predators have evolved mechanisms for overcoming the irritation caused by the hairs and their toxins. Foremost among these is the European Cuckoo, *Cuculus canorus* L., but also the Pheasant, *Phasianus colchicus*, and various other birds, though to a lesser degree. Cuckoos, in fact, feed largely on hairy larvae during their short stay in Britain from April to early July when they tend to congregate for feeding in places where such larvae are at a high density. In the reed beds and ditches *Cosmotriche potatoria* is a favourite choice of diet. Cuckoos are common on the Yorkshire heather-moors at a time when the larvae of ssp. *callunae* are actively feeding in their third instar, and large numbers are eaten.

Here there may well be an advantage in being cryptic and I have found the melanic form of the larva on Rombalds Moor is much more difficult to recognize on its background than f. *typica*. I have on many occasions noted that a black larva, when placed alongside a yellow one, takes cover and hides beneath vegetation more rapidly than the yellow one. In Caithness, where Cuckoos are uncommon, it is unlikely that there have been similar selective pressures for larval crypsis, but see footnote on p. 251.

f. *olivacea* and f. *olivaceo-fasciata* in Yorkshire

The proportion of f. *olivaceo-fasciata* to that of f. *olivacea* occurring in the wild population nowhere bears the relationship of a heterozygote to a homozygote recessive. There would be a significant deficiency of f.

olivaceo-fasciata if this form was, in fact, the heterozytgote ($\chi^2_{(1)} = 56\cdot44$, $P < 0\cdot001$, Cadbury, unpubl.). Nor do breeding experiments from either form confirm such a hypothesis. More likely is it that different genes are involved or that f. *olivaceo-fasciata* expresses itself differently in the presence or absence of modifiers.

A further observation by Collinson is of interest in connection with this. He states that during the 40 years he has studied ssp. *callunae* on Rombalds Moor, f. *typica* females have, to some small but recognizable degree, become progressively darker. Such polygenic effects are comparable to those occasionally found in the typical form of industrial melanic species; for example, *Gonodontis bidentata* in central Birmingham, where the black f. *nigra* does not yet occur.

There is, in fact, no evidence so far that the heterozygotes of the black eggars have been evolving dominance during the last 90 years since they first appeared on the Yorkshire moors; nor is there, in fact, in Caithness, which has a far longer history.

(iv) *L. quercus* ssp. *quercus*: the Cheshire and Lancashire sand-hill polymorphism

For several reasons, black eggars under this heading are of particular interest. In the first place, the race in which they occur is a different one from that of the previous two polymorphisms. Secondly, though of local distribution, the Wirral Peninsula (Cheshire) population must for a long time have had little or no contact with that of Lancashire. This is because of the width of the estuary of the River Mersey and the extent of the City of Liverpool to the immediate north. Yet black eggars occur as a polymorphism in each locality.

To the south-west, the Wirral Peninsula is separated from the sand-hills of North Wales by the River Dee which has a broad estuary; black eggars do not occur there either.

A further difference is that, with one possible doubtful exception, all black eggars have been reared from dark larvae and none from typical brown ones.

Fifteen F_2 broods (heterozygous f. *olivacea* × heterozygous f. *olivacea* from Wallasey) produced 77 f. *olivacea*, all from melanic larvae, whilst 217 f. *typica* came from brown larvae. No crossing-over occurred as it did in the Yorkshire 'chocolate' form but the larvae are not as dark as 'black silky'. Here again there is linkage, and a single gene, or maybe a super-gene, is likely to be responsible for both larval and imaginal melanism.

Cuckoos were, until recently, common throughout the area and the larvae of ssp. *quercus* may be more vulnerable than those of ssp. *callunae* because they are fully grown earlier in the season—in May and June—at the height of this bird's abundance in this country.

The imaginal frequency of the black eggar is today low in both the Cheshire and the Lancashire communities. The total population in the Wirrall Peninsula has diminished in the last 30 years to a point of near extinction. This is due to the various indiscriminate activities of Man. In most places the species manages to exist, breeding in heavily polluted hawthorn hedges surrounding allotments or bordering lanes.

Cadbury recently assembled only 45 males in 13 days at Caldy and one of these was a black eggar. I can confirm his findings in regard to the population density here. In 1954 I spent five days beating and searching for larvae in May and June, both by night and by day but only obtained one larva from a hawthorn hedge. In a week of assembling at Wallasey golf course in 1955, I took five f. *typica* only. No early records exist of random sampling, the sole interest being in collecting the black eggar, but at that time this species was more common. The records show, however, that such forms regularly occurred there.

To the north of Liverpool the Lancashire sand dunes have been somewhat less affected by human interference, except by industrial pollution. Here the black eggar has been known to occur since 1880 (at Crosby). At this time (from 1873 to 1938), there was a colony of Black-headed Gulls which bred regularly on the sand dunes near Ainsdale, though we have no record of their numbers.

The total population of the eggar here is still high and 557 males were assembled by Cadbury in ten days in 1959 of which 7 ($1 \cdot 3 \pm 0 \cdot 5$ per cent) were f. *olivacea-fasciata* and 2 ($0 \cdot 4 \pm 0 \cdot 3$ per cent) f. *olivacea*. In 1968, Harper assembled 457 males at the same locality of which 1·6 per cent were of the black form; this does not differ from Cadbury's 1959 figures.

(v) A theory for a recessive melanic polymorphism

Recessive Melanism is always rare. That the frequency of such a melanic morph can be maintained in a population at a level great enough to constitute a polymorphism suggests a most unusual situation. Yet in *L. quercus*, in three separate areas in Britain, this has arisen independently, and also in several other regions in Europe.

Though the phenotypes appear similar in each locality, we have obtained direct evidence in two instances that they are controlled by separate genes; in the third (Cheshire–Lancashire) the melanic polymorphism occurs in the one-year life-history ssp. *quercus*. Yet here again we have shown recently, but indirectly, that the gene responsible for f. *olivacea* is different from those controlling the two other polymorphisms. In 1968 Harper paired females, heterozygous for f. *olivacea* from Rombalds, to wild f. *olivacea* males from Lancashire. As the Yorkshire females were from f. *typica* larvae, themselves heterozygous for 'chocolate', and as we have accepted that all Lancashire f. *olivacea* come from melanic larvae, 50 per cent of larvae and 50 per cent of imagines could be expected to be melanic if the same genes were involved: neither black larvae nor black moths appeared in the offspring. We can therefore state that these forms, though phenotypically similar in each of the three polymorphisms, are genetically distinct.

The ecologies in the three localities bear little relationship to one another, though there are elements common to all.

Because of the clearly defined clines it is likely that the oldest polymorphism is that in Caithness. It is here, therefore, that one would anticipate evidence of a tendency to evolve dominance. Yet there is none.

In this respect, *L. quercus* (and possibly a few other species, for instance *Meristis trigrammica* f. *obscura* in Britain and *Endromis versicolora* f. *lapponica* in north Sweden (Newman, 1943) must be considered in a different category from the majority of species which have melanic polymorphisms. It is important, therefore, to try to assess their age and also the previous history of *L. quercus* in each locality.

Caithness

It is my belief that, as is the case for Industrial Melanism today, Man is indirectly responsible for the melanism of *L. quercus*—present-day Man for the industrial forms and earlier Man for the polymorphism found in the eggar in Caithness. In the last three to four thousand years he has produced a hitherto unknown biotope—moorland—throughout Caithness and also elsewhere. Previously, and ample evidence remains to be seen today in the peat-hags, the country there was covered with Hazel scrub, and also Birch and Aspen; moreover pine forests existed there towards the Sutherland border. We have a fair picture of this earlier biotope in the few remaining Caledonian Pine forests remaining today, such as the Black Wood of Rannoch, and here heather and *Vaccinium* grow between naturally regenerated trees, which may be widely separated.

L. quercus extended its range to such ecologies long ago when it took to feeding on heather, as it does today in the Black Wood where I have found the larvae. The change of diet to heather from deciduous leaves, along with other conditions imposed by its new habitat, necessitated a two-year life-history.

The earliest known traces of Man in Scotland are *c*. 4000 B.P. and traces of Neolithic farms are recorded from Orkney and Shetland early in the second millennium B.P., when flocks or herds of sheep and goats were introduced. No attempt at the cultivation of barley was undertaken then, nor until the Beaker people from central Europe arrived there *c*. 1800 B.P., and the plough was not introduced until a much later date.

At this period there must also have been a clearance of scrub and trees in some valleys and replacement by grasslands.

Evidence of the Bronze Age shellfish-eating people has been found at Keiss *c*. 1000 B.C. Also the highest known concentration of Brochs in Britain exists around Wick, eight miles distant.

The overall picture we get is that considerable activity took place in Caithness two to three thousand years ago and that timber was cleared and regeneration was limited because of grazing by the introduced herds. This has continued to the present time with an exacerbation between the ninth and thirteenth centuries A.D. due to Norse settlers, and again in the middle of the nineteenth century when much 'waste land' was reclaimed for farming.

This change in the countryside led to other effects—carnivores largely disappeared and probably for the first time the Common Gull, *L. canus*, was able to nest on the mainland in comparative safety. This was the earliest occasion when eggars and gulls came face to face, each having extended its range for different reasons on to moorlands similar to those found there today. More recently, because of the emphasis on grouse-moors, controlled burning of the heather took place annually. Bracken, growing amongst heather, disappeared and, as few Birch trees grew there either, the only background was one of olive-green heather amongst peat. This is the position today throughout Caithness; nor does it vary until one reaches the mountains to the south and to the west.

The black eggar may, then, have had cryptic advantage in some places, locally, for at least 2000 years. These places would, at first, be in those areas where sheep or goats grazed, which would never be far from Brochs because of the need for protection. The main centre would have been around Wick, as judged from the distribution map of Brochs at that time.

If regular burning, however, is necessary to give the black eggar the the cryptic advantages it enjoys today, the recessive polymorphism found here today must be of more recent origin.

Yorkshire

A history of the Yorkshire population is more definite. Though here again open moorland came into existence at the time of the Iron Age when trees were felled for fuel, we have direct evidence that the melanic morphs were rarities at the end of the last century. The increase since 1930 coincides with the appearance of at least six nesting colonies of Black-headed Gulls, *L. ridibundus*, which previously did not exist here. Nor has the Common Gull, *L. canus*, any place in maintaining this polymorphism. It is likely, therefore, to be under 100 years old and its origin thus coincides with the period of heavy pollution of the moorlands which has been taking place during this time. It is impossible to assess the relative importance of the three contributing factors—air pollution, the regular burning of the moors, or the arrival of the gulls.

Cheshire–Lancashire

The Cheshire–Lancashire eggars are completely undocumented prior to the beginning of this century. Nor have we a record of their frequency then at Wallasey or elsewhere for any one year. We do know, however, that the whole coastal belt along which this species is found has been heavily polluted for the last one hundred years and that the *Salix repens*, on the stems of which the larvae and imagines sit, are blackened. I have regularly observed large numbers of gulls resting on the sand dunes in recent years, chiefly the Herring Gull but also Black-headed, though an earlier breeding colony which existed near Ainsdale disappeared in 1938. I have seen no sign of gull predation here, nor would one expect it today when the eggar is at so low a density. In my opinion the Cheshire–Lancashire polymorphism is the most recent of the three and it has been brought about by the indirect effects of air pollution. Black eggars are not to be found (except as very rare mutants) on similar but unpolluted sand hills to the south-west (for example Anglesey) where f. *typica* is common.

We can make an intelligent guess that the Yorkshire and Lancashire polymorphisms are not less than 100 years old but that in Caithness the polymorphism may be 3000 years old, though it is unlikely to be older. Yet nowhere is there evidence of dominance modification. This is in contrast to the majority of cryptic species affected by industrialization where polymorphisms have arisen with melanic forms showing full dominance at inception.

I think that the most likely explanation is that we are witnessing in *L. quercus* the first occasion on which melanic forms have given such overwhelming cryptic advantage to a species. The gene-complex was unprepared for such a situation and maintained no dominance-modifiers but depended on mutation for their presence and this must be an extremely slow process. In the majority of cryptic species this occurred many hundreds of thousands of years ago and the gene-complex has maintained such palaeogenes ensuring full heterozygote expression at each mutation.

One further point must be considered. The black eggar polymorphisms occur in three separated districts, each associated with gulls—without doubt the main predators—yet *L. quercus* f. *typica* occurs throughout Britain in most suitable places. The males fly violently and if f. *typica* is more efficient in flight than f. *olivacea* it means that the areas in which the polymorphisms exist are constantly receiving genes from f. *typica* by migration from outside.

In such a situation, dominance of the melanic forms may be unachievable.

I have dwelt at considerable length on Recessive Melanism and in particular on recessive melanic polymorphisms because I believe that a study of such exceptions to a rule can throw light on more usual situations: in this particular context, on industrial and ancient melanic polymorphisms.

Part VI. Miscellaneous melanisms

15. Miscellaneous melanisms in moths and butterflies

Introduction

WE have so far considered melanisms in the Heterocera only and these have fallen under three separate headings, Industrial Melanism, Ancient or Relict Melanism, and Recessive Melanism. For the most part, such melanisms have referred to morphs, or, less frequently, races. Yet, by our definition of melanism in the Lepidoptera, we must include other examples, not so far mentioned. These are, first, *Aposematic Melanism*, a distinct entity, secondly *Melanism in the Rhopalocera*, and thirdly those coming under the heading of *Exceptional Melanism* (Chapter 16). I want to draw attention to a number of unusual, but highly interesting situations occurring amongst moths which do not conform and fail to fall into the previous classification. Because of the significance of Industrial Melanism, too many biologists have come to think of melanism in the Lepidoptera as being limited to industrial polymorphisms maintaining dark individuals.

Both the usage of the term 'melanism' and its distribution in Lepidoptera must be recognized as being wider than this. I give some examples only under each of the three headings and they are in no way intended to be comprehensive.

(i) Aposematic Melanism in the Heterocera

Extreme blackness and complete whiteness are under certain circumstances recognized as warning deterrents and serve to remind a predator of a previous unpleasant encounter. This does not necessarily imply the use of specific repellent chemicals, but rather a degree of unpalatability.

Earlier in this book I have put forward the suggestion that black and white must have established a contribution to aposematic defence before the recognition by predators of red and yellow as colour signals. Whilst these are today more frequently met, yet black and white are still symbols of distastefulness.

In the Lepidoptera (and also the Trichoptera and other orders) this is often enhanced by specialized day-time behaviour, such as gregarious flight.

Two good examples of this, occurring in Lepidoptera in Britain, are found in *Atolmis rubricollis* L. (Arctiidae) and *Odezia atrata* L. (Brephidae). The Red-necked Footman is jet black with an orange tail and thin crimson collar. It flies in numbers in the afternoon sun, usually around the top of a particular Oak tree. I have not yet tested this moth by presenting it to birds, but I have tasted a squashed one and it was bitter. *O. atrata*, the Chimney Sweep, is another black moth, but with a white silhouette around the tips of the forewings. This frail insect flies gregariously in the sunshine in local untilled grasslands where the food plants, *Conopodium* or *Chaerophyllum* species, exist. I have given this species to a wild Robin, *Erithacus rubecula*. The first was pecked and mutilated, but not eaten. No interest was shown in subsequent offerings.

Gregarious day-flying Lepidoptera with black colouration are commoner, I think, in the tropics; for example, the Syntomiidae of South America. Little work has been done on their behaviour or toxicology.

(ii) Melanism in the Rhopalocera

Melanism amongst butterflies bears little relationship to that in the majority of moths; nor is this surprising when one considers the fundamental differences in habits. Yet some of the melanism is comparable.

Thermal Melanism is represented by several high-mountain species (for example, *Erebia* see Chapter 10 (**vii**)). It is also found in high-latitude species. A good example is the Fritillary, *Argynnis improba* ssp. *improbula*, which occurs in northern Sweden at a latitude of 69°N and at a height of around 3000 feet. This butterfly is heavily pigmented with black, and the normal dark patterning is greatly suffused (G. Howard personal communication). I shall refer to other species in this family which have melanic forms in Britain later in this chapter. In the same northern localities as ssp. *improbula* there is found a strange monomorphic Blue Butterfly (Lycaenidae) (G. Howard, personal communication). It will be appreciated that species in this family are normally sexually distinct, the male having bright blue colouration. In *Lycaena glandon* both sexes are of a muddy brown-blue colour and both sexes take flight in the brief periods of fitful sunshine.

Cryptic Melanism in butterflies is less frequent (except in Batesian mimics). There is, however, in Britain one species which maintains a dark morph limited to the female sex, which undoubtedly makes it less conspicuous when in flight in shade and shadow. This is *Argynnis paphia* L., the Silver-washed Fritillary, and its dark f. *valesina* which is discussed later in this chapter.

Aposematic Melanism occurs in several genera of Rhopalocera (for example *Battus*) and confers protection when they are conspicuous in flight or when feeding.

Of great interest is the North American Black Swallowtail complex recently studied by J. van Z. Brower (1958b, 1960) and L. P. Brower (1959a, 1959b) in the field and whose genetics have been elucidated by Clarke (1971) and Clarke and Sheppard (1955, 1957, 1962a).

Battus philenor L., the Pipe-vine Swallowtail (Plate 15.1, left 1), is a highly unpalatable species occurring in the eastern United States; its larva feeds on the toxic plant *Aristolochia*. Both sexes have black colouration and J. van Z. Brower has shown that this melanism is aposematic. She found that captive Florida Scrub Jays, *Cyanositta coerulescens*, refused to touch this species after a single encounter. *Battus* is the universal model to several edible species in widely separated genera which mimic it. All have forms with black colouration which are frequently limited to the female sex. In certain localities as many as four different species fly together with the model, and being black gain protection from their resemblance. These then, are the so-called Batesian mimics. Three of the commoner ones are *Limenitis arthemis astyanax* Fab. (the Red-spotted Purple), the female of a Fritillary, *Speyeria diana* Cram., and the female of *Papilio polyxenes asterius* Fab., the Parsnip Swallowtail (Plate 15.1, right 1). Though both sexes are black, but with different patterning, the female only is considered mimetic. More interesting is the black morph of *P. glaucus*, the Tiger Swallowtail (Plate 15.1, right 2 and 3), which occurs as a polymorphism amongst the females of the species in the south-eastern districts of the United States only, but at different frequencies.

J. van Z. and L. Brower sampled 3387 black swallowtails from four localities in the south-eastern United States. They showed that the highest frequencies of black *P. glaucus* coincide with areas where *Battus philenor* was most abundant. In other places where the model does not occur, such as Canada, the *P. glaucus* population is monomorphic yellow.

The situation must be more complicated than this, however, for Batesian mimicry demands that the population of the model must usually be in excess of that of the edible mimics, the proportion varying according to the number of predators and the degree of unpleasantness experienced on eating a model. An excess of models is not always necessary for all Batesian model–mimic balances, however. The highly original experiments of J. van Z. Brower (1960) demonstrate this admirably. She used meal worms which had been coloured according to whether they had or had not been treated with a solution (66 per cent) of quinine dihydrochlorine, which is bitter. In 160 trials both treated and control meal worms were presented to nine uninitiated Starlings (*Sturnus vulgaris*) in differing orders and proportions. The experiments demonstrated that if the experience of eating a highly distasteful insect was sufficiently unpleasant, protection was given to other species which were mimicking it even though such mimics were edible and outnumbered the models. From these experiments it can be argued that the more toxic a species is, the more important is it for the mimic to copy the model perfectly in order to attain full protection. Because several palatable species are involved, each edible, and each superficially of similar melanic appearance, such Batesian mimetic species must demand considerable inter-relationship one with another.

J. van Z. Brower showed (1955a, 1958b) that Scrub Jays which had not previously experienced *Battus* would readily eat the Batesian mimics. Furthermore, she demonstrated that after a single encounter with *Battus*, certain individual Jays were able to recognize a difference between some mimetic species and the model; others left all black swallowtails severely alone. The 'I.Q.' and experience of each Jay must vary (L. P. Brower, Alcock, and J. van Z. Brower, 1971).

But selective predation, depending entirely on the distribution of the model *Battus*, could not alone maintain the dimorphism of female *P. glaucus*. Burns (1966, 1967) has claimed that preferential mating takes place in which the yellow males choose yellow rather than black females and has suggested that this contributes to the situation. He states that 'reproductive advantage (of the light form) and mimetic advantage (of the dark form) are opposed forces that may sustain female dimorphism'.

Prout (1967) pointed out that two fitness-components alone could not maintain a balanced polymorphism, nor was this implied by Burns. The mating advantage of yellow *P. glaucus* is in some ways similar to that of *Argynnis paphia* (see below), in which f. *valesina*, the dark female, has a cryptic advantage on the wing but fails in direct competition to attract the male first (Magnus, 1958).

The size of the *Battus* population, the number of Batesian species occurring locally, the number and the individual intelligences of predators, and selective mating must each be a component which contributes to the mimicry-complex.

The genetics of melanic mimetic *Papilio* species has been investigated by the large-scale breeding experiment of Clarke and Sheppard (1962a, 1962b). In this, using a hand-mating technique (Clarke and Sheppard, 1956b), they were able to procure hybrid individuals between melanic and non-melanic species and forms. By these means they showed that in the F_2 hybrids of *P. polyxenes* × *P. machaon* f. *typica* (Plate 15.1, left 3) of European origin (which are yellow in both sexes) the melanism is unifactorial and dominant to typical yellow and is controlled by a pair of autosomal genes. In addition to this they showed that the sexual dimorphism of the black female is also unifactorial and dominant, and finally that though this sexual difference cannot be detected in the yellow ground-colour of *P. machaon* and *P. zelicaon* yet they carry the gene for it.

This is in contrast to the extremely rare melanic *P. machaon* ab. *nigra* Reutti (Plate 15.1, left 2), the British Black Swallowtail, which occurs in both sexes and has been bred or taken in the wild near Horning, Norfolk on rare occasions. Here the melanic is recessive, and occurs in both sexes (Cockayne and Newman, 1931).

More complicated is the inheritance of the melanic female of *P. glaucus* which appears not to conform to any known genetic principle. Though a clear-cut switch-mechanism exists which provides either black or yellow females, it has so far eluded satisfactory analysis even though Clarke and Sheppard bred 70 broods. From these it is clear that melanic female *P. glaucus* always have more black offspring than yellow females, and vice versa. The gene controlling melanism is probably situated on the Y-chromosome. Yet normal female inheritance, be it carried by this gene or in the cytoplasm, is not followed. Also, and of importance, the Browers have shown that the frequency of black *P. glaucus* is to a large degree dependent on the numbers of the model *Battus philenor* at any particular locality. This situation can only be brought about by intensive natural selection which suggests that selective predation is largely responsible. (Latitude, with the attending changes in temperature and humidity, is ruled out.) There must, I think, be another and quite separate gene, possibly autosomal, which effects the position and expression of the one on the Y-chromosome; it is this gene which must be subject to differential selection.

In North America then, we have a situation in which a number of melanic Batesian *Papilio* spp. copy the distasteful *Battus* which is a sufficient deterrent to give satisfactory protection. These various black mimics maintain their melanism, either in both sexes or as a polymorphism in the female, by various genetic switch-mechanisms; none is recessive to yellow.

By contrast, the only European black swallowtail is excessively rare and it is recessive. It occurs in both sexes. Here there are no appropriate models. Sheppard has suggested (personal communication) that it is tempting to compare this situation with the case of the rare recessive melanics in moths and the dominant melanisms which are so successful when a dark background is present, and that the North American *Papilio* spp. (and other species) have been able to make use of melanism due to the presence of *Battus* and have become dominant, whilst the European *P. machaon* has no such aposematic model and has had no such opportunity, though such a gene is present at a low frequency in the population.

Such occasional melanism, which is probably also recessive, has been recorded in other species: for example *Melanargia galathea* L., the Marbled White, and *Pieris rapae*, the Small White Butterfly, both in Europe and North America. A gauge of the rarity of such specimens can be appreciated from the fact that the only known melanic of the first-named species obtained the highest price on record for a British butterfly, £100 at auction over twenty years ago. It is now in the National Museum of Wales.

Polymorphism in butterflies is common but in Britain not one which is maintained largely by cryptic advantages and disadvantages. Yet this appears to be the case in *Argynnis paphia* L. (Nymphalidae). Form *typica* is normally yellow-brown in both sexes and the species is found in most suitable open woodlands throughout southern England. The melanic form, *valesina* (the Greenish Silver-Washed Fritillary), occurs in the New Forest, Hampshire where it occupies 10 to 15 per cent of the female population, but it has a considerably lower frequency elsewhere in England.

Goldschmidt and Fischer (1922) bred this species on a large scale and demonstrated that this form is controlled unifactorially by an autosomal gene which is sex-limited. The form *valesina* is dominant to f. *typica*, and is carried by the male (in which it does not express itself) as well as by the female.

Combined back-cross broods from f. *valesina* females produced equal numbers of each form (n of females $= 710$). Yet back-crosses from known heterozygous males gave a significant deficiency of f. *valesina* females. The story is, therefore, more complicated than the earlier interpretation that f. *valesina* was linked with a recessive lethal (Ford, 1965).

It is difficult to test the cryptic advantages of the two female forms in flight, nor am I able to suggest an experimental approach. 'Collectors' in the New Forest, however, where the species was, until recently, common, will state that f. *valesina* is much more difficult to see and follow within a wood than f. *typica*, and I, myself, subscribe to this. They also say that f. *valesina* frequent woodland shade rather than open rides between trees.

Magnus (1958) has worked on the mating behaviour of this species. He used a mechanical rotator comparable to that employed for the electric hare in greyhound racing. Living females, paper models, and super-models (i.e. models larger than life-size) were tested. He showed that the males reacted to brown super-models and normal brown females in that order, but not to the dark green form. In fact, it is likely that copulation with such forms only takes place fortuitously when they are in juxtaposition, as for instance, when both sexes are feeding on bramble flowers. This could be tested by dissecting wild samples for spermatophores as demonstrated by Burns (1966, 1967) in *Papilio glaucus*. If this were true, it would imply that f. *valesina* could only survive in areas of high density population, and my own observations confirm this. However, it is stated that in the Far East (southern China) one hundred per cent of the female population is f. *valesina*. A comprehensive account of the biology of this polymorphism is given by Ford (1971).

This species serves to demonstrate an advantage and a disadvantage in a butterfly both of which help to maintain two morphs in its population at varying frequencies.

Argynnis paphia has, however, more to contribute to melanism, though in this instance we have no idea of the selective advantages and little of the genetic inheritance. There is a f. *melaena* D'Aldin in which all four wings are clouded with black markings. It occurs in both sexes and in both forms of the female, but of particular interest is it that f. *melaena* occurs on certain years only. It is tempting to put forward a theory that environmental effects at a critical period are responsible, and this may in part be true. We know, however, that the late L. W. Newman had a brood of *A. paphia* from a normal female from the New Forest in which f. *melaena* segregated in a 1:3 ratio (personal communication). More likely is it that this form is a product of both inheritance and environment; by which I mean that a recessive character, controlled genetically, is only able to manifest itself under some environmental stimulus imposed at a critical period of larval or pupal life (see Chapter 18).

Other Rhopalocera species behave similarly, I suspect—for example f. *nigrina* and f. *seminigrina* of *Limenitis camilla* L. (Ford. 1957. Plate xxiii) and probably many Vanessids. Such forms can today have little survival value though they may have had in the past under different environmental conditions. They certainly demand more thorough study, and urgently, before they disappear together with our other woodland butterflies.

The *Argynnis* spp. also demonstrate Geographic Melanism, though I would hesitate to interpret the causes, except to suggest that there may be thermal advantages. The females of *A. aglaia* L., throughout the range of the species, are brown with small, black markings. In its northernmost location in Britain, the western isles of Scotland, and occasionally elsewhere, these markings are extended and specimens heavily suffused with black occur. Such dark individuals may gain advantage in certain years only, when cloud prevails during late July and early August, for then this butterfly must take advantage of every gleam of sunshine.

Species in other genera may show similar darkening at high latitude and altitude. Thus *P. napi* L. the Green-veined White has increased black markings in Caithness, Scotland, where the females may be dusky (and also in Ireland). A similar form (a subspecies), which is even more dusky, occurs in the Alps and the Carpathians, where it is referred to as ssp. *bryoniae* Hueb. (Bowden, 1962), and I have recently seen specimens collected by G. Howard from Lapland in which the females are smoky-brown but the males exceptionally white. Studies of the genetics of such melanic races and their hybrids have been undertaken by Bowden (1962, 1966) and Lorkovic (1962).

Melanism in the Rhopalocera, though having somewhat different uses from that in the Heterocera, can therefore be classified under similar headings: cryptic (in flight), aposematic, geographical, and thermal. Industrial Melanism plays no part whatsoever, nor should we expect it to do so.

16. Exceptional melanism in the Heterocera

WHENEVER one finds a species or mutant form which, in one way or another, does not conform to others in the same genus, it must become a subject of intense interest and therefore be studied. Such exceptions in the melanism of the Lepidoptera can be classified loosely under four headings, two of which constitute the next chapter.

(i) Incomplete dominance with distinct recognizable heterozygotes occurring in cryptic species.
Polia nebulosa Hufn.

(ii) Incomplete dominance occurring in aposematic species.
Spilosoma lutea Hufn. f. *zatima* Stoll.
Arctia caja L. f. *brunnescens* Stattermeyer, f. *fumosa* Hörhammer
Panaxia dominula L. ssp. *persona* Hübn., f. *italica* Standfuss, f. *nigradonna* Kettlewell
Panaxia dominula L. f. *medionigra* Cockayne, f. *bimacula* Cockayne
Aglia tau L.

(i) Incomplete dominance in cryptic species

Only rarely are the heterozygotes of melanic forms quite distinct from either homozygote, when usually they have been given separate aberrational names. It frequently happens, however, that certain individuals only can be recognized as being heterozygotes, because of their less extreme melanism. This happens even in those species which come under the heading of Ancient or Relict Melanism. Here, in theory, sufficient time has elapsed for full dominance to have been achieved if the selective forces demanded it. Many species such as *Triphaena comes* f. *curtisii* (Bacot, 1905*a*, 1905*b*) in Scotland or *Amathes glareosa* in Shetland have a range of dark individuals, all of which are quite distinct from f. *typica*, but which are not always recognizable in a 1:2:1 ratio (Ford, 1955*c*).

In Shetland, we have investigated *Amathes glareosa* intensively. On Unst, the northernmost island, 97 per cent of the population are f. *edda* and here we can therefore anticipate from the Hardy–Weinberg Law, assuming equal viabilities, a heterozygote frequency of 29 per cent. Yet in a sample of 2539, we were unable to score 12 per cent (=300) as 'light f. *edda*'. By contrast, in South Shetland, Dunrossness, where f. *edda* is maintained at a frequency of 2·5 per cent in the local population, the majority here were lighter than any found on Unst (Table 11.3). This is similar to a small sample I had from the Orkneys (R. I. Lorimer) the only other place in the British Isles where f. *edda* occurs. Here, though the frequency is 15 per cent, all the individuals have been classified as 'light'.†

There can be several interpretations of this. The gene controlling f. *edda* in Orkney and South Shetland may be a different one from that on Unst though this is unlikely; genes for dominance modification may not have been selected in the more southern localities where f. *edda* occurs, there may be modifying genes which increase the blackness of heterozygote f. *edda* present in Unst, northern Shetland (Kettlewell and Berry, 1969).

Under the present heading of Incomplete Dominance, I do not refer, however, to such situations. Rarely in cryptic species the heterozygote is quite distinct from either homozygote. This occurs in *Polia nebulosa* (Plate 16.1), which is found in Delamere Forest in Cheshire, and its history is an enlightening one. I refer to it in this chapter under the heading of 'Exceptional Melanism' because of its rarity, though it could equally well have appeared under 'Industrial Melanism'. The imago depends for survival on camouflage when at rest by day on tree trunks. Over the width of Britain there is a discontinuous cline from Northern Ireland and western Scotland where it is whitish to eastern Britain where it is grey. Delamere Forest, however, has been heavily polluted from the nearby cities of Manchester and Liverpool for over 100 years. Here, earlier in the century, jet-black specimens were bred in small numbers from wild larvae, collected in the spring, but such individuals had a snow-white fringe silhouetting their outline. This is the homozygote, f. *thompsoni*. Another melanic and a much commoner one was of an intermediate form, being grey without white fringes and this was shown to be the heterozygote, f. *robsoni*. The homozygote therefore, whilst possibly having an advantage in blackness was conspicuous, because of its fringes. More recent sampling by Ashworth and myself in which we collected over 200 larvae produced no f. *thompsoni* nor f. *robsoni*, nor f. *typica*. The whole population had been replaced by another melanic f. *plumbosa*, similar to the heterozygote f. *robsoni* but recognizably different from it by the uniform colouration of its scapulae. The F_1 generation from these produced no f. *thompsoni*. This then is another example in which one melanic form replaces another. It is dis-

† As forecast we have this year (August 1972) recorded f. *edda* for the first time on the British mainland. C. Gibson and I sampled *A. glareosa* at three localities in northern Caithness and 44·4 per cent were f. *edda* (n = 45). Here the majority were of a light silvery-brown colour, but a few were darker' (Kettlewell and Gibson, in press).

cussed under the heading of the natural history of Industrial Melanism in Chapter 19.

Polia nebulosa f. *robsoni*, a recognizable heterozygote of an industrial melanic, is a rare example of incomplete dominance in a cryptic species. That it should now be extinct (or extremely rare) and replaced by another melanic form which may be under multifactorial control is highly significant. It suggests that the homozygous melanic, f. *thompsoni*, may have recently been at a disadvantage, possibly because of the white fringes, and the heterozygous f. *robsoni* because of the white scapulae.

(ii) Incomplete dominance in aposematic species

Introduction

I have already pointed out in the previous chapter the difficulties imposed on an aposematic species becoming cryptic. In situations in which the chemical repellent has proved to be ineffective to a particular predator, however, and the warning colouration therefore of no consequence, this can take place *locally*. I give one example of this, *Spilosoma lutea* Hufn. f. *zatima* Stoll. Because such a melanic cryptic form occurs peripherally in each of its known localities, it must constantly be receiving genes by migration from the main gene-pool which favours warning colouration. Under such conditions of semi-isolation full dominance would be unlikely to have been achieved.

More difficult is it to understand the partial dominance of melanism in two species of aposematic Arctiidae though one of these is the best documented polymorphism in Lepidoptera in the world; namely *Panaxia dominula* and its *medionigra* and *bimacula* forms. The frequency of these has been recorded each year for over 30 years by E. B. Ford (1971). I give, in some detail, the work I have done on the homozygote f. *bimacula*. The second species which supports a form with incomplete dominance and to which I shall refer is *Arctia caja* L., the Garden Tiger, and its melanic f. *fumosa*.

Even more difficult to account for, is the melanism which exists in *Aglia tau* which I discuss; but of this species I have no personal experience in the field, and I quote the breeding experiments on its various so-called 'melanic forms'.

Spilosoma lutea Hufn. f. *zatima* Stoll (Plate 16.2)

In Britain there are three species of *Spilosoma* (Arctiidae) which hatch around midsummer and are distinguished by having white or buff aposematic colouration, *S. lutea* Hufn., *S. lubricipeda* L., and *S. urticae* L.

In certain places *S. lutea* maintains a melanic polymorphism (f. *zatima* Stoll.) though probably at a low frequency. *S. lubricipeda* in northern Britain and also in Ireland, has to a varying degree the snow-white of the forewings replaced by brown, in a high proportion of the population. A similar or identical form occurs as an extremely rare mutant in London. I have taken two in 6000 (approximately) near Oxford. I am unable to indicate the method of inheritance, or the advantages accrued from such a colour change. It is a general rule, however, that many northern and western cryptic species, both in Britain and Scandinavia, maintain melanic forms or polymorphisms which are darker than their southern confrères. That this should extend to such aposematic moths as *S. lubricipeda* and also *Phragmatobia fuliginosa* L. f. *borealis* is a fact which I cannot explain. The third species of *Spilosoma* in Britain, *S. urticae*, which is confined to water meadows locally in the southern half of England, has, as far I know, no such forms.

S. lutea f. *zatima* is clearly distinct from f. *typica* and, except for a buff central area, the forewings are rayed with black. This form is controlled by a single gene which has incomplete dominance; and it offers an alternative method of survival—by crypsis. The homozygous f. *zatima* are more extreme than the heterozygyotes and they range imperceptibly to f. *deschangei* which usually represents the homozygote. Federley (1920, 1923) thought that three dominant modifiers effected f. *zatima*; the darkest, with black fringes, he designated ZZ AA BB CC. Goldschmidt (1924) obtained his f. *zatima* from a different source and he came to the conclusion that they might be controlled by a different allelomorphic series.

The form *zatima* is rare in Britain and is probably limited to the coastal sand hills of east Yorkshire and Lincolnshire. Yet a more recent spread into industrial areas is possible, for example into Barnsley, Yorkshire, and Ottershaw, Surrey (recorded by R. Bretherton, in 1962) but this may be another and separate mutant. It has been referred to as ab. *totinigra*.

Seitz (1906–32) states that on the Continent f. *zatima* is confined to the Channel coast and the islands off Germany, in particular Heligoland. In each locality the biotope is sand-hill. I understand that more recently it has been found regularly in such localities on the Yorkshire and Lincolnshire coastline, where the moth rests by day on fallen Elder (*Sambucus*) on the dead boughs of which it is highly cryptic and on whose foliage the larvae feed. It is likely that gulls once again are responsible for such a polymorphism, though I have no evidence of this.

Though *S. lutea* is unpalatable it is not as toxic as *S. lubricipeda*, which it resembles and which usually hatches at a slightly earlier date. Because of this, *S. lutea* may gain some degree of protection (Rothschild, 1963).

Form *zatima* provides an example of a rare phenomenon; the forfeiting of an aposematic colouration and its substitution by a cryptic melanic form having near-dominance. It is comparable to the black forms of the Two-spot Ladybird, *Adalia bipunctata* (Coleoptera), which Creed (1963, 1966) and Lusis (1961) have shown to be dominant to the typical red form, though Creed considers that other factors more important than colouration maintain this polymorphism.

Arctia caja f. *brunnescens* Stattermayer., f. *fumosa* Hörhammer (frontispiece)

It is difficult to write about this species without first referring to the present methods of establishing nomenclature in aberrational forms of Lepidoptera. This same statement holds good for *Abraxas grossulariata* and other aposematic species which show great variability. On the one hand the indiscriminate naming of all abnormal specimens, be they of genetic or environmental origin, a combination of two separate forms, or varying heterozygous expression, has led to a state of the greatest confusion. On the other hand, similar phenotypes, but of different genetic origin, have the same name.

I have already pointed out that the amount of variability which occurs among aposematic Lepidoptera in nature is surprising (chapter 13); it is likely that these aberrational forms will be increasingly studied: not with the idea of discovering more variation, but in an attempt to elucidate the mechanisms which control their expression.

If scientific names have to be erected for such forms, and I for one believe they have, let us stop at the point where every combination of them has to receive yet another name. The breeders of *Drosophila* recognize some hundreds of characters in *D. melanogaster* and are able to converse on these universally without mis-identification, yet this is accomplished without the use of a single latin name. A simplification and standardization is necessary for dealing with variation in such species as the one under discussion. Hours of wasted discussion on the priorities and synonyms of names would also be avoided.

There are two particular forms of *A. caja* which I wish to discuss, and each has several synonyms. The first which I shall refer to as f. *brunnescens*, has red-brown hindwings and the forewings vary from normal to smoky. It is the heterozygote of the second, f. *fumosa*, the homozygote, which is an insect with dark chocolate-brown forewings and hindwings, with the normal markings obscured, but visible beneath; the overall impression is that of an all-black insect. The genetical situation is, therefore, one of incomplete dominance of the dark form.

The frequency with which f. *brunnescens* has appeared on inbreeding from wild stock is remarkable, certainly the gene cannot be rare in the southern English population. Smith, S. G. (1955) and Wright and Smith (1956) have contributed substantially to our knowledge of the *brunnescens–fumosa* situation.

Wright, who was breeding from a wild-caught light-coloured female (ab. *schultzii*) noticed that 'the last five to emerge from 50 individuals in the F_1 generation had "smoky brown-pink" hindwings'. We know now that of the 50 offspring he had already hatched, another 20 must have been heterozygote f. *brunnescens* (assuming equal viability). The first appearance of this form in inbred stock has frequently shown no simple Mendelian relationship. The fact that visible f. *brunnescens* occurred only in the last five hatchings does suggest that a component of their environment had played a part in the expression of this form. Wright obtained pairings from these and records an F_1 consisting of 11 f. *typica*, 24 f. *brunnescens*, and 17 f. *fumosa*. He appears to have had no difficulty in recognizing the heterozygote f. *brunnescens* in this brood.

Gordon Smith and others have had the same experience in the offspring of a first out-cross of f. *fumosa*, which contained a number of individuals not recognizable as f. *brunnescens*, and he suggests that 'if similar specimens had been captured in the wild state they would probably have been classified as normal'. Yet when these were paired together, one quarter were homozygotes (f. *fumosa*).

This mutant form would, therefore, be ideal material for demonstrating selection for dominance and recessiveness in laboratory experiments. It is likely that only a few modifiers which are dominant are responsible for heterozygous expression, but the effect of environment must also be taken into account. This incomplete dominance is comparable to that found in the *medionigra–bimacula* relationship in *Panaxia dominula*.

There is one major difference between the two species, however. The enzyme responsible for the melanism in the imagines of *A. caja* also produces melanism in the larvae. Wright first noted in a brood that the 17 f. *fumosa* had arisen from 17 melanic larvae which he had segregated. Subsequently he and Gordon Smith showed conclusively that all larvae of the homozygote f. *fumosa* are melanic and have total absence of the usual lateral orange hair-tufts. Wright was also able to differentiate the heterozygous larvae in his stock though I failed to do this in my own which he had given me; but he was an artist and as such was able to recognize minute colour differences which I could not. The f. *fumosa* larvae are a clear entity and no crossing over has been reported. This then is the second example referred to in which the same, or a closely linked gene, controls both larval and imaginal melanism.

Panaxia dominula L., and its melanic subspecies and forms

Panaxia dominula f. *typica* has iridescent green-black forewings with a constant pattern of white spots. The hindwings are crimson with black markings. It is a highly distasteful species and on attack it secretes droplets of fluid from its cervical glands. In Britain it flies by day in the sun, but in southern Europe it has a regular night-flight. When disturbed at rest in wet weather it feigns death, at the same time exposing its crimson hindwings.

P. dominula lives in circumscribed colonies and never ventures far afield from the centre, which is usually focused on its favourite food-plant in Britain, Comfrey, *Symphytum*. Such colonies, separated by a distance of two miles of arable land, retain characters distinct to each. This species, therefore, offers exceptional opportunities for population studies, both in the laboratory and in the field.

Panaxia dominula f. *typica* occurs locally throughout northern and central Europe, also in a few localities in Norway and Sweden and as far east as Leningrad (Standfuss, 1896). Though subspecies occur with yellow, red, and intermediate coloured hindwings in Portugal to the west (ssp. *lusitanica* Stdgr) and a closely related species *P. rossica* Kolehati in Russia replaces it to the east, only in Italy is melanism found as a recurring character. This is in ssp. *persona* Hübn.

A somewhat similar form, with greatly increased areas of black, is found as a polymorphism, but at a low frequency, in one locality only in Britain. It is f. *bimacula* Cockayne. The heterozygote f. *medionigra* Cockayne is darker than f. *typica* and is intermediate in appearance (Plate 16.3).

I discuss both situations in which melanism occurs in some detail. I believe that pattern changes to the extent involved, in particular in ssp. *persona*, must themselves contribute to species survival, quite apart from the pleiotropic effects which the gene controlling them may produce. It must be noted that all these colour changes occur in areas on the periphery of the range of species.

P. dominula ssp. *persona* Hübner f. *italica* Standfuss f. *nigradonna* Kettlewell

The situation in Italy is of exceptional interest (Kettlewell, 1942–3, with figures). Only in the north-west of that country (Piedmont) is *P. dominula* f. *typica* found. Elsewhere, the populations to the south as far as Sicily form subspecies which are in part melanic and maintain combinations or modifications of the following characters which are features of ssp. *persona*: (*a*) the red on the hindwing is replaced by yellow; (*b*) the body is always steely blue-black; (*c*) there is a diminution of the white markings on the forewings, and the hindwings may be largely black, but thinly rayed with yellow. Specimens with all degrees of melanism fly together.

Subspecies *persona* differs from f. *typica*, occurring in the rest of Europe, by many genes. Apart from colour-pattern dissimilarities, though the genitalia of each are identical (Kettlewell, 1942–3, 1943–4, 1945, 1946*a*), the following character changes have been recorded:

(*a*) According to Passerini the larval patternings are recognizable distinct in the two (Boisduval, 1834).
(*b*) There are mating and assembling scent differences—male *P. dominula* f. *typica* pair with female ssp. *persona* with difficulty and always choose female f. *typica* when presented with both at the same time (Standfuss, 1896).
(*c*) Such pairings are frequently infertile (Goldschmidt, 1924).
(*d*) There is a high mortality rate in hybrid larvae (only 12 per cent, 40 per cent, and 22 per cent were reared in such crosses (Standfuss, 1896).
(*e*) The shape of the forewings in ssp. *persona* is always narrow and more pointed than in *P. dominula* f. *typica*.

Genetics of ssp. *persona*. 'Yellow hindwings' is an incomplete dominant, heterozygotes being intermediate. The gene responsible is therefore a different one from that which controls the rare yellow mutant found throughout the range of f. *typica*. Here yellow is recessive.

The genetics of the melanic components. Black body-colour appears to be dominant to the normal red and controlled separately from wing melanism. Goldschmidt (1924) has shown that two pairs of genes are responsible for the degree of black pigmentation, each having incomplete dominance. A range of forms fly together in many localities in Tuscany, Campagna, and Calabria. The lightest, f. *italica*, with reduced forewing markings and increased black patterning on the yellow hindwings, he designates aa bb. The darkest, AA BB, is largely black on all four wings (f. *nigradonna*). Intermediates contain many individuals with black banded hindwings (Aa, Bb, etc.).

To the east of the Apennines in southern Abruzzi there is a more or less uniform ssp. *mejellica* Dannehl, characterized by small size and diminished white markings on the forewings and wide black bands on the outer margin of the yellow hindwings. We have then, throughout central and southern Italy, various races of *P. dominula* all of which differ from f. *typica* by increased black colouration.

I have no knowledge of these subspecies in the wild and have failed to find ssp. *persona* on several occasions. The great change of colouration and pattern must, I think, contribute direct advantages derived apart from any physiological ones.

P. dominula f. *bimacula*, f. *medionigra* Cockayne (1928)

These two forms have, in fact, the best documented history of any Lepidopteran on record; this is the result of the sampling of a colony at Cothill, Berkshire, since 1939 by E. B. Ford, (1955a) and regular mark–release–recapture experiments by him, R. A. Fisher (Fisher and Ford, 1947), P. M. Sheppard (1951, Sheppard and Cook, 1962), and others, from 1941 up to the present time. The mutant form *bimacula*, was found at Cothill in 1926; in this form most of the white markings on the forewings, with the exception of the two basal spots, are absent. Because the overall impression is that of a black insect I include it, and without hesitation, in a book on melanism. I have worked on this particular form for over 30 years.

Shortly after the discovery of f. *bimacula*, breeding experiments showed that it was a homozygote and that the heterozygote could be recognized by either one of two characters: the absence of the central dot on the forewing frequently with a reduction in other white markings, or the presence of an extra black spot on the hindwing. In fact the heterozygote f. *medionigra* usually displays both.

Many thousands of words have been written on the Cothill *P. dominula* and these two forms in particular, in regard to their frequency changes. My main interest, however, has been the homozygote f. *bimacula*. As only 28 of these have been caught at Cothill in 23 years (1939 to 1970) in yearly samples totalling 18,385 (Ford, 1965, 1966, 1971), one would expect this form not to have contributed materially to the gene-frequency fluctuations which vary from 1·1 per cent to 11·1 per cent.

Sheppard has shown that there is non-random mating of heterozygotes in which unlike genotypes are preferred by the females (Sheppard, 1952), though not by the males. The heterozygotes are likely, in fact, to be responsible, in various ways, for the Cothill polymorphism though they have a 30 per cent viability disadvantage to f. *typica* (Ford and Sheppard, 1969). Because the frequency of f. *bimacula* does not depart from that expected from the number of wild-caught f. *medionigra*, the homozygote appears not to be at a physiological disadvantage, though the smallness of the yearly samples of f. *bimacula* precludes certainty on this point. It would be unusual for a homozygote mutant, in what must surely be a mutation of recent origin, to be so well adapted to a gene-complex.

Laboratory and field investigations on f. *bimacula*. *P. dominula* is, in common with many gregarious and semi-gregarious species, relatively immune from polyhedrosis, a virus disease, which most assuredly renders the laboratory breeding of Lepidoptera a highly hazardous occupation. On rare occasions only have we had nuclear polyhedral disease confirmed in certain broods from inbred stock of this species.

A second attribute is the ease with which artificial colonies can be established when once the ecological requirements are satisfied. For over 30 years I have founded such colonies whose genetic origins have previously been studied, frequently from pure-line recessive mutants.

I have used two different methods: first by introducing the species into suitable habitats which were already available, and secondly, but more recently, I have made artificial ones in urban gardens. Though the ecological requirements in both instances were correct, the one type behaved entirely differently from the other. Whilst the founders introduced into unoccupied habitats survived at normal population levels, the artificial urban ones, feeding on transplanted food plants, usually built up a population explosion within three to five years. This has on one occasion been followed by extinction the following year, owing, I think, to destruction of the food plants. Sheppard (1956, 1961) has recorded this colony which I had introduced in 1949 into the grounds of the British Museum, Tring, Herts. Only brief mention is therefore necessary.

Prior to my departure to S. Africa in 1949, I released approximately 600 larvae of this species which were the outcome of eleven years of intensive inbreeding of three major gene-character differences and their modifiers: ab. *albomarginata* (and ab. *flavomarginata*), ab. *juncta*, and ab. *brunnescens*. Details of each have been recorded by Sheppard; the majority of the *P. dominula* in the three broods released were individuals with a greatly increased amount of white (or yellow) on the forewings and spotless hindwings (except for the central dot). Others had crimson on the hindwings replaced by red-brown. The f. *bimacula* played no part in the colony whatever. Nevertheless, it is appropriate to record here two important, but unexpected results. First, by 1953, after only four years, the larvae had assumed plague proportions and devastated the Comfrey plants I had originally introduced, and most of the nettle and bramble. The caterpillars covered the ground in large numbers feeding on previously unrecorded food-plants and even grass. At least two Cuckoos were daily devouring numbers of larvae, as could be observed from the Museum windows. The following year, in 1954, imaginal numbers bore no relationship to the larval density of the previous year. From that year onwards no *P. dominula* were to be found whatever, and the species became extinct.

The second observation is that following the release of the highly abnormally coloured founders in 1949 such phenotypes rapidly disappeared from the population in subsequent generations, the majority of individuals being f. *typica* in 1952, 1953, and 1954, the last year. A random sample of more than 50 taken in my absence showed no mutant forms except ab. *juncta* which has a varying expression and was present in 5 per cent of the later samples.

This experiment, whose only design was that of keeping my stock available for use on my return to Britain, has convincingly demonstrated the heavy selection pressures which must be brought to bear on the gene-complex in order to eliminate abnormal forms; this could not be caused by genetic drift only. It is unlikely to be due to visible changes, but provides evidence that genes which control normal patterning are also optimal for the physiological and other requirements of a species.

The interesting results of this experiment suggested that introduced colonies of this kind, particularly in artificially made habitats, would help considerably in studying the selective advantages of other mutant forms of this species under natural conditions.

Laboratory breeding. Fisher and Ford (1947) selected heterozygous f. *medionigra* for absence and presence of expression of the mutant gene. The two lines which evolved showed the development of recessiveness on the one hand, and dominance on the other.

In 1943 I showed that temperature also affected the degree of expression. That portion of back-cross broods whose larvae and pupae had been subjected to a constant temperature of 70°F (21°C) showed a greater degree of black hindwing markings than those individuals given normal varying day and night temperature (32° (0°C)–70°F (21°C) Kettlewell 1943–4). Some f. *medionigra* were extreme, even tending to a f. *bimacula* phenotype with banded hindwings.

My work, however, has largely been concerned with the homozygote and at this time it was accepted that pairings of *bimacula* × *bimacula* were always infertile. This may, in fact, have been true in certain stocks, but not in others. After four generations of pure-cross f. *bimacula* in which I selected for darkening factors I obtained a strain in which (*a*) the forewings were totally black, except for the basal spots, (*b*) the hindwings were black except for small lines or spots of crimson, and (*c*) the abdomen, normally crimson with small black dorsal spotting, was totally black. This was in contrast to wild caught f. *bimacula*, and others bred from larval samples from Cothill, which sometimes had remnants of the normal white markings on the forewings.

Considerable variation in the patterning of f. *bimacula* therefore existed, and, unlike f. *medionigra*, appeared little affected by environmental conditions and was therefore the direct result of genetic modification.

Artificial colonies of f. *bimacula.* I made my first artificial colony of this species in 1939 in the garden of my home at Cranleigh, Surrey. This was no experiment; it was in fact, referred to as the 'dump' because it was constituted from individuals not needed for breeding stock. I very soon came to realize how much there was to be gained from living alongside a species on which one is working. Observations at all times of the day and night are essential—and throughout the year. It was here I first discovered that this sun-loving species indulged in an occasional mass flight, one to two hours before midnight—a crazy one in which both sexes tore around the top of the highest tree in the neighbourhood. It only happened once or twice a year and always on the hottest nights. It had never before been recorded though it flies by night on the Continent. I now know that this apparently aimless exercise (for they neither mate nor feed then) is of importance in that it serves to 'call in' the individuals to form the centre of the colony for the following generation. This may vary within 200 yards from year to year and there may be several epicentres near by, but always separated by distances of several hundred yards of apparently suitable habitat from another colony. This Cranleigh colony, largely consisting of f. *typica*, has continued to thrive for 30 years until the present time, but unfortunately it has been impossible to keep record here of the f. *medionigra* frequencies.

Sheppard (Sheppard and Cook, 1962) has later made use of introduced colonies in which the f. *medionigra* gene has been released at different frequencies. He has recorded the subsequent fluctuations in each of these.

The objects of the present experiments. The colonies of this species which I wish to record here are of pure f. *bimacula*. These were designed for an entirely different purpose to previous ones. First I wanted to see whether a pure strain of f. *bimacula*, so different in appearance from f. *typica*, was viable in the wild. Secondly I needed to observe whether there were behaviour differences between this form and f. *typica*. Thirdly, I wished to know whether insectivors reacted in the same way to each. Lastly, my main purpose was to see whether selective pressures would rebuild the white patterning of the forewings, similar to that of f. *typica*. This could only be brought about by modification of the gene-complex, because in the majority of wild-caught f. *bimacula* only small traces of such markings remain, if any.

The introduced colony of f. *bimacula.* In 1961, after three years' preparation, I commenced the first artificial colony of this form. Approximately 3000 newly hatched larvae of f. *bimacula* of two different strains were released in the autumn onto Comfrey (*Symphytum* spp.). The site was in the two acres of garden, river banks, and marshland which surround my home at Steeple Barton, Oxfordshire, where this species did not previously occur.

Viability in pure colony: population size of imagines. The estimated population of f. *bimacula* in the first year (1962) was approximately 250. This was assessed as four times the number observed at rest during a three-hour period each day; this was a rough and ready method, suggested by Sheppard, based on his experience in many mark–release–recapture experiments. When several colonies have to be examined each year, more time-consuming methods are impossible, and, I think, unwarranted.

In 1963 the population was smaller (200 approximately), and thereafter it increased yearly with a peak in 1966. During these five years, considerable adjustments in the gene-complex must have been taking place. Pure-line homozygous f. *bimacula* have therefore been demonstrated to be capable of maintaining themselves in nature in these completely rural surroundings, and also in a second introduced (but not artificial) colony in Blenheim Park, Woodstock. Here the colony, also founded in 1962 from the same strains, continues to thrive, but in smaller numbers than at Steeple Barton.

This is probably due to the fact that its centre coincides with the maximum population of pheasants which feed on the larvae of this species.

Behavioural differences. Both sexes of f. *bimacula* indulge in less flight activity by day than f. *typica*, the females in particular. Not until 1968 had I ever witnessed the midnight flight in this form, though I suspect that it may have occurred previously, because of the larval distribution in this colony which is always centred around certain trees only, and these vary from year to year. Nor, though I have used a mercury-vapour trap in the centre of the colony for six years, have I taken this species in this way. On the Continent *P. dominula* f. *typica* is attracted to light regularly. (Rothschild, personal communication.)

In flight, f. *bimacula* is difficult to follow, except in the case of the male immediately before copulation when it may hover over the female. At rest on the upper side of green leaves it is easier to see than f. *typica*, having lost its disruptive patterning.

In both f. *bimacula* and f. *typica* the males hatch before the females, but for the first three years the discrepancy was more evident in the ab. *bimacula* colony than in wild colonies of f. *typica*. This led to an interesting result.

The Steeple Barton colony lies on a gentle slope, at the bottom of which, and some 30 feet lower, is the River Dorn. The temperature here at night is several degrees colder than at the top of the slope. Originally I had released the larvae over the whole area. The River *P. dominula* hatched 7–14 days later than those higher up. The result was that the river males left the area and paired with later females from the upper zone. Finally, none of the river females appeared to have found a mate. All eleven I picked up at river level late in the season in 1962 laid infertile ova. However, six out of seven females picked up at night from the higher level laid fertile eggs. From that time the river part of the colony was extinct and the following year, though the larvae were in large numbers higher up, none were observed by the river, nor have they reappeared there since.

I am fully aware that some of this evidence is circumstantial, but it does suggest that differences in temperature lead to differences in time of emergence and that this is another factor (apart from the night flight) which influences the centering of a colony for the following year. As I noted this in a colony of f. *bimacula* and it has not been recorded in f. *typica*, I mention it here, though a similar situation may exist in the normal form.

Predation of imagines. Several birds have been recorded as taking *P. dominula* f. *typica* on the wing, including Swallows, *Hirundo rustica*, and in particular the Spotted Fly-catcher, *Muscicapa striata*. It is of interest that the only example of 'beak marks', in this species in the Hale Carpenter Collection of these 'marks' in the Hope Museum, Oxford, is a f. *bimacula*. Dragonflies (*Odonata* spp.) have also been seen on numerous occasions pursuing and catching *P. dominula* in flight.

In the ten years of my living alongside my Surrey colony which maintained various forms, including f. *bimacula*, but largely f. *typica*, I never observed any bird predation, though each year the species was flying daily in numbers.

Yet in the first hatch of f. *bimacula* at Steeple Barton in 1962 I picked up 15 partially eaten or mutilated specimens. The first five of these were wing remnants only. All the later ones, however, were with abdomens, but they had their heads nipped off. I think that the Spotted Flycatcher was responsible for most of these as I witnessed attacks on flying f. *bimacula* on several occasions. In most of these it was directed on assembling males —and successfully, though 6 of the 15 corpses were sexed as females. The local birds had previously had little experience of day flying aposematic species, and certainly none of f. *bimacula*.

Since 1963, though the population size has risen, I have found less and less evidence of bird attacks on f. *bimacula*. In 1966 I picked up four sets of wings and two decapitated f. *bimacula* between 24 June and 16 July amongst the 374 insects I saw (total population approximately 1500). This contrasts with 15 in 1962 in a population estimated at 250. The predation rate fell from 6·4 per cent in 1962 to 0·4 per cent in 1966. The local birds, I think, had recognized their relative unpalatability by the same process as they have in nature in colonies of f. *typica*.

Larval predation. In spite of their spines and warning black and yellow colouration, bird predation at times is responsible for the devastation of larval colonies. Previously it was thought that Cuckoos, *Cuculus canorus*, and Pheasants, *Phasianus colchicus*, were the only predators, in particular Cuckoos; I have seen as many as half a dozen at the same time gorging themselves on *P. dominula* on the River Itchen, and 23 have been recorded simultaneously feeding on a colony of this species by a water-bailiff on this same river.

In the course of checking the ultimate fate of radioactive *P. dominula* larvae which I had returned to a colony from which I had extracted them (Kettlewell, 1952*a*), it was, on one occasion, necessary for me to shoot a

Cuckoo which, with others, had been feeding on them. I extracted the remains of over 60 *P. dominula* larvae (determined by head segments) from this one bird. The Cuckoo, however, is not the only predator. I have seen the Song Thrush, *Turdus musicus*, feeding its young on fully grown larvae, the Blackbird, *Turdus merula*, the Great Tit, *Parus major*, and the Starling, *Sturnus vulgaris*, eat them.

In 1967, a year of great larval abundance in the Steeple Barton colony, I noted for the first time between 10 and 15 Starlings dropping into the Comfrey areas where f. *bimacula* larvae were feeding. There were 7 pairs with nearly fully fledged young and each bird visited on an average, once in under 10 minutes. I shot one with two last-instar *P. dominula* larvae in its beak and recovered one from which I bred a f. *bimacula*. As this behaviour continued for several days and was extended to the whole area of the colony, I am certain that many thousand of fully grown larvae were destroyed. This may be reflected in the lowering of the imaginal population that year (estimated at 500).

In another colony a number of Great Tits were seen in 1968 to be carrying away *P. dominula* larvae, presumably to feed their young. It would appear, therefore, that in spite of their spines and colouration, many different kinds of birds attack them and that, because the full growth of the larvae coincides with the growth of the nearly fully fledged young of the predators, large numbers may be taken.

Though the larvae of *P. dominula* show warning black and yellow colouration at close quarters they can be, from a distance, cryptic against a fragmented background because of the disruptive pattern. For this reason, no doubt, larvae about to undergo ecdysis, usually wander away from green foliage and take up position on dead herbage where they are not easy to see. Larvae when feeding exposed on leaves are conspicuous by day.

I have observed that larvae in some colonies feed and expose themselves only by day, larvae from other colonies only by night and in some both day and night feeding takes place.

It is surprising that under the pressure of bird predation all *P. dominula* larvae are not nocturnal feeders in the same way as those of *Callimorpha quadripunctaria* Poda, the Jersey Tiger.

Larval variability. In certain localities the markings of the larvae are dimorphic (Kettlewell, 1942–3). Some have white spotting between the yellow areas and others have not. In a colony near Deal, south Kent, it appears that 'white spotting' is absent in the whole population. As the frequencies of the two forms vary widely on different years it is possible that this difference could be an environmental effect. Also, on one occasion, following an extremely hard winter, when I had a high mortality rate in my outdoor cages, many surviving larvae in all broods were nearly completely black; both yellow and white markings were absent. On no occasion have I found this form in the wild.

An estimate of larval–pupal mortality. As it was important to obtain a measurement of such predation, I carried out extensive laboratory work on techniques for marking larvae with radioactive isotopes. This was, in fact, an extension of previous work which I had undertaken in Africa on marking locusts (*Locusta migratoria* and *L. pardalina*) (Kettlewell, 1955e).

In 1953 I extracted a large sample of larvae from the *P. dominula* colony at Sheepstead Hurst, Berkshire, where it was abundant that year. These were fed on Dead Nettle (*Lamium* spp.) grown in water culture containing ^{35}S. After 24 hours of feeding I released 1227 of these into the colony. Two months later I undertook a full mark–release–recapture investigation on the wild imagines, at the same time scoring each for their radioactivity. By these means I was able to calculate the late-larval (and pupal) death rate. This was shown to be in the region of 90 per cent and it is likely that when population levels are high, maximum predation takes place during this period of their life history (Kettlewell and Cook, 1960).

Wing pattern changes. The main feature of f. *bimacula* is the disappearance of the light spotting on the forewings with the exception of the two inner spots, along with the presence of black banding on the hindwings. Occasionally, however, residual dots remain on the outer areas of the forewings. On the evidence provided it appeared that, if the normal patterning of f. *typica* was of adaptive significance, its light spotting might be reconstituted by other genes in f. *bimacula* when subjected to natural selection under natural conditions. In the other direction, as previously stated, I had shown in laboratory broods that darker f. *bimacula* could be produced in four generations of selective breeding. It could be of interest to see whether the patterning was affected in one direction or the other in artificial colonies. We were, in fact, testing in the field in pure culture, a new form, homozygous and laboratory produced.

Results in the introduced colony of f. bimacula. A random sample of the sibs of the original releases (1962) consisted of 22 specimens of which (excluding basal spots) 8 (36·4 per cent) had all-black forewings, 10 (45·4 per cent) had one to two minute white dots on the outer margin, and 4

(18·2 per cent) had only 3 or 4 small dots. None had more extensive white markings.

In 1967 in a sample of 100 examined alive, only 16 per cent had spotless outer forewings, 81 per cent small dots varying in number from one to three, and 3 per cent had more extensive white patterning, a form not previously seen by me (Plate 16.3, right 1–4). This could be in part of environmental origin, but I think this unlikely. Though it is too early to attach any importance to this, it does suggest that some of the white spotting is being reconstituted, even after only five generations in the wild. The genes controlling these must be different ones to those which are normally responsible in f. *typica*.

More interesting is a quite unexpected development. In 1964 a hitherto unknown form of *P. dominula* appeared in the colony, and most of its characters were diametrically opposed to those of f. *bimacula* (see Plate 16.3, right, 1–4). The forewings have longitudinal yellow and white streaks and spots: the orange-red hindwings, except for the central dot, are immaculate. In flight the overall impression is one of an orange-yellow insect. One or two others have occurred yearly in my f. *bimacula* colony, eight in all. (I have referred to these as 'f. *pseudojuncta*' (Plate 16.3, left 4).)

We have bred from each and on no occasion, amongst the thousands of offspring we have reared, has this form reappeared in the laboratory. All have been f. *bimacula*. This applies also to the subsequent F_1 and F_2 generations of inbred stock.

The origin of this variable form is, therefore, unknown though we have imposed temperature, humidity, and other environmental variants on both larvae and pupae. If it spreads, an entirely new polymorphism will have arisen.

I have discussed f. *bimacula* at length because I believe that similar opportunities are to be found in other species in which a polymorphism, or a single morph contributing to it, can be studied in the wild by the use of artificial colonies.

We have, in fact, attempted this with a number of species:
   (i) *Lasiocampa quercus* ssp. *callunae* and its melanic form, f. *olivacea*, using a heather moor on an island on which this species does not occur.
   (ii) *Ectropis consonaria* and its melanic, f. *nigra*, in isolated Beech woods in semi-urban areas.
   (iii) *Lycia hirtaria*, and its recessive f. *nigra*, in London squares.
In each instance the founders were released at a known gene frequency but in none of the three species have the results so far been worth the con-

PLATE 16.3 (page 277). *Panaxia dominula* L. (× 1) and its two forms *medionigra* and *bimacula*.

Left

1. *Panaxia dominula* f. *typica*.
2. *Panaxia dominula* f. *medionigra*.
3. *Panaxia dominula* f. *bimacula*.
4. *Panaxia dominula* new mutant form 'f. *pseudojuncta*', occurring in a colony of pure *bimacula*.

Right

1–4. Various expressions of f. *bimacula* taken from a colony of pure *bimacula*.

(Photograph by Robin Tanner).

PLATE 17.2 (page 294). *Hepialus humuli* L. (× 1).

Left
1. *Hepialus humuli* L. ♂ f. *typica* (British mainland).
2. *Hepialus humuli* L. ♀ f. *typica* (British mainland).
3. *Hepialus humuli* L. ♀ from Shetland.

Right
1–3 *Hepialus humuli* L. ♂♂ from Shetland, race *thulensis* Newman.

(Photograph by John Hayward).

PLATE 17.4 (page 251). *Lasiocampa quercus* ssp. *callunae*. Wild larvae from Caithness (typical and melanic), 1972. Compare with larvae from Yorkshire, Plate 14.5.
(Photograph by Simon Whalley) (× ¾).

siderable efforts of attempting to create circumscribed populations in the wild. This is in contrast to the introduced colonies of *P. dominula*—in particular f. *bimacula*.

The unknown selective advantage of 'melanistic' P. dominula. If one believes, as I do, that a major colour change in a species either in a polymorphism, or still more so in a subspecies, implies a colour advantage in a particular biotope (apart from other effects of the genes controlling it) it is reasonable to compare two forms which, though geographically separated, are phenotypically similar. Many of Standfuss' crosses of subspecies *persona* and red *typica* of German origin (=ab. *romanovi* Standfuss) are, in fact, remarkably like f. *bimacula*. Furthermore, similar forms, with red hindwings, occur in the valleys of the Dolomites and Southern Tyrol (=ssp. *pompalis* Nitsche) (Kettlewell, 1942–3). As yet, I have been unable to attempt crosses of ssp. *persona* and f. *bimacula*.

The black components of ssp. *persona* have in north-eastern Italy become dissociated from the yellow, in the same way that in the west (Piedmont) the yellow is dissociated from the black. The survival contributions of 'yellow' and 'black' must therefore be distinct and separate. We have no idea what these may be, though this could be tested in the wild in north-west and north-east Italy respectively. Nor do we know the reasons why f. *bimacula* and f. *medionigra* maintain a polymorphism in one locality in Britain except for the disassortative mating of the heterozygotes.

Artificial colonies have demonstrated the following:
 (i) f. *bimacula* is viable in pure line in introduced colonies.
 (ii) Colonies of f. *medionigra*, the heterozygote, founded at a known frequency, fluctuate yearly both in population-size and gene-frequency.
 (iii) Larvae vary in diurnal and nocturnal feeding and are subjected to greater visual predation than we had originally anticipated in an aposematically coloured insect.
 (iv) The warning colouration of f. *bimacula* has been shown to be effective as a deterrent in an artificial colony after only three years of experience by bird predators.

Field studies on ssp. *persona* need to be undertaken before any theories on the causes of the melanism are postulated.

Aglia tau L.

Aglia tau f. *typica* is a yellow-brown moth with a well-developed 'eye-mark' situated centrally on the upper surface of each of its four wings. Though present on the underside of the forewings, it is normally not recognizable on the underside of the hindwings.

We know a little about the genetic control of such 'eye-marks' from another species, *Saturnia pavonia* L., where each 'eye' is made up of at least four components: a pupil with light reflex, an iris, and a margin signifying the outer limits of the eye. In rare cases any one of these markings can be absent separately; this suggests that cross-over or deletion has taken place. Still more rarely the 'eye-mark' is totally absent. It is likely that the whole is controlled by a super-gene with fairly tight linkage.

Such 'eye-marks' must be the end product of intensive selection. But selection for what? We do not know. Ford (1955a) has pointed out, and I have observed, that during eclosion, a most vulnerable moment in the life of any day-hatching moth, the 'eye-marks' are particularly conspicuous on the bulging wings and this may signal threat to prospective predators. Such an eye-mark must be enhanced on a light, rather than a dark, background.

It is remarkable therefore to find melanism in such species. Yet on the industrial moors of Yorkshire, dark or smoky forms of *S. pavonia* occur though the normal eye patterns are visible.

Melanic *A. tau* are widespread over much of the continent at a 3 per cent level of the population, and at a higher frequency around industrial areas (Hasebroëk), particularly in Czechoslovakia.

That an eye-spot is important to survival as a threat deterrent in the imago seems indisputable because of the complexities of its origin. Yet *A. tau* is likely to depend on camouflage also because it follows the pattern of all other industrial melanics which most assuredly so do. Standfuss showed that two separate genes control the melanism; in each the dark form is dominant, but incomplete. In contrast to the melanic *dominula* ssp. *persona* of Italy where the factors contributing to melanic forms are at separate loci and therefore additive, the two genes controlling melanism in *A. tau* occur at the same locus and are allelomorphic. Two alleles for melanism must lead to six different phenotypes—typical *tau* mm, f. *melaina* Gross M^1m (one heterozygote) M^2m, f. *fere-nigra* Th.-Mieg (the other), with the two homozygotes M^1M^1 and M^2M^2 darker (Standfuss, 1910a, 1910b, 1914). A combination of these last two, M^1M^2, produces another recognizably different form *weismanni* Standfuss. When paired *inter se* all three melanic forms are produced and as they must be homozygotes, *tau* f. *typica* must be absent.

It would be interesting to found colonies from released f. *weismanni* in isolated industrial areas where this species does not occur. Though the first generation would, in theory, have 50 per cent of this form and 25 per cent of each of the other melanics (as homozygotes), the frequencies would, no doubt, be subsequently changed as a result of mating preferences, viability and other differences between the melanic forms about which we are at present in ignorance. *A. tau* f. *typica* could only occur by mutation.

17. Melanism influenced by sex

(i) Sex-limited melanism

Cycnia mendica Clerck f. *rustica* Hueb. (Arctiidae)

Sexual dimorphism in the Rhopalocera is common. Here the male is frequently monomorphic, even where the females exploit a range of colours and patterns. In mating, which is always by day, the females usually react to a uniform sex-signal which they recognize by a standardized male colouration.

By contrast, in the Heterocera, this is rarely so (yet the same method is used in part in the Hepialidae). Sexual dimorphism in moths must therefore have a different origin and purpose and frequently this reflects entirely different habits between the two sexes; a common one being nocturnal flight in one sex versus diurnal in the other.

In England, as in Europe generally, *Cycnia mendica* has white females with small black dots similar to those in *Spilosoma* spp. which we have already discussed (p. 273). Males are melanic and are either black or, less frequently, dark brown. Yet in Ireland and certain other places, both sexes are white, the male colouration being similar to that of the female, though less commonly they may be pale buff. This is race *rustica* which is distinct (Plate 17.1). On crossing the two races a range of forms occur in the F_2 generation, and the males vary from white, through buff-brown (f. *standfussi*, the heterozygote), to black as in the British race. This is due, not to multifactorial inheritance as was suggested by Cockayne (1919), but to a simple major gene with a variable heterozygote. Black colouration in the male is therefore sex-limited and has no dominance (Onslow, 1921*b*, Adkin, 1927*a*, 1927*b*).

This species offers rewards for field and laboratory research in many directions. In particular, why is the male melanic in England but white in Ireland? If what I have already said in the two opening paragraphs is correct, it must be assumed, *a priori*, that English males behave in a different way from Irish ones and furthermore that this behaviour difference is unlikely to be connected directly with mating. We have evidence of this from three separate observations:

    (1) From sampling at mercury-vapour traps we know that *C. mendica* males in England fly late at night. The females never do so ($n = 500+$, Oxfordshire).

(2) Females in England fly around midsummer in sunshine during the late afternoon when they are conspicuous. I have never observed males by day, except at rest in grass tufts.

(3) Pairs can be found copulating at the first light of dawn (P. Feeney in Oxford, personal communication).

The black males and white females in England may therefore reflect a habit difference between the sexes. In England and much of the northern Continent males fly by night only, whereas the white females fly by day and depend then probably on aposematic colouration in the same way that the related *Spilosoma* species do. I have no knowledge of the toxicology of female *C. mendica* though this is being investigated at the present time. Without this knowledge, however, we can predict that English and Irish males of this species are likely to have behavioural differences. I have no evidence that Irish males fly by day.

Ford (1955a) states that race *rustica* in Ireland is an interglacial relict, yet there are other areas where white males are found, as in south-east Europe (Onslow, 1921b). If similar race characters are to be found in the Iberian Peninsula this is likely to be correct; if not, it is more probable that the Irish race has arisen more recently and separately.

There is considerable genetic variability in *C. mendica* throughout its range. On the one hand a black male is recorded from Ireland and on the other whitish males from Kent, in south-east England (Barrett). In Oxfordshire I have sampled an average of 70–100 males per annum for twelve years: 5 per cent of these are paler than the normal black f. *typica*.

I think that in this species the gene-complex is poised for expression of melanism in the male or its repression according to the local selective pressures which demand fundamental habit or behavioural changes in that sex. As yet, we have no idea what these may be.

Hepialus humuli L. race *thulensis* Newman (*Plate* 17.2)

Hepialus humuli is sexually dimorphic throughout its range and, with the exception of those on the two groups of islands, Shetland and Faroes, the male is china-white on the upper surface of all four wings but black on the underside. The female is entirely different and has a patterned brown surface. In Shetland and the Faroes, which lie to the north of Britain, the majority of the males have a dark patterning on their forewings, with black hindwings, and it is difficult in some individuals to differentiate them from the females (=race *thulensis*). Approximately 3 per cent in our samples from there are yellow-white on the upper surface of the forewings, similar to but not so vivid a white as their mainland confreres. Furthermore, in the majority the hindwings are black. I give in Table 17.1 the results of random samples from north and south Shetland taken in 1964 (Cadbury, unpublished).

TABLE 17.1
Hepialus humuli—♂ phenotype frequencies in
Shetland, 1964 (Cadbury, unpublished).

Locality (northernmost at the top)	Phenotypes				Total
	Ochreous-brown as in ♀	Cream with dark pattern	Immaculate cream	China-white	
Unst					
Burrafirth	7	0	1	0	8
Baltasound	28	6	1	0	35
	35 81·5(±5·9)%	6 13·9%	2 4·6%	0	43
South Mainland					
Kergord (light trap)	1	0	0	0	1
Tingwall	17	9	0	0	26
Easter Quarff	2	2	0	0	4
Cunningsburgh	2	0	0	0	2
Channerwick	2	1	1	0	4
	24 64·9(±7·0)%	12 32·4%	1 2·7%	0	37

The problem therefore is, why is there in the male this surprising change in colouration? The story is likely to be an interesting one though it is as yet incomplete and unpublished. The males fly during mid-summer at dusk when they 'lek' in groups of 5–50 in limited areas of certain meadows only. Here they 'pendulate', hovering over the same spot for up to 3 minutes. They position themselves several yards apart and are immensely conspicuous. The female, which is not easy to see on the wing, flies at a male and strikes him in flight; they fall to the ground and copulation takes place immediately, with the male hanging from the female in a state of catalepsy. Alternatively, she alights near a hovering male, when she is difficult to see. Cadbury has witnessed this behaviour pattern on several occasions and was convinced that the final attraction of the male to the female must be a scent pheromone because the female was not visible to him and frequently hidden by vegetation. Harper in 1960 also recorded a female dropping within a few yards of three pendulating males and witnessed that she took up position and held her body as in a normal female which is 'calling'. One of the males slowly found its way through the undergrowth and mated with her.

We have elucidated part of this complicated story and it appears to be this.

Mating depends on two components, a visual one and an olfactory one. The sequence is that the males lek over suitable breeding-areas which are natural untilled hay-meadows with an abundance of plant species as well as grass. Such an aggregation is probably decided by scent in the first instance but spacing is determined by vision. A white 'super-model', when interposed, will cause a male to move away a short distance and then to follow it. Females are attracted by both scent and vision. Females in Caithness, north Scotland, settle on white muslin bags containing Caithness males and on one night in July 1963, Cadbury watched 'at least a dozen females fly up to a white cage which contained fifteen freshly captured males'. By contrast, during the hundreds of experiments we have conducted in southern England, only once has a female been attracted in this way (Oxford, 1966), though there is record of a female in Hampshire (S. England) fluttering around a box whose lid was slightly open and contained a male completely hidden from view (Michael, 1949).

Of equal interest is an observation I made in 1966. Two Shetland males attracted three Oxford males which fluttered up and down the outside of a white muslin bag containing them (though the inmates were difficult to see). This had never happened previously when we were using English males under identical conditions; nor did males take notice of empty white bags used as controls.

The assembling of the Oxford males only took place when the Shetland males commenced to fly within their muslin bag. I can only assume that race *thulensis* emits a pheromone when in flight, and that this emphasis on scent is sufficient to compensate for the loss of normal visual signals of both sexes, the normal china-white males being equally conspicuous to gulls as to their own species.

We have observed no crepuscular predation whatsoever in England, though this has been recorded by Reid from Aberdeenshire in Scotland (Reid, 1893). By contrast in Shetland, two species of gull, *Larus canus*, the Common Gull, and *Larus ridibundus*, the Black-headed Gull, appear nightly in their hundreds and scout this species in the meadows during the long northern twilight. White moths must be at a great disadvantage from predation, but at an advantage for mating. It would seem that in Shetland for this reason the scent component has become all important. Cadbury (unpublished) found that the scent attractant arose in a brush-gland of the male which is situated on the tarsus of the third pair of legs. Cadbury also showed that such 'brushes' when dissected out from Caithness males and placed in a white muslin bag attracted both male and female *H. humuli*.

It appears then that, in Shetland, crepuscular predation by gulls is responsible for a change of emphasis in sexual attraction from *visual* to *olfactory* recognition. This is because at a latitude of 60°N there is no darkness at midsummer and all light-coloured moths are conspicuous.

The increased pigmentation and patterning of race *thulensis* raises a point which I am so frequently asked. What constitutes a melanic form? The answer cannot be made clearer than in race *thulensis*. So-called melanic, melanistic, and melanochroic forms refer to heterogeneous and genetically quite indeterminate groups. In all species where an increase in dark pigmentation occurs, though not necessarily producing an all-black insect, such instances must be recorded as coming under the heading of melanism.

The genetics of H. humuli, race thulensis. Because of the difficulty in breeding root-feeding larvae, and in particular *H. humuli*, no broods have been successfully reared. Also, it is certain that race *thulensis* takes more than one year to complete its life cycle, possibly three. It is likely that the majority of *H. humuli* from southern England have a two-year life-history. We have dug up half-grown larvae of race *thulensis* in Shetland in the month of July at the same time that the moth was flying.

In the last few years small numbers have been bred in the laboratory by Aitkenhead and Baker (1964), feeding the larvae on carrot. They completed their cycle in about eighteen months. More recently we have succeeded in rearing the larvae to full growth from the egg by feeding them on an artificial pabulum, containing 14 organic chemicals. The technique, which necessitated the keeping of each larva in a separate box, was devised by D. R. Lees. The larvae, which lived in cocoons, appeared to go into aestivation on full growth, but none reached the imaginal state. Any knowledge of the genetic origin of race *thulensis* must therefore, for the present, be deductive.

There are two or more distinct characters which contribute separately to this race, one being dark patterning, another black hindwings, and a third cream ground-colour. Cadbury and I found the last of these at a frequency of just under 3 per cent ($n = 67$, near Lybster, Caithness, Scottish mainland) in 1965. I doubt whether china-white males occur in Shetland. The patterning of race *thulensis* is remarkable in that it can be coloured black, pink, or brown, but all serve to darken the male when in flight. Such markings could be the result of polygenic inheritance.

There is certain evidence, however, against this, as occasional individuals with a well-developed pattern similar to that of race *thulensis* have been captured in widely separated areas far from Shetland. A single specimen was taken by D. Cunningham in Dumfries-shire (Scottish mainland) in 1946; another by W. Collinson near the industrial area of Halifax, Yorkshire, whilst, on the Continent, Lempke and Sneller record 'seven males with red forewing markings' from near Rotterdam. The distribution of race *thulensis* and of forms similar to it suggests that a single major gene is responsible for male patterning and that this character has incomplete dominance; furthermore, that in Shetland it is likely that it occurs in a gene-complex which favours the expression of a variety of recombinations.

By this means, a sensitive system is permitted in which the advantage of lighter coloured males for attracting females visually is free to operate locally within an isolated community where such conspicuousness generally confers disadvantages from predation. Such a system could account for the considerable local frequency-differences of the various male phenotypes recorded in Shetland.

It may be that the laboratory experiments on the hybrid English × Irish *Cycnia mendica* throw light on the genetic origins of the male melanism of *H. humuli*, race *thulensis*.

(ii) Monomorphic melanism in a normally sexually dimorphic cryptic species

Xylomyges conspicillaris L. (Caradrinidae)

This species is confined in Britain to two adjoining districts—Devon and Somerset, and Gloucestershire and Worcestershire which are continuous over 100 miles in a south–north direction. Only rarely is it found elsewhere. (I took a ♀ f. *melaleuca* at Steeple Barton, Oxfordshire in 1970 which was infertile.) What the specific ecological requirements are we do not know. It also occurs widely in Europe.

The usual form in Britain is f. *melaleuca* View which is black but with small, light disruptive markings which make it exceedingly difficult to find as it rests by day on posts and also on Elm trunks.

Two other forms, however, occur in Britain, f. *typica* and f. *intermedia* Tutt (Plate 17.3). Both are lighter in colour than f. *melaleuca* and are the more usual forms on the Continent of Europe. In Britain these two forms are rare and only occur regularly in the more northern of the two districts.

Prior to 1937, most collections of British wild-caught *X. conspicillaris* consisted of f. *melaleuca* with a few f. *typica* and f. *intermedia* but with no obvious genetic relationship, and at that time many of us assumed that f. *intermedia* was the heterozygote of f. *melaleuca*.

I happened to visit my friend and co-worker, Dr E. A. Cockayne, at a particularly lucid moment in regard to this species, the genetics of whose forms had preoccupied him for some time, and I will give a circumstantial account of what happened. He told me that he was suffering from a chill because whilst in a bath a few days previously it suddenly occurred to him that f. *typica* and f. *intermedia* seemed to occur in wild samples at about the same frequencies. He had immediately leapt out of his bath and with towel around him, sexed these two forms in his own collection. He found that all of the f. *intermedia* were females and all f. *typica* were males and they were in approximately equal numbers. Subsequently he scored the two forms which were bred from a female f. *melaleuca* by G. B. Coney in 1933. Of the 146 *X. conspicillaris* which he bred, 107 were f. *melaleuca*, 22 f. *intermedia* (all females), and 17 f. *typica* (all males) (Cockayne, 1937–8). From this it is clear that *X. conspicillaris* is usually a sexually dimorphic species on the Continent but that in Britain, particularly in the southern part of its range here, it is replaced by the monomorphic melanic f. *melaleuca* which has dominance in both sexes to the dimorphic f. *typica*.

It is strange that the frequency of the forms should vary so greatly in two adjoining districts. There is a higher rainfall, however, in the more southern of the two districts, Devon and Somerset, where it is between 30 and 40 inches per annum. To the north of the River Severn, in Worcestershire, it is only 24–27 inches. The moth normally rests on dead wood by day. Maybe the increased humidity and cloud in the south-west lead to darker backgrounds, favouring the melanic form of the species.

18. Environmental melanism

IN the present stage of our knowledge it is not possible to define environmental melanism with any precision. Nevertheless, this heading represents at least two situations where dark forms occur as a result of a physical impact. First, a temperature variation occurring at a critical time may bring into effect one of two alternative gene-mechanisms thereby switching one pattern to another (Standfuss, 1896, Goldschmidt, 1938). Such an example of temperature effects is seen in the seasonal forms of *Araschnia levana* L. (Süffert, 1924) in Europe, or again *Precis sesamus* in Africa. Intermediates (f. *porina*) are rare, except under planned laboratory conditions. Nor, as far as I know, does an example exist of such a switch to an undisputed melanic form.

Though this may be true of imagines, it certainly is not of larvae. Here it is a commonplace, but the 'switch' is decided by 'background' (and not by temperature) at a particular and sensitive moment in the larval life, thereby enabling the caterpillars of many species to undertake polyphagous habits (see Chapter 4).

The second situation is provided by a somewhat different mechanism; yet again it depends on a particular gene-complex for its expression. Here the general premise is that in imagines of several species from widely different families, low temperatures lead to darkening. There is a gradient of pigment directly correlated with time/temperature and this is decided during a short critical period in the pupal stage which affects the speed of metamorphosis. Such environmental darkening occurs more commonly in those species which regularly migrate from Africa and the southern part of the Iberian Peninsula.

It is necessary here to refer to the methods of the pigment determination of the imago more precisely. It demands the timing of three components: the availability of the chemical substances, the relative rate of development of the scales, and the flooding of the wings by 'activators' at a particular time (Ford, 1945). All are, in part, under genetic control.

Flooding implies fluid and during the metamorphosis the fluid–plasma (blood) distributes the enzyme tyrosinase throughout the insect's body and, by way of the nervures, the wings. The chemical precursors of pattern in the form of the amino acid tyrosin are laid down earlier within the scales themselves and the order in which these scales develop is usually determined genetically. Whether the embryo-pattern is ultimately fulfilled or not depends on the timing of the tyrosin–tyrosinase combination. This is in contrast to pigment formation in mammals. For here the tyrosinase is not in the blood, but laid down in these parts of the pelt which will become black.

In my experimental work on *Heliothis peltigera* I shall show that in this species the pattern is decided at an earlier stage than ground-colour, and that subsequent environmental changes have no effect on a form which is indelibly fixed at that time.

The conflict of an environmental and a genetical explanation

During the second half of the last century melanic Lepidoptera were found to occur in many diverse habitats. Not only were black forms, previously unknown, being discovered for the first time in the little visited northern and western isles, but Industrial Melanism was rapidly spreading throughout urban central England. The entomological literature for the 50 years between 1880 and 1930, is filled with records of new melanic forms, and the spread of previous rare ones, and latterly speculations on the causes of this hitherto unknown phenomenon (see Chapter 5). Emphasis was always on environmental causes and culminated in the dogmatic assertions of Heslop Harrison that Industrial Melanism was due to direct effect of air pollution on individual larvae, and furthermore that such melanism was thereafter inherited.

Heslop Harrison was a zoologist of considerable repute and a Fellow of the Royal Society, and not surprisingly he greatly influenced a body of opinion. When the bubble was pricked by Hughes (1932), Fisher (1933), Haldane (1935), Ford (1937), and others, the effects of external influences were largely discounted.

I think that as a result of this the contribution of the environment to the expression of colour and pattern has been largely under-rated since then and it has been assumed that genetical variants were alone responsible for them. This, of course, may be correct in the different forms of a majority of species. In others, however, insects of identical genetical constitution can vary from light to dark depending on, for example, the temperature during the pupal period. It is these I wish to discuss.

Migrant species

Many regular migrating species vary in their tone of colour much more than the indigenous species they come to join: for example, the early (=primary) migrants of *Nomophila noctuella* L., *Heliothis peltigera* Schiff., *Rhodometra sacraria* L., and *Agrotis ypsilon* Van Rott. Spring captures of these are usually of a pale, sandy, straw colour. It is noticeable that later in the season the majority of the imagines of these same species are darker, frequently conspicuously so. It was tempting to theorize on their origins. Because it was known that these species (with the exception of *A. ypsilon*) could not maintain themselves during the winter months in northern Europe, it seemed likely that the early specimens had flown from areas of North Africa or southern Spain.

Evidence of origin. Accidental radioactive labelling of N. noctuella L.
On 13 Febuary 1960, France reported that the first of two atomic tests had been carried out successfully in the deserts of North Africa. As I had then been recently working on the radioactive labelling of larvae with the object of studying population dynamics, it appeared to me that large areas of desert country would have received radioactive fall out and that there would be an uptake of this into plants and subsequently into insects.

I therefore advertised in scientific journals, requesting that any examples of early Lepidoptera migrants should be sent to me and specifically naming the four species already referred to. I received a number of Lepidoptera, the majority of no consequence and only three *N. noctuella*. Each insect was tested under a Geiger counter. Only one, a pale-straw coloured *N. noctuella*, which I caught in a trap at Steeple Barton, Oxford, on 10 March of that year was radioactive and this doubtfully so (15 to 20 per cent above background). An autoradiograph showed, surprisingly, not an outline of the insect, but a single spheroid point-source (Kettlewell and Heard, 1961). This was dissected out at Harwell and it was found to be resting in the scales external to the thorax. It was a sphere with a diameter of 9μ and was identified as being typical of previous ones observed after test explosions: it consisted of fused silica coloured by traces of metal. Considering the back-logging of the atomic cloud, the weight of the particle, and its disintegration rate, we could say with some certainty that the moth must have flown into the cloud, probably near its source in North Africa. I have recounted this in detail as I believe that it is the first proof we have had as to the origin of such immigrants.

It is likely that the other pale coloured Lepidoptera which so frequently arrive with *N. noctuella* have a similar origin and are carried upwards and northwards from Africa on convection currents.

The mean day temperature at this time of the year is high, so that rapid metamorphosis is ensured. These conditions, I believe, are responsible for the pale colouration of the early immigrants. Experimental confirmation of this was obtained (Kettlewell, 1943–4) from work on the

Caradrinid moth, *Heliothis peltigera* Schiff., which so frequently accompanies *N. noctuella* on its early excursions.

Temperature experiments on Heliothis peltigera *Schiff*

This sub-tropical species frequently arrives in northern Europe under certain meteorological conditions from the Mediterranean countries and such invasions have been described as occurring in Britain from the earliest records (Barrett, 1895–1902). Specimens bred in captivity vary greatly in colour from the lightest yellow to dark chocolate-brown. The early immigrants (March to May) are always paler. Autumn specimens bred in Britain tend towards dark chocolate-brown. I have netted *H. peltigera* both by night (at the flowers of *Scrophularia*) and by day over clover, when it was conspicuous; I have never found it at rest. One of the problems raised by this colour variability is whether it has any adaptive significance.

The main larval food plant in Britain was, until 1928, considered to be Restharrow (*Ononis*) and not until that year did we realize that another very different one, *Senecio viscosus*, which is confined to shingle beaches on the south coast of England (for example, Dungeness, Kent, and the Crumbles, Eastbourne) was a more favoured choice. The Garden Marigold (*Calendula* spp.) also gave it an opportunity for extending its range inland and was at times made use of. Since that date large numbers were bred in captivity.

Cockayne (1930) stated that as a result of his experiments he had 'very little doubt that light ground-colour and reduced markings were the direct result of heat applied to the pupae'. My own results in 1931 conflicted with his as I bred uniformly dark examples from a batch of pupae that I had forced in heat. They had, however, been lying in their cocoons in the sand previously for a period of 2–3 weeks when they must have been subjected to varying temperatures. The mean hatching time was 43 days. Wightman (1931) allowed 20 days to elapse between the disappearance of the larvae underground and the commencement of treatment at 38°C (100°F) and, as in my own no doubt, the metamorphosis had already started in many. He produced medium-dark to dark specimens. W. G. Wynn (1933), studying the effects of humidity on variation, obtained opposite results in 1931 to those he did in 1933, almost certainly because his pupae were four weeks old before treatment.

At that time, 40 years ago, I was working as a medical student at St. Bartholomew's Hospital, London and thanks to Dr. Garrod, the Senior Pathologist, I had his laboratory incubators at my disposal. Consequently I (Kettlewell, 1943–4) was able to carry out temperature experiments on

H. peltigera whilst at the same time complying with a fixed rota for the accouchement of the mothers of Islington. The two responsibilities fitted perfectly.

Pupal treatment. Each pre-pupal larva was placed separately in a chipette box and the treatment, heat at 30°C (86°F) or cold at 12°C (54°F), commenced *within 24 hours of the act of pupation.*

Experimental results. The uniformity of the results under these controlled conditions was surprising in view of the fact that the stock was heterogeneous, being from wild-collected larvae from Dungeness, Kent. At 30°C (86°F), regardless of humidity, the pupae hatched rapidly and produced pale-straw coloured imagines. Only one per cent did not conform and these were refractory to continuing heat. By contrast, pupae which were given $5\frac{1}{2}$ hours at 30°C (86°F) in order to stimulate the commencement of metamorphosis, and then placed at a constant 12°C (54°F), produced dark-brown imagines and took three times as long to hatch. More precisely it appeared that the amount of pigment laid down was correlated with the time of development and this was decided by the temperatures imposed at critical periods in the pupal life.

Pupal reaction
Because of the uniformity of their behaviour and also because of their translucency, the pupae of *Heliothis peltigera* appeared to offer special opportunities for studying pigment deposition.

At this point it may be of advantage to refer to the pupal state in general. In univoltine species, there is a period of inactivity following pupation of from three months to three years prior to the commencement of the imaginal metamorphosis. In multivoltine species kept at high temperatures there is no such resting period and eclosion follows rapidly.

If P = day of pupation, C = commencement of metamorphosis, and H = day of hatching, PC represents the dormant period found in the majority of univoltine species, and CH the period of increased metabolism leading to eclosion. The two are quite distinct. In 1944 I gave then what I now regard as unfortunate designations. The first ($=PC$) I called the 'passiphase' and the second ($=CH$) the 'actiphase'. Though I now dislike these terms greatly I have found they served a useful purpose. In some species, such as *Eriogaster lanestris* L. and *Brachionycha nubeculosa* Esp., however, this method of analysis is not applicable for here the imagines develop but do not hatch for two to six years, remaining dormant within their pupal shells, but this type of behaviour is exceptional. It is one of the ways in which species, especially northern ones, escape recurring bad conditions if they emerge early in the year.

Pigment deposition in H. peltigera

We are here concerned with the laying down of the pigments in the pupae of *H. peltigera* which, when kept in continuous heat, can be considered as in the actiphase throughout: there is no resting period. At 30°C (86°F) the pupal state (CH) lasted only $8\frac{1}{2}$ days (200 hours) (except for the one per cent which deviated widely in their reaction) and all the imagines were a pale-straw colour. Nevertheless, exactly the same colour and behaviour was obtained from the deviants following treatment by 21 days in the cool at 12°C (54°F).

The first visible sign of development within the pupae is the darkening of the eyes and, at 30°C (86°F), this took place consistently in 5 days (117 hours). The actiphase can therefore be divided in to two parts, CE and EH (where E refers to eye-darkening). In order to find out more precisely the exact time of the laying down of the pigments, I subjected pupae in each part to a constant 12°C (54°F) or 30°C (86°F). Experimentally this led to the prolonging of each period but quite separately. The following is a summary of the results:

(1) Temperature during the passiphase has no effect on the subsequent pigmentation, which is entirely decided in the actiphase.
(2) Prolonging the CE period resulted in an increase in the normal patterning.
(3) Prolonging the EH period produced insects with a dark ground-colour.
(4) Prolonging both periods (=CH) to a total of 30 to 40 days led to the darkest *H. peltigera*, which were of a deep chocolate colour with pronounced patterning. This group also had the highest mortality rate.

The significance of environmental melanism

H. peltigera is therefore a species which regularly migrates north early in the year, has several broods in succession, and, because of lowered autumnal temperatures, the later imagines are dark instead of light. I have found a comparable situation in the migratory moth *Rhodometra sacraria* L. (Hydriomenidae), but here instead of cold and prolonged pupation producing chocolate colouration they give rise to the crimson (ab. *rosea*

Oberth) the dark-veined (ab. *sanguinaria*), or the smoky (ab. *fumosa* Prout) form, according to the exact treatment (Kettlewell, unpublished). The same phenomenon is likely, I think, to be found in many migratory species whose populations have to conform to hot, dry deserts and cool, wet environments in the course of a year, for instance *Utetheisa pulchella* L. (Kettlewell, 1963c, 1963d, 1964a).

The all-important question here must surely be whether autumnal darkening can provide survival advantages. I have found *R. sacraria* at rest in stubble fields on several occasions and its disruptive stripe and the positioning of the wings around the base of the stems undoubtedly make it highly cryptic. Although we now know that a southern migration is undertaken in the autumn by many migrant species, we have no knowledge of whether *H. peltigera* and *R. sacraria* act in this way. If they do, and dark colouration offers advantages during the return of such individuals to the southern reservoir, the increased pigmentation can be said to have adaptive significance. If not, it appears that both the act of migration and the colouration are fortuitous and of no consequence to the survival of the species.

'Buffering'

The colour and patterning of every species of Lepidoptera is finely geared to its well-integrated ontogenic development and to its habits, particularly in cryptic species. But this may be affected by physiological requirements. For those same genes responsible for one are usually also responsible for the other. It is noticeable that even extreme environmental conditions affect cryptic patterns but rarely (as they do in *H. peltigera* and *R. sacraria*), and the majority of cryptic species remain identical whether bred in heat or cold. I would suggest that in these a precise patterning is all-essential for survival and that the genes responsible for this are buffered against environmental effects. By this I mean that the gene-complex has been selected for colour stability and that this has priority over other considerations.

In aposematic species the situation is different. Pattern in particular is affected by the temperatures experienced during pupal life. I have already suggested that the reason for the existence of such variability is that warning colouration does not demand exact copying and that provided that it is exhibited on attack by a predator at the right moment, pattern is of little consequence. Hence the large number of aberrational forms in such aposematic species as *Abraxas grossulariata* and *Arctia caja,* Mutant genes affecting pattern have not been buffered by selection because this has not been 'required' for efficient defence.

Environmental effects in aposematic species

The technical difficulties in attempting to demonstrate environmental effects are typified in *Arctia caja*. So great is the latent variability that it is rare to breed from a wild female whose offspring, when treated under different temperatures during the pupal period, do not show extreme divergence. Yet that part of a brood, which is subjected to the normal range of temperatures they experience as pupae in nature may show little if any variability. Further, the number of aberrational forms varies with the degree of temperature abnormality. I quote the results of one experiment only, though this is typical of many: 635 pupae of an F_2 generation were given one of three treatments—(1) 'normal' (continuous (14°–21°C (60°–70°F)), (2) eight hours at 21°C (70°F) and 16 hours 'cool' at 4°C (40°F), (3) four hours at 21°C (70°F), 20 hours at 4°C (40°F). In batch 1, of the 161 bred, 3·8 per cent were classified as aberrational. In batch 2, 10·7 per cent of 228 pupae, and in batch 3, 18 per cent of 228. The corresponding mortality rate was 3 per cent, 1 per cent, and 4·4 per cent respectively. The aberrational forms varied from those with nearly white forewings to an example in batch 3 which had totally black ones. The simplest explanation for such results is that a number of genes, undisclosed under normal conditions, express their characters when given periods of heat or cool which they do not normally experience in the wild. Imagines in batch 1 hatched as always three to five weeks earlier than those in batch 3. All the aberrational forms which appeared experimentally are found at times in nature but offspring from these do not necessarily produce the same phenotypes in subsequent generations.

Environmental effects on single-gene expression

I have already referred to a switch-mechanism responsible for the complete change of pattern in the butterflies *Araschnia levana* in Europe and *Precis sesamus* in Africa (p. 301). We have no evidence of how many genes are involved in these but it is likely that there are several and one hundred per cent of the populations react uniformly to a seasonal change. Phenotypic variation in the expression of single genes under the influence of different environments has been demonstrated elsewhere: in plants, in Barley where an albino strain only manifests this character below 6·5°C (Collins, 1927), or again in the arthropod *Gammarus chevreuxi*, but not before 1944 in a British Lepidopteron (Kettlewell, 1944). In that year I recorded that I had bred a brood of *Panaxia dominula* from a typical female taken in Kent whose offspring behaved differently when reared in the normal varying temperatures between 7° and 27°C (45°–80°F) from those given constant heat at 21°C (70°F). Of the twenty bred under normal conditions all were f. *typica*. The 23 insects, which hatched earlier in the year, whose pupae had experienced continual warmth, were entirely different however. Five had nearly all-black forewings but with minute tracings of the white patterning, and the black markings of the hindwings accentuated, six were normal f. *typica*, and 12 were intermediates which phenotypically resembled a well-known form, *paucimacula* Schultz which from time to time appeared in wild stock from the same locality (Deal, Kent). Previously I had bred from these and had produced f. *typica* only under normal outdoor temperatures. Because of the 1:2:1 ratio in my brood (though I was unable to continue an F_1 since, as so often in the Arctiidae and Hysidae, the pairings were infertile), this single brood has demonstrated that pattern variability can be maintained in a population and this may remain undisclosed except under particular external influences. I regard such instances as being of the greatest interest because it is possible, indeed likely, that such patterns were built up in the past when the climatic conditions were entirely different. Under those prevailing today the appearance of such forms is successfully buffered. Ford (1957) has dawn attention to the phenotypic resemblance between certain individuals of *Vanessa urticae* of British origin when subjected to heat treatment and the Corsican race *ichnusa*. With cold treatment the same stock will produce dark forms similar to f. *polaris* found in north Scandinavia. I believe that many so-called pheno-copies (Goldschmidt, 1938) are relict patterns from the past.

Conclusion on environmental melanism

The concept of melanism in this chapter is one used in its broadest sense, that of darkening; be it the chocolate-brown of *H. peltigera* or the dark grey or crimson of *Rhodometra sacraria*. In both there is increased pigmentation.

The examples I have referred to can be divided into two groups. First, species which regularly migrate from North Africa, where they have a pale sandy colouration, assume dark colouration when bred in the cooler climes of northern Europe. Whether this has any adaptive advantage we at present do not know. Secondly, and in aposematic species only, the typical pattern comes to vary greatly when abnormal environmental conditions are imposed. In *Arctia caja* these forms may be white or black amongst sibs. In *P. dominula* a single gene can alter phenotypic expression according to the temperature treatment imposed during the pupal period. Because such genetic variability exists but is only exposed under abnormal environmental conditions I have suggested that this type of environmental melanism is a relict of the past.

Résumé of Part VI

'Miscellaneous Melanism' was originally designed as a single chapter in order to include a number of melanisms which did not conform. The fact that I have had to break such a unit into four separate chapters is, in itself, evidence of the diverse paths which melanic species, subspecies, and forms of species have taken.

These four chapters demonstrate widely different genetical and evolutionary processes contributing to melanism and each example awaits a great deal more research. One of my objects has been to draw attention to such rewarding projects.

Part VII. The Synthesis

19. A theory of melanism

I HAVE attempted to describe and account for Industrial Melanism. On the one hand I have provided evidence that certain specialized biotopes have acquired dark-coloured Lepidoptera in the past and that these non-industrial forms are the forebears of the Industrial Melanisms of today. On the other, I have shown that Recessive Melanism but rarely contributes to this group. The reason for the dominance of the industrial melanics, even at inception, remains to be explained. Each of these three situations needs fuller detail.

The contribution of non-industrial polymorphisms and new mutations respectively

Ancient Melanism occurs today in the seven specialized categories which I have denoted (Chapter 10) and the melanic forms found there, usually at low frequencies, have from time to time served as reservoirs for the recent industrial melanic explosion we are now witnessing. In Britain we have evidence that only in a few species have the melanics been able to take advantage of such readily available sources; these include such species as *Allophyes oxyacanthae*, *Phigalia pilosaria*, and *Cleora repandata*. This is because Non-industrial Melanism has usually not occurred in proximity to industrialization. For this reason many industrial melanic species have had to wait for rare mutations to take place prior to spreading.

This accounts, I think, for the lag of one hundred years and more before the first record of a black form in such cryptic Lepidoptera as *Brachionycha sphinx* ab. *fusca* Cockayne and *Xylocampa areola*. Melanism has still not appeared in *Graptolitha ornitopus* in Britain even though in North America all the related species in this genus have distinct dark forms. This whitish moth is conspicuous at rest on present-day trunks and palings; not surprisingly it has become rare or absent from large areas of England though it is still common in parts of southern Ireland.

The history of Industrial Melanism on the North American continent is, I think, entirely different. For here non-industrial melanic forms have been freely available. To some extent this may have been a result of the presence of fire-resistant trees throughout most of North America, in particular certain Pines which are virtually indestructible (Chapter 10 (**v**)). In such coniferous forests melanic forms have continued to thrive even at a low frequency. More important is it that such forests may frequently exist close to industrial areas; hence the almost simultaneous occurrence of melanism in over 100 species of cryptic moths in Pittsburg and elsewhere. Melanism is found throughout many states of North America; one kind fades into another.

The significance of the difference between recessive and dominant melanism in Britain

With few exceptions both in Europe and in North America, industrial melanic forms are dominant or incomplete dominants; but rarely are the heterozygotes variable or indeed visibly recognizable. So little has been recorded from Canada and America on Recessive Melanism, or indeed, about the genetics of melanism itself in the Lepidoptera, that I must exclude the North American continent from the tentative theories I now put forward on the position of Recessive Melanism.

In Britain under one per cent of species maintaining melanic forms which can have any claim as being due to the effects of industrialization, are recessive. In particular, we have worked extensively on one, *Lasiocampa quercus*, and here three melanic forms, phenotypically identical, but genetically separate, occur at frequencies of 1 to 70 per cent in three widely separated localities. All are recessive. In two of these localities we have witnessed and filmed heavy predation by gulls. In Caithness we showed by release experiments that this predation was selective and that the melanic form had a 90 per cent cryptic advantage. This is even greater than that I determined in *Biston betularia* f. *carbonaria* which is typical of industrial melanic species with dominant inheritance.

With these data it is pertinent to inquire as to why so few recessive melanic polymorphisms occur: it must represent a most unusual situation.

I have suggested that the reason for this is that only in the last few thousand years has *L. quercus* extended its range into heather moorland (or onto exposed *Salix repens*, another dwarf shrub species). Here, outside the deciduous woodlands which had previously been its normal heritage, for the first time this species has come into conflict with predators of large size and capacity; in particular gulls. When yellow, the females are cryptic at rest on dead deciduous foliage, but are eliminated when on heather-moors because of their conspicuousness. Disadvantageous characters involving colour which in the past have been rapidly eliminated, and which have had little effect on the gene-complex, are usually recessive when they recur as mutations today.

The melanic forms of *L. quercus* are different therefore, for here, though recessive, they may contribute up to 70 per cent of a population as homozygotes. Yet even after a period of several thousand years dominance has not evolved, and the majority of heterozygotes are indistinguishable from f. *typica*. The achievement of dominance in response to a changed environment must always, as in the past, be a slow process in species in which the gene-complex has been previously 'unadapted'.

I believe that what we are witnessing today in *L. quercus* and its subspecies *callunae* is the first occasion on which melanic crypsis has presented advantages to this species. I believe that the same genetic metamorphosis has taken place in the past history of each industrial melanic form which today has full dominance.

The control of dominance

I have attempted to explain the universal dominance of the industrial melanics. I have also suggested that their presence today has arisen from two separate sources—the local migration of earlier non-industrial melanic forms and fresh mutations. In each, the black segregate clearly from the light; but rarely are there intermediates. In Britain a considerable body of literature goes to show that the majority of industrial melanic forms were unknown prior to industralization, in spite of extensive collecting from the eighteenth century onwards. How else, therefore, could our melanic forms have arisen in Britain, except by mutation? There is contributory evidence to this premise. Within a single species different centres of melanism have occurred in isolation widely separated from each other; thus melanic *Cleora rhomboidaria* Schiff. have at least two centres of origin, Norwich in the east and Portsmouth in the south.

The problem posed is how, after extinction of a melanic form for a number of years, has dominance of such a form been maintained within a gene-complex. As early as 1955 I attempted to test how the gene responsible for f. *carbonaria* of *B. betularia* from industrial Birmingham stock would express itself when placed in a complex where this melanic form did not occur (Cornwall, south-west England). Clear-cut segregation occurred in all the F_1 broods. Using random f. *carbonaria* from these and outcrossing them to Cornish f. *typica* for three generations, the expression of f. *carbonaria* changed, for the black forewings were uniformly peppered with white scales (Kettlewell, 1965e). On the one hand they differed from the parents of F_1, on the other they had little resemblance to the earliest f. *carbonaria* which had been taken in the middle of the last century. Such specimens frequently had small wedges of white markings on both forewings and hindwings (Plate 7.1). This suggested that the earliest f. *car*-

bonaria might be genetically different from present day forms occurring in Birmingham; also, that the gene-complex in Cornwall was different from that in Birmingham.

The breakdown and buildup of dominance

The breakdown. The results of the 1955–9 experiments encouraged me to repeat the out-crossing of *B. betularia* f. *carbonaria* into populations farther afield; namely to the North American Peppered Moth, *Biston* (*Amphidasis*) *cognataria* Gn. (Plate 9.1). The two species have been separated from each other for several hundred thousand years and though they have deviated in characters the crosses *B. betularia* × *B. cognataria* (of Canadian origin) are fertile and very little upset in sex valency takes place. The North American species, *B. cognataria*, has a comparable melanic form (f. *swettaria* Bns. and McDngh.) to f. *carbonaria* of *B. betularia*, and this is common in many industrial areas in the eastern United States, but as far as I know is rare or absent from most of central Canada. It was from such a locality that I obtained my stock.

Heterozygous f. *carbonaria* of industrial (Birmingham) origin were paired with Canadian wild type. Fifty per cent of the F_1 were recognizable in all broods as being f. *carbonaria* (or modified f. *carbonaria*) but after only three generations there was no clear-cut segregation (Plate 19.1). For this reason it was necessary to extract from the offspring of each brood all the darker individuals which appeared phenotypically to contain the f. *carbonaria* gene. From these, parents were chosen at random for crossing with Canadian wild type. In the F_4 a complete gradation of insects from black to light occurred. At this stage 94 per cent of the gene-complex must have been of Canadian origin. In this short time the dominance of f. *carbonaria* had been dissipated and clear-cut segregation had vanished in every brood.

The buildup of dominance. Random specimens of the intermediate f. *carbonaria* obtained in this way were crossed with *B. betularia* f. *typica* of industrial origin. It was here my intention to reconstitute dominance by reintroducing the British gene-complex over a period of years. It came as a surprise to discover that in the first generation clear-cut segregation took place (Plate 19.2). From this the following two points can be stated with confidence. (1) There are genes present in British *B. betularia* of industrial origin that ensure the expression of near-full dominance. (2) These modifying genes are themselves dominant but not linked with that for f. *carbonaria*. It would be of the greatest interest to find out whether these genes were universal throughout Britain or had been selected more recently since the advent of industrialization.

I repeated, therefore, crosses between intermediate f. *carbonaria* (ex F_4 *B. betularia* f. *carbonaria* × Canadian f. *typica*) and British f. *typica*, but on this occasion they came from Cornwall, south-west England, where melanism in this species does not at the present time occur (Plate 19.3). Clear-cut segregation took place in all broods in the first generation but the f. *carbonaria* were not as extreme as those I had obtained when using industrial stock. It is clear from this first that there are genes present in the complex of British *B. betularia*, even in the absence of Industrial Melanism, which ensure the full expression of f. *carbonaria* of each mutation; secondly that the selection for the blackest f. *carbonaria* has taken place in industrial areas. By contrast no such modifiers are present in Canadian stock (Plate 19.4).

We have no reason to believe that *B. betularia* f. *carbonaria* is in any way different from the majority of other cryptic species which have melanic forms. If this is true it must mean that the local gene-complex of each species is geared in such a way that at mutation each melanic form is ensured fairly complete dominance.

The unresolved question

How dominance-modifiers, which are themselves today dominant, can be maintained in a population in the absence of melanic polymorphisms, except by mutation, is indeed perplexing, though if one accepts them as palaeogenes, as in certain blood groups, it becomes somewhat less so. The fact that they are so preserved in *Biston betularia* ensures that the blackest form, f. *carbonaria* is available on mutation and that such extreme melanism does not have to be built up afresh on each occasion which would, indeed, be a lengthy process.

The recurring need throughout the ages and the important contributions of melanism to species survival suggest that palaeogenes controlling it may play a special part analogous in some ways to the supergene controlling the different mimetic forms of *Papilio dardanus* (Clarke and Sheppard, 1960c). It is more likely that the different alleles which developed during different periods in the past, lie within a cistron. Recent work by Clarke and Sheppard (1964) and also Lees (1968) on *Biston betularia* has shown that one member of the '*insularia* complex' is allelic to f. *carbonaria*, and in *Phigalia pilosaria* Lees (1971) has demonstrated that two of a number of different melanic forms are also allelic. Similar situations are likely to be found commonly in other species.

The natural history of Industrial Melanism

That industrialization should have led to conditions favouring melanism is entirely fortuitous. Its success is a result of the effects of three components, a darkening of the backgrounds, diminished light intensity due to haze, and physical superiority due to a tendency to evolve heterozygous advantage.

As smoke began to drift across England at the end of the eighteenth century, at first the more extreme melanic mutations were not selected. It is a fact that in many cryptic species these blacker forms only replaced early lighter melanics at a later date. Thus in *Biston betularia*, f. *insularia* has usually appeared in advance of f. *carbonaria*. In *Cleora rhomboidaria* in some localities one melanic population has been entirely replaced by another and more extreme one, f. *perfumaria* by f. *nigra*. In industrial Yorkshire, every lepidopterist knows that the blackest forms of many species are found around their industrial centres. This, then, is the recurring substitution of one gene by another.

There is, however, a second means of becoming darker which has also been taking place in the last 100 years. This is brought about by an entirely different mechanism which is identical to that I have already described in the buildup of dominance in the *B. betularia* × *B. cognataria* hybrids—one controlled by modifiers. Such an example is seen in the local populations of *B. betularia* in Cheshire, Yorkshire, and elsewhere, for here the white dots generally remaining around the thorax of f. *carbonaria* are now completely absent in a majority of individuals. Corroboration of polygenic modification has been demonstrated in the laboratory where industrial f. *carbonaria* outcrossed to Cornish f. *typica* produced a form of *carbonaria* with light speckling (see p. 315).

These two distinct mechanisms, substitution and modification, account for the melanic phenotypes which are found today. The first is the result of millennia of evolution on melanism; the second depends on the laws governing gene/chromosome mechanics such as recombinations, which take place today as in the past. Melanism, like sex, is an ancient and fundamental part of nature.

APPENDIX A: Breeding techniques

THE Lepidoptera offer, I think, exceptional opportunities for both laboratory and field experiments: this is because of their particular qualities—they are small, easily bred (but only with specialized knowledge), their genotypes are usually recognizable, and they offer rewards to research into their genetical complexities, their biochemistry, and their survival mechanisms in the field, developed under a million years of intensive selection.

For over 50 years I have bred Lepidoptera. For the first 30, chiefly from mutant forms taken in the wild, with the object of finding out their modes of inheritance (Kettlewell, 1963b). I am now discussing the entirely different requirements which we have had to develop in the last 15 years, namely the mass-breeding of imagines for various experimental uses: for example for mark–release–recapture in the field, or for testing viability differences in the back-cross broods, and for a host of other uses. The demand for living material from dealers, of such species as *B. betularia*, *G. bidentata*, and *P. pilosaria*, which demonstrate the genetics of melanisms and their advantages, and disadvantages, has increased greatly in the last ten years, this chiefly from universities, schools, and other centres of learning. It is useless to attempt to undertake breeding experiments without some prior knowledge of large-scale breeding, and for this reason, I give a résumé of the various techniques we have developed at Oxford in the last 15 years. Many of these are due to the enthusiasm, originality and ingenuity of Gillian Brooks, who has tested, and indeed pioneered, successful short-cut methods in the mass breeding of a number of species.

The limitation imposed on breeding larvae is chiefly from that complexity of diseases known as polyhedrosis. Usually it occurs in one of two forms: acute nuclear, or chronic cytoplasmic (Smith and Rivers, 1956). I have, on one occasion, experienced a situation in which every individual in an apparently healthy brood of *Pontia daplidice* (Rhopalocera) swelled up, burst, and died within a period of 48 hours, due to a particularly virulent form of virus. I have had similar catastrophes at times with *B. betularia*. This can be avoided by taking certain precautions, and by following strict rules. First, by taking into account the fact that polyhedrosis is passed from one generation to another in the egg (possibly also in the sperm). Pairings should not be taken therefore from infected broods, and if they have to be by necessity, use only those individuals which were the earliest to pupate, and males rather than females. Secondly, because the disease is not only highly contagious but also because the virus can live away from its hosts for long periods embedded in crystals, all containers, be they tins, cages, or sleeves, must be sterilized in an autoclave; in fact the whole breeding programme must be carried out having regard to asepsis as in an operating theatre or a maternity ward. Wooden cages must be washed out with strong soda (which dissolves the polyhedral crystals). Thirdly, the earlier that young larvae are given maximum room, as when moved into large sleeves, the less likely are they to develop this disease by passage (as envisaged by Louis Pasteur) (Hamilton, 1885).

I have shown, by marking full grown larvae of *Panaxia dominula* with ^{35}S, that a 90 per cent mortality rate takes place in nature during the late larval and pupal stages (Kettlewell and Cook, 1960). In the laboratory breeding of these insects, this can largely be avoided as I believe that (at least in *P. dominula*), this is due to predation, particularly by voles. Nevertheless, pupae in the laboratory can have a high mortality rate depending on their treatment, which varies according to the species. I must therefore attempt to give a guide for methods of mass-breeding throughout the various stages of ovum, larva, pupa, and imago.

Ova

The requirements of the Rhopalocera for ovipositing are entirely different from those of moths. Usually, they have to be provided with growing food plants and they lay only in the sun. Muslin cages (containing potted plants) or sleeving (as for example with *Papilio* spp. on greenhouse *Citrus* or *Choisya*) are to be recommended. A further consideration is that suitable flowers must be introduced, or alternatively females can be hand-fed by holding the wings dorsiflexed and extending the proboscis onto a pad of cotton wool soaked with honey and water. Hovanitz has stated (personal communication) that females so fed (for instance *Colias* spp.) lay more eggs than those merely sprinkled with water. The ova should be collected daily as they have many predators, not least the 'red spider' mite, and also certain Tachinid flies, and Ichneumonids, which parasitize Lepidoptera eggs (and they can traverse the muslin mesh with ease).

Most Heterocera lay well in glass-topped tins if provided with their specific requirements. Thus *B. betularia*, *A. cognataria*, *L. hirtaria*, and other species which have long ovipositors lay their 'cakes' of eggs freely in cotton wool which has been 'fluffed up' (nylon wool is not accepted) or alternatively muslin 'balls' (not nylon). Species which drop non-adherent eggs at random such as *L. quercus*, *P. dominula*, and *H. humuli* lay freely in the same type of container, but it is of advantage to provide a small amount of the larval food-plant. We make use of this in the laboratory by placing a funnel beneath a battery of caged females which directs the eggs into a test-tube. Other more specialized species will only oviposit on the edges of paper or twigs, (for example *C. elinguaria*). The majority of species however, lay eggs or batches of eggs regardless of their environment (for example *A. caja*, *S. lutea*, and many Sphingids).

In most species eggs must be examined daily for signs of darkening prior to hatching. With few exceptions (and these are of the greatest interest, for example *Euplagia quadripunctaria* or *Argynnis paphia*) young larvae die within 24 hours unless they have their correct food-plant immediately. Over-wintering ova (for example *B. sphinx* and *C. elinguaria*) should be kept in a refrigerator at around freezing point so that the larvae hatch simultaneously when given warmth in the late spring.

Larvae

Young larvae should be segregated from unhatched eggs as soon as they have commenced to wander. In Geometrids and other species, which hang from a silken thread, individuals are most easily picked up by means of a camel-hair paint-brush. *First-instar* larvae of most species (for example *B. betularia*,

PLATE 19.1 (p. 316). The breakdown of dominance: third generation from P_1 *B. betularia* f. *carbonaria* (of Birmingham origin) crossed to Canadian (Wicklow Bay) *B. cognataria* f. *typica* ($\times \frac{1}{2}$).

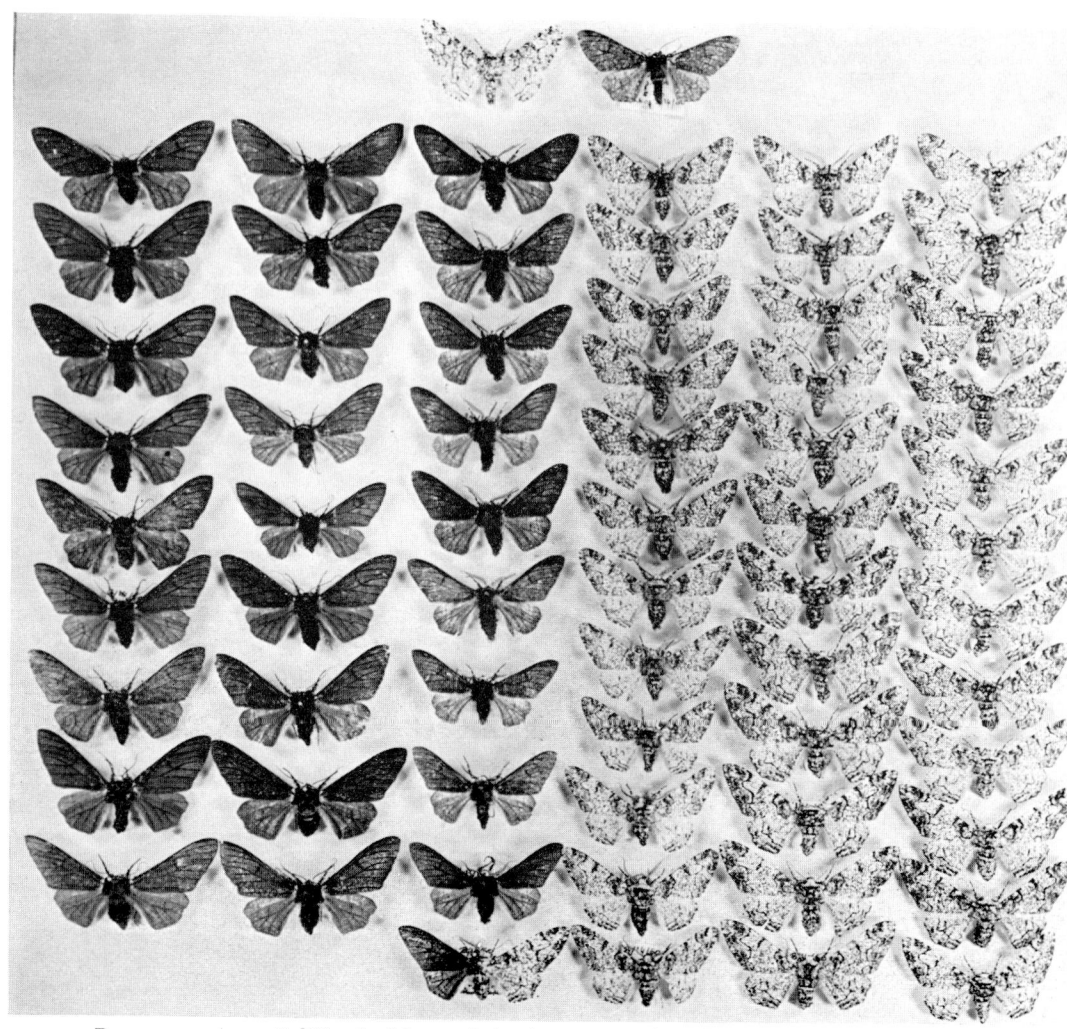

PLATE 19.2 (p. 316). The build-up of dominance: random 'f. *carbonaria* hybrid' (third generation) × *B. betularia* f. *typica* from industrial Birmingham, England (first generation) ($\times \frac{1}{2}$).

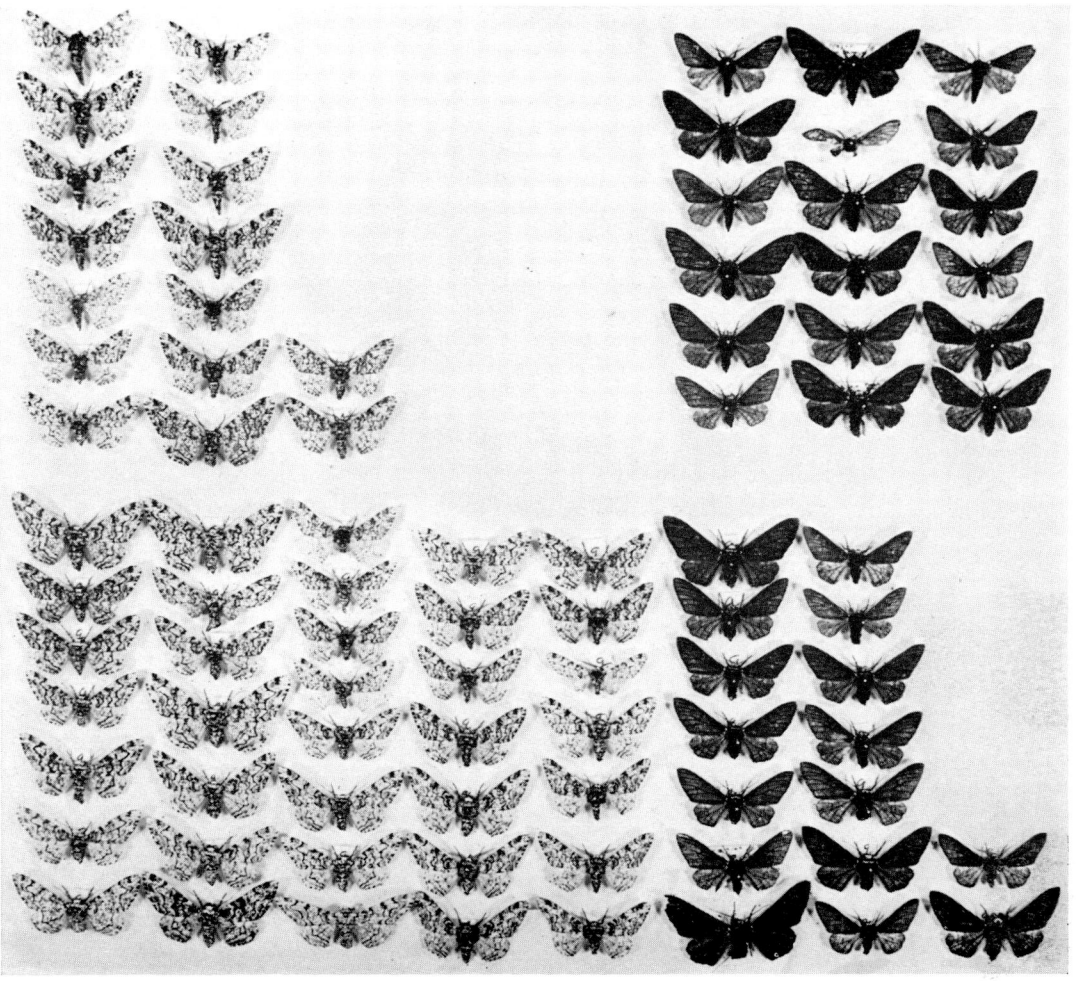

PLATE 19.3 (p. 317). The build-up of dominance: random 'f. *carbonaria* hybrid' (fourth generation) × B. *betularia* f. *typica* of Cornish origin, where f. *carbonaria* does not occur (first generation) ($\times \frac{1}{3}$).

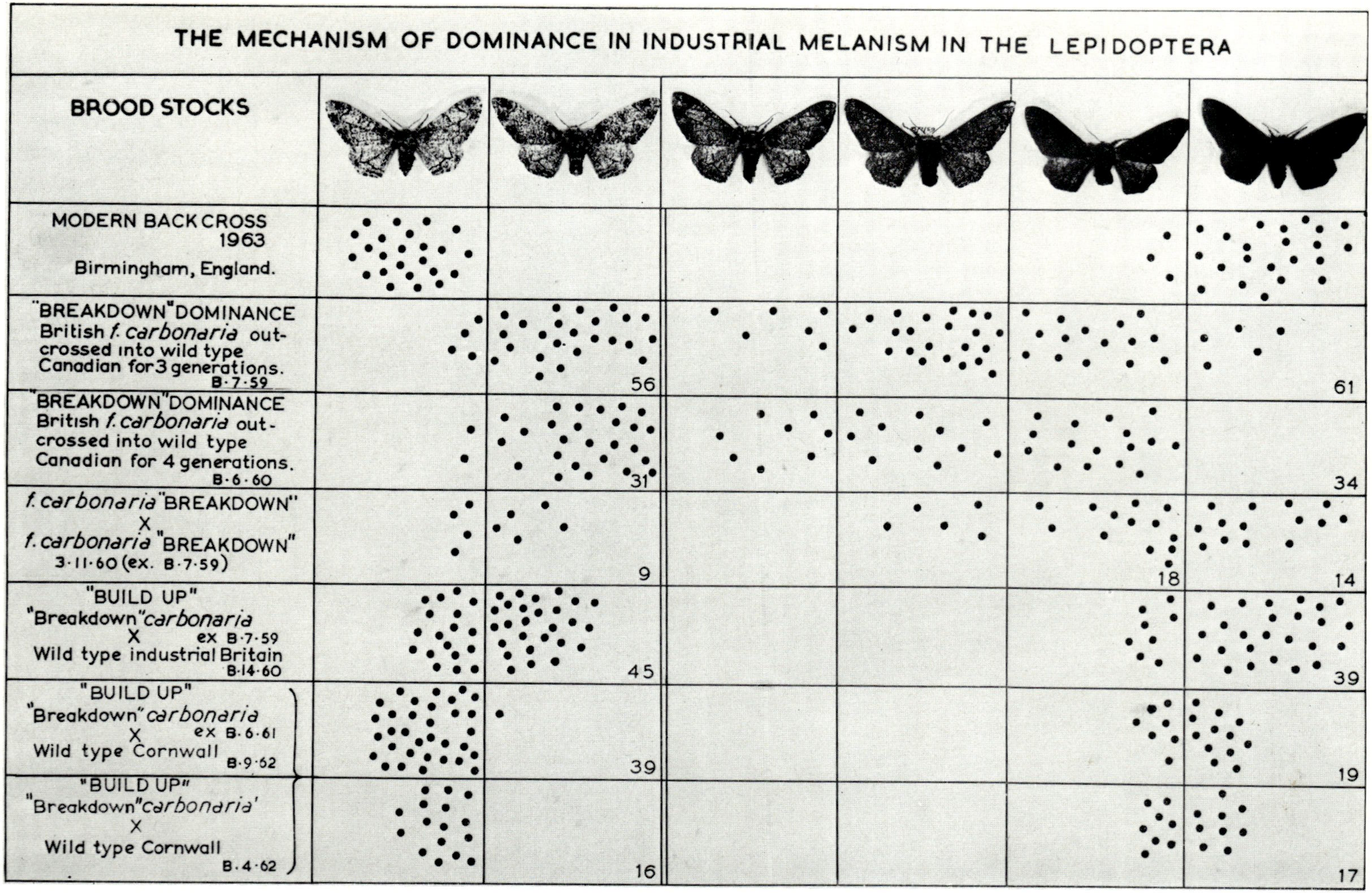

THE MECHANISM OF DOMINANCE IN INDUSTRIAL MELANISM IN THE LEPIDOPTERA

BROOD STOCKS						
MODERN BACKCROSS 1963 Birmingham, England.						
"BREAKDOWN" DOMINANCE British *f. carbonaria* outcrossed into wild type Canadian for 3 generations. B·7·59		56		61		
"BREAKDOWN" DOMINANCE British *f. carbonaria* outcrossed into wild type Canadian for 4 generations. B·6·60		31			34	
f. carbonaria "BREAKDOWN" × *f. carbonaria* "BREAKDOWN" 3·11·60 (ex. B·7·59)		9			18	14
"BUILD UP" "Breakdown"*carbonaria* × ex B·7·59 Wild type industrial Britain B·14·60		45				39
"BUILD UP" "Breakdown"*carbonaria* × ex B·6·61 Wild type Cornwall B·9·62		39				19
"BUILD UP" "Breakdown"*carbonaria'* × Wild type Cornwall B·4·62		16				17

PLATE 19.4 (p. 317). Distribution chart of hybrid phenotypes.

A. cognataria, L. hirtaria, P. pilosaria, and also those of *A. caja* and *L. quercus*) are best kept in air-tight cylindrical tins with glass tops (diameter 3″, height 2″) or alternatively in plastic sandwich-boxes. It is essential that a filter paper which can be changed regularly covers the bottom. At this stage I consider 100 larvae per tin the limit for the first four species and certainly not more than 50 for *A. caja* and *L. quercus*, which are larger. On replenishing food do not remove leaves unless mouldy, merely mix fresh ones with the old and change the filter paper. Never introduce foliage which is wet and never place tins in a position where sun can strike; one minute only of midday sun usually kills 100 per cent of young larvae in closed containers. Never disturb during ecdysis which frequently takes place on spun silk on the glass top. *Second-instar* larvae need a continuation of the same treatment but 'diluting' must be carried out by dividing the contents of one tin into two. Alternatively such small larvae can be directly sleeved onto *Salix* spp., particularly *S. caprea* (Sallow). It is essential that the sleeve material is of extremely small mesh (georgette or terylene silver-seal). As soon as they are large enough, all larvae can be reared in large sleeves (6 ft × $2\frac{1}{2}$ ft.) of wider mesh. They will usually contain sufficient food for 100 larvae of *B. betularia, B. cognataria,* or *L. hirtaria*, till full growth.

As a labour-saving technique, as the time of pupation approaches, a large flower-pot, containing bulb fibre and moss, is introduced at the lowest level of the sleeve and is either supported on a stake or rested on ground level. In practice we merely collect the pots at the end of the summer and extract the pupae which from October onwards are laid out in dry fibre in shallow tins (Kodak ciné are ideal). These should be examined weekly for the extraction of any dead individuals.

The breeding of *A. caja* is somewhat different from that of the univoltine Geometrids for here it is possible to obtain four generations a year. Using a day-length of $17\frac{1}{2}$ hours (by Venner switch control of a mercury-vapour bulb) and a temperature of 27°C (80°F), larvae reach full growth in 20–30 days. They should be moved into large perforated-zinc cages (3 ft × 2 ft × 2 ft) for their last two instars and they will make their cocoons satisfactorily in the interstices of egg trays which have been loosely tied together. Pupae should be extracted and laid out on coconut fibre. The larvae feed well on cabbage in the winter months.

The breeding of *L. quercus* ssp. *callunae* demands further techniques for here we have a larva which in nature hibernates the first winter and spends the second in the pupal state. As it is essential to keep the two-year cycle for outdoor experiments (imagines only occur in odd years) larvae are best sleeved the first winter on evergreen Privet, **Ligustrum vulgare**, and changed onto *Salix* in the spring.

The diapause can however be broken and this has advantages when laboratory work into the genetics of melanic forms is the main consideration. This can be accomplished by placing first-instar larvae into a $17\frac{1}{2}$ hour daylight and 27°C (80°F) regime from commencement. We have reduced the 24-month cycle to 12 months by this means, but in only a proportion of each brood.

Larvae, when near to full growth in July or August, must be removed to large, well-ventilated containers and hand-fed prior to pupation. (Note: the larval hairs and the cocoons are highly toxic, producing local urticaria, and severe eye conditions if rubbed onto the cornea.)

Pupae

The pupae of Rhopalocera must never be moved from their attachment sites. By contrast the pupae of most Heterocera should be extracted a month or so after pupation, in particular those which incarcerate themselves in soil. They should be kept in bulb fibre in air-tight containers until immediately prior to hatching, when accommodation for wing-drying must be considered. Thus the Kodak film tins, so excellent for overwintering pupae, are fitted with a 6" high collar of stapled cardboard on which the tin lid rests. Improvisations and improved techniques must be developed with each species according to their special requirements.

Imagines

Bred moths are usually needed for one of three objects: for mark–release experiments in the field, for recording genetic variability, or for obtaining selected pairings.

For the first it is necessary to accumulate numbers and it is important to remove each insect (especially males) as soon as possible after they have dried their wings; they should then be placed in a refrigerator at about 7°C. I find that lower temperatures than this may affect the subsequent pairing behaviour and the fertility, particularly of females. In regard to the second, that of recording variability and its genetics, it is essential to conserve each specimen permanently as no power of description is as satisfactory as a visual record. For obtaining special pairings it is usually sufficient to cage a selected pair in a small muslin bag; it is entirely erroneous to think that larger cages give better results. Each species has its own particular requirements for assembling and copulation to take place: thus *B. betularia* and *B. cognataria* mate between 10 p.m. and midnight, and *L. quercus* in sunshine only, usually in the afternoon.

The important rule here is never to leave males in close proximity to the females if pairing does not take place on the first occasion, for they rapidly go into a state of fatigue when continually subjected to female scent. Always extract and isolate the males and introduce them again at the appropriate hour on the following day.

None of the imagines I have mentioned in Appendix A need feeding because in each the proboscis is rudimentary. In others, however, such as Rhopalocera, or in moths like *Panaxia dominula*, *Xylomyges conspicillaris*, and *Amathes glareosa*, and in all Noctuids in which the proboscis is functional, feeding with water and glucose or honey is essential for maximum ovipositing. Only by trial at each stage can successful large-scale breeding be accomplished.

APPENDIX B: Melanism in British Heterocera

This resumé does not attempt to be comprehensive, rather it calls attention to some of the species, and their localities, in which melanism has been recorded. Specimens can be seen in the National (Rothschild–Cockayne–Kettlewell) Collection in the British Museum of Natural History, London; the species referred to follow the order in which they are arranged in this Collection. When possible I also give their genetics, the class of melanism into which each falls (Industrial, Ancient, or Recessive), and on occasions I give the more important references.

The prefix 'ab,' or 'f.' is of little significance other than to record whether a mutant is a rare aberration or a recurring form; S represents sample-size. The nomenclature is that of Kloet and Hincks (1945) and also of Goodson and Read in *Aberrational and subspecific forms of British lepidoptera*, Vols. 1–12 (unpublished, copies in British Museum of Natural History, Oxford University, and elsewhere).

Laothoe populi L.
    ab. *suffusa* Tutt. Frequent in Scotland.
Herse convolvuli L.
    ab. *obscura* Tutt.
Sphinx ligustri L.
    ab. *brunnea* Tutt.
Hyloicus pinastri L. (see text)
    ab. *nigrescens* Lempke. Industrial melanic in the Netherlands, not yet established in Britain.
    ab. *ferrea* Cless. Industrial melanic on the continent. ? dominant.
    ab. *unicolor* Tutt. Suffolk (Cockayne 1926c). Recessive.
    ab. *nigra* Walther.
Deilephila porcellus L.
    ab. *nov* (R.–C.–K. Collection). Kendal, Westmorland.
Deilephila elpenor L.
    ab. *daubi* Niepelt.
Macroglossa stellatarum L.
    ab. *subnubila* Schultz.
Cerura furcula Cle.
    f. *borealis* Boheman. Northern melanic race.
Cerura vinula L.
    f. *phantoma* Dalman (see text).
    f. *arctica* Zetteredt. Northern (Scandinavian) melanic race. ? rare aberration in Britain.
Stauropus fagi L.
    ab. *obscura* Rebel. Industrial melanic, dominant to f. *typica*, centred around London. To the north-east, Epping (where over 90 per cent are melanic) and Cambridgeshire. To the south, Kent, and to the west, Surrey, Hampshire, Gloucestershire and as far as Somerset.

324 Appendix B

Sample figures (showing percentage of melanics):

		melanics %	
Surrey	Ottershaw	93	($S = 175$, R.F.B. 1949–59)
Hampshire	Winchester	5·1	($S = 39$, D.W.H.F. 1956)
	Rowland's Castle	8·0	($S = 25$, F.C.S. 1956)
	Chandler's Ford	6·8	($S = 59$, B.G. 1956)
Berkshire	Kingsclere	7·4	($S = 54$, R.S. 1956)
	Windsor	95	($S = 20$, A.H.H.[4] 1971–2)
Worcestershire	Worcester	11·8	($S = 17$, R.H.C. 1956)
Bedfordshire	Woburn	12·5	($S = 16$, S.H.K. 1956)
Kent	Ashford	16·6	($S = 18$, C.A.W.D. 1956)

Melanic *fagi* have also been recorded (but with no sample numbers) from Cambridgeshire (B.O.C.G.), Dorset (Lyme Regis) (C.J.L., one in 1957), Kent (Westwell and Ham Street, A.M.M.), also Orpington, Pett's Wood (R.G.C.), Somerset (Weston-super-mare, C.S.H.B.), Sussex (Arundel, approximately 5 per cent G.W.H.).

Nearer London it appears that the melanic frequency is much higher, near 90 per cent, at Herts (Totteridge), (R.I.L.), Middlesex (Pinner, 98 per cent W.E.M.), and it is probably of the same order in Epping Forest. In 1971, it appeared to be spreading.

Drymonia dodonaea Schiff. (syn. *trimacula* Esp.)
  ab. *peripurescens* Cockayne. Ancient melanic.
Notodonta dromedarius L.
  f. *hibernica* Cara. Northern race. Ancient.
  ab. *nigra* Cockayne.
Notodonta anceps Goeze. (syn. *trepida* Esp.)
  f. *fusca* Cockayne. Genetics: ♀ f. *typica* ex-Westmorland gave 45 f. *typica* and 14 melanic.
    ? Recessive heterozygote parents (A.R.). Ancient melanic.
    Melanic forms occasionally in Surrey (Witley, J.L.M.).
    ? Industrial melanic.
Lophopteryx capucina L. (syn. *camelina* L.)
  f. *nigra* Riesen. Ancient melanic.
Clostera curtula L.
  ab. *webbiana* Rebel.
Tethea octogesima Hueb. (syn. *ocularis* Guen.)
  f. *fusca* Cockayne. Variable in expression, widespread in the Netherlands. Appeared for the first time in southern and eastern England 1945, increased rapidly since then.

Sample figures (showing percentage of melanics);

		melanics %
Cambridgeshire	Cambridge	'... I should assess that over half were of the dark form.' ($S = ?$, B.O.C.G. 1955)
,,	,,	100 ($S = 7$, C.F.R. 1955)

Essex	Westcliff-on-sea	33·3 (S = 30, H.C.H. 1956)
,,	,,	31·6 (S = 38, H.C.H. 1957)
Kent	Ashford	33·3 (S = 12, C.A.W.D. 1951–3)
,,	Folkestone	
	Romney Marsh	first appeared 1952.
,,	Dymchurch	'a fair number' (A.M.M. 1952)
London area	Totteridge	first appeared 1958.
,,	,,	50 (R.I.L. 1969)

f. *frankii* Lempke. Industrial melanic, dominant, first recorded in south Netherlands 1925, not very common there (Lempke). A series of about two dozen *octogesima* are in the R.–C.–K. Collection labelled 'ab. *nov*. Cambridge'. I think these are f. *frankii*.

This species is the only example I know of, where a definite migration (and not mutation or local spread) has led to widespread Industrial Melanism.

Tethea or Schiff.
  f. *albingensis* Warnecke.
    Industrial melanic, dominant, Durham (J. D. 1914), Scarborough, York, Forge Valley, ? authenticity. This form arose in Hamburg in 1910 (Gerschler 1915).
  f. *permarginata* Hasebke. Industrial melanic.
Tethea duplaris L.
  f. *obscura* Tutt. Ancient melanic in Scotland, industrial elsewhere, as in Yorkshire. In Inverness-shire, Scotland, 'all dark' ($S = 100+$, G.W.H.). Hampshire, (Rowland's Castle), in 1957, melanics were 82·35 per cent ($S = 17$, F.C.S.). Surrey, (Witley), 'all melanic' (J.L.M. 1965).
Tethea fluctuosa Hueb.
  f. *unicolor* Lempke. Rare, Kent (1971). (de Worms (1971). *Entomologist's Rec. J. Var.* **83**, 324)
  f. *concolor* Lempke. Continental.
Asphalia diluta Schiff.
  f. *nubilata* Robson. Ancient melanic in Scotland, and industrial in Yorkshire and elsewhere. ? dominant.
Achlya flavicornis L.
  race finmarchica Schoeyen. In Scandinavia only.
  race scotica Tutt. Scotland.
  f. *pseudoalbingensis* Franz. Industrial melanic in Germany and Netherlands (Plate 6.1). ? authenticity of Durham records.
  ab. *atrescens* Cockayne.
Polyploca ridens Fab.
  f. *unicolor* Cockayne. Recurring form around London (Surrey and Hertfordshire). 'Probably all melanic at Bushey' (B.G.).
Dasychira pudibunda L.
  f. *concolor* Stgr. (Plate 5.7, left 5 and 6) Incomplete dominant, industrial (Betrem 1929).
  f. *obscura* Lempke. Industrial melanic, possibly the heterozygote of f. *concolor* may have spread from the continent. Melanic forms common in Netherlands and southern Jutland (C. de W.). So far not found north of

London and the Thames, but is increasing in Surrey (R.F.B.) (see text). Sussex, Arundel '1 in 20' (G.W.H.) London (Totteridge), 4 per cent ($S = 214$, R.I.L.) ,Pinner, occasional melanic specimens (W.E.M.). Kent (Cuckmere), 2 out of 3 (H.B.D.K. 1972).

Lymantria monacha L.
Three separate melanic forms each having dominance (one sex-linked) combine to produce the blackest, f. *atra* Linstow (see text).
f. *atra* Linstow. In Britain only obtained by laboratory breeding.
f. *erimata* Hübner.
f. *transiens* Thierry-Mieg. Industrial melanic in northern Germany, but probably ancient in pine forests.

Malacosoma castrensis L.
ab. *nov.* (R.-C.-K. Collection).

Lasiocampa quercus L. (Plate 14.1)
f. *olivaceo-fasciata* Cockerell, recessive, ancient in Caithness, ? industrial in Yorkshire and Lancashire.
f. *olivacea* Tutt. Recessive, ancient in Caithness, industrial in Yorkshire and Lancashire. So-called 'f. *olivacea*' is genetically distinct in each of its localities.
f. *lurida* Cockayne. Recessive.
(See text.)

Endromis versicolora L.
f. *lapponica* Bau. ? Race in northern Sweden. Aberration in Scotland (Aviemore). Recessive to f. *typica* (Newman 1943).

Saturnia pavonia L.
f. *nigrescens* Cockayne. Industrial melanic.

Nola cucullatella L.
ab. *fuliginalis* Steph. Dominant. Industrial. Epping, 10 per cent ($S = ?$) Welling, north Kent (R.L.E.F.).

Roesilia (syn. *Nola*) *confusalis* Herr.-Sch.
ab. *columbina* Image. Industrial melanic, Epping and elsewhere around London, ? dominant.

Setina irrorella Clerck.
ab. *brunnea* Vorbrodt. ? Thermal melanic (see text).
ab. *melaina* Tutt.

Eilema deplana Esp.
ab. *plumbia* Cockayne.
ab. *violagrisescens* Daniel. ((1952). *NachrBl. Bayer Ent.* **1**, 2; (1952). *Entomoligist's Rec.* J. Var. **64**, 10, 273). Continental melanic, ? Britain.

Spilosoma lubricipeda L.
f. *brunnea* Oberth. Recurring form in northern and western Scotland (for example Elgin), also Northern Ireland. Geographic melanic.
ab. *nigrescens* Cockayne. Specimens from Walthamstow.

Spilosoma lutea Huf.
f. *zatima* Stoll. Incomplete dominant with variable expression (Federley 1920; Goldschmidt 1924), Lancashire and Yorkshire sandhills (type locality, Heligoland (see text)).

ab. *totinigra* Seitz. ? Industrial melanic, Sheffield.

Cycnia mendica Clerck. English males black, Irish males white (as females) = race rustica Hueb. which is dominant to *typica* but has variable expression (see text; Onslow 1921*b*; Cockayne 1919).

Parasemia plantaginis L.
    ab. *matronalis* Freyer (syn. *melaselegans*). Scotland.
    ab. *brunnescens* Schawerda. Occurs in southern England, rare.

Phragmatobia fuliginosa L.
    f. *borealis* Staudinger. Scottish race, though this form occurs occasionally on moors in southern England (see text). Harper (per. com.) has suggested that temperature may play a part: 'All bred and caught specimens in northern Scotland are f. *borealis*. Offspring forced at 70°F. emerge in the winter and are markedly redder, approaching the English form.'

Arctia caja L.
    f. *brunnescens* Statt. The heterozygote of f. *fumosa* Horhammer with variable expression, seems to be of widespread occurrence (see text and frontispiece).
    f. *fumosa* Hörhammer. Homozygote, incomplete dominant (see text and frontispiece).
    ab. *nigropennalis* Statt.
    ab. *obscura* Cockerell.
    ab. *melanozoster* Cockayne.
    ab. *clarki* Tutt. Genetics of these melanics are not fully understood, some are rare recessives, others appear only to express themselves under given environmental conditions.

Arctia villica L.
    ab. *nigrella* Fettig.
    ab. *caliginosa* Schultz.
    Genetics not understood ? environmental plus genetic.

Hypocrita jacobaeae L. (syn. *Callimorpha*)
    ab. *grisescens*. Spuler.
    ab. *nigrana* Cabeau.
    ab. *totonigra* Richter.
    All of extreme rarity.

Euplagia quadripunctaria Poda. (syn. *hera*).
    ab. *nigricans* Kempny. This form occurred in one brood only, probably of genetic origin.

Panaxia dominula L.
    ab. *nigra* Sp.-Hof. (syn. ab. *nigroviridis* Thierry-Mieg.)
    The dozen in Britain are all from Kent (Deal), ? subviable recessive.
    f. *bimacula* Cockayne. Incomplete dominant, originally occurred at Cothill, Berkshire only; now introduced at varying frequencies elsewhere (see text).
    f. *medionigra* Cockayne. Heterozygote of *bimacula*.
    ab. *paradoxa* Reich. Bred from numerous localities; I think this is not genetically determined and probably the result of a virus infection.

Euxoa nigricans L.
    ab. *fuliginea* Hübner.

Euxoa tritici L.
  f. *nigra* Tutt. Ancient melanic and possibly industrial in urban areas.
  f. *nov.* (R.-C.-K. Collection). Irish race, Sligo.
† *Agrotis segetum* Schiff.
  ab. *subatratus* Haworth. In female.
  ab. *nigricornis* Villes. Gloucestershire (Cheltenham) *typica*, 55·44 per cent; intermediate, 32·61 per cent; melanic 11·95 per cent ($S = 92$, R.S.J. 1959). Bretherton (per. com.) records that at Ottershaw, Surrey, he examined 1234 examples, completely black forms occurred in both sexes (usually females), but complete gradation from light to black occurred.
Agrotis vestigalis Hufn.
  f. *nigra* Tutt. Ancient and geographic local race as on Surrey heaths.
Agrotis clavis Hufn. (syn. *corticea* Hueb.).
  f. *subfuscens* Haworth. In males.
  f. *nigra* Tutt. In females.
Agrotis cinerea Hueb.
  f. *obscura* Hübner. Females usually blackish, both sexes dark in Oxfordshire (Steeple Barton).
Agrotis puta Hueb. Females are melanic and males light.
  ab. *nov.* Male melanic from Scilly Isles (Exh. Br. Ent. Soc. 1972).
Agrotis exclamationis L.
  f. *picea* Haworth. Ancient and geographic as in the Isle of Man (A.H.), and recently industrial which is spreading rapidly in some localities, for example Penrith, Cumberland (W.F.D.). Sample Gloucestershire (Cheltenham) gave 0·36 per cent melanics, 0·88 per cent intermediates, and 98·76 per cent f. *typica* ($S = 3115$, R.S.J. 1959); Oxfordshire (Steeple Barton) melanics 1 per cent ($S = 368$, H.B.D.K. 1972).
Ammogrotis lucernea L.
  f. *kerronsis* Kane. Geographic local race in Ireland.
  The species is also melanic in Orkney (R.I.L.) and Caithness (H.B.D.K. 1972).
Rhyacia simulans Hufn.
  f. *suffusa* Tutt. Local race in the Hebrides (R.C.K.). One hundred per cent of this form on the Isle of Uist (A.R. 1967). More recently I have seen a series from Scorradale, Orkney, taken by Pelham-Clinton. Ten of the eleven specimens were black (i.e. 91 per cent), the single insect was distinct and pale. This suggests that f. *suffusa* in Orkney may be controlled by a single gene, in the same way as *Amathes glareosa* f. *edda* and *T. comes* f. *curtisii* are, and not multifactorially.
Spaelotis ravida Schiff.
  f. *suffusa* Tutt.
Diarsia festiva Schiff.
  f. *nigra* Byt. Salz. In race thulei Staud. in Shetland and race conflua Treit. in north Scotland, very dark forms occur, particularly in thulei when individuals can even be confused with f. *edda* of *Amathes glareosa* Esp. (see Plate 10.2, nos. 1 and 2). Ancient and geographic. Cadbury outcrossed

† In many Agrotid species, the females are melanic and the males less so.

thulei to *festiva* from Scilly Isles, the few insects bred were intermediate, therefore probably polygenic.

Diarsia dahlii Hueb.
   f. *perfusa* Kane. Ancient and geographic in Ireland.

Amathes glareosa Esp. (see text)
   f. *edda* Staudinger. Incomplete dominant, ancient, and geographic in Britain, limited to Shetland (including Fair Isle), Orkney (see Chapters 10, 11, 12), and the extreme north of the Scottish mainland (1972).

Amathes agathina Dup.
   f. *scopariae* Millière. Geographic form in Yorkshire and the north.
   f. *infuscata* Culot.

Amathes xanthographa Sch.
   f. *obsoleta-nigra* Tutt. Geographic melanic; but recently I think becoming an industrial melanic around London and elsewhere. Dark forms occur occasionally throughout its range, particularly in Scotland. In Shetland, varying degrees of blackish forms are common, but unlike *Amathes glareosa*, there appears to be no cline.

	intermediate %	melanics %	
Scotland Kingussie	11·09	0·22	($S = 1804$, G.W.H.)
Shetland, Hillswick	17·5	62·5	($S = 40$, H.B.D.K. 1961–2)
Mangaster	23	39·5	($S = 43$, H.B.D.K. 1962)
Tingwall Valley	21	78·4	($S = 172$, H.B.D.K. 1961)
Dunrossness, S. Shetland	—— 94·1 ——		($S = 353$, H.B.D.K. 1960)

Eurois occulta L.
   f. *passettii* Thierry-Mieg. Ancient and geographic melanic (probably northern) occurs throughout central Scottish highlands as far as Newtonmore (G.W.H.). I think a grey/black polymorphism is found here at varying frequencies.
   In southern Britain only f. *typica* and their offspring occur, for example, Kent (Dungeness) (H.B.D.K. 1939) and Oxfordshire (Steeple Barton) (H.B.D.K. 1955); in Shetland and Aberdeen f. *typica* only. I think all these were primary immigrants from Scandinavia (Bretherton, R.F. (1972). *Proc. Br. ent. nat. Hist. Soc.* **5**, 95–119).

Triphaena comes Hueb.
   f. *curtisii* Newman. Ancient and geographic, incomplete dominant whose expression is determined by the local gene-complex (Ford 1953b, 1955c). Widespread from Perth northwards. Sample from Kingussie, Inverness-shire, melanics 5·3 per cent ($S = 38$, G.W.H.). Melanic forms occasionally found in Southern England.? Industrial Melanism, also on Lancashire sandhills.

Triphaena pronuba L.
   f. *brunnea* Tutt. Males only.
   f. *postnigra* Turner.

Lampra fimbriata Von. Sch.
   f. *nigrescens* Busse. Males only.

Polia nebulosa Huf.
   f. *thompsoni* Arkle. The homozygote, incomplete dominant.

f. *robsoni* Collins. The heterozygote, industrial. Both these forms were found in Cheshire near Delamere (Bowater 1914a) (see text). Now rare and in a more recent sample of over 200 collected by myself and Ashworth, there were none of either form: all were of another melanic, f. *bimaculosa* Esper (syn. *plumbosa* Mansbridge). Widespread throughout Britain (Mansbridge 1917). Sample from London (Totteridge), gave 8 per cent ($S = 51$, R.I.L. 1958). Surrey (Ottershaw), 0·24 per cent ($S = 414$, R.F.B. 1946–59).

Hada nana Hufn.
  f. *proxima* Freyer. Geographic, dark specimens in Scotland and Ireland.

Hadena caesia Schiff.
  All British specimens are darker than continental examples, darkest forms occur in south-west Ireland (E.S.A.B.).

Hadena conspersa Schiff.
  f. *hethlandica* Staud.
  f. *obliterae* Robson and Gardner (syn. *obscura* Gregson). Geographic, probably polygenic with one major gene contributing in each locality. Melanic forms occur in three main areas:
    (1) Shetland. I have bred large numbers from Unst where all individuals are dark or very dark. Also from Hillswick on Shetland Mainland where they are melanistic, but not so extreme. In Orkney they have increased pigment, but to a lesser degree than Shetland.
    (2) In the Hebrides, Isle of Lewis.
    (3) In north Devon (Clovelly) where such forms are found on the coastline.

Hadena lepida Esp. (syn. *carpophaga* Bork.)
  ssp. *capsophila* Dup. The Irish race (darkest in the south-west. A smaller form (though no doubt genetically different) occurs in the Isle of Man. There is a west-east cline, blackest in Ireland to uniform white individuals in Kent (Dungeness) and uniform pale yellow in Suffolk. Coastal specimens from Wales are intermediate.
  f. *suffusa* Tutt. This should refer to dull-black specimens from the Blasket Islands and the Tearach, west of Ireland.

Xylomyges conspicillaris L.
  f. *melaleuca* Vieweg. Ancient melanic, occurs in both sexes. This is the usual form in south-west England (see text).

Bombycia viminalis Fab.
  f. *obscura* Staud. (syn. *suffusa* Warren).
  f. *unicolor* Tutt. The blackest example. ? homozygote. Ancient and geographic, widespread in Scotland. Aberdeenshire (Strathdon), three melanics in sample of four (H.B.D.K. 1968). Industrial melanic in many places in northern and southern England. Surrey (Ottershaw), melanics were 10 per cent ($S = 59$, R.F.B. 1946–59). Intermediates occur.

Aporophyla lutulenta Schiff.
  Race lunebergensis Freyer. Scottish race, smaller and blacker (its corresponding pale form in this race is f. *sedi*).

Aporophyla australis Bois.
  f. *ingenua* Freyer. Occurs locally, probably genetically distinct in each locality. Dungeness, Kent, also Cornwall and Ireland.

Lithophane socia Hufn.
  f. *nigricans* Klem. Local and geographic. This form can make this species difficult to separate from its twin species *L. semibrunnea* Haw. I have taken it in Worcestershire, near Malvern, also found in many western localities.

Allophyes oxyacanthae L.
  f. *capucina* Millière. Geographic and ancient, in Britain; also industrial. Absent from certain areas of Britain and also from the Continent including Netherlands (Lempke per. com.).

   Genetics. The commonest melanic f. *capucina* is controlled by a single gene. Lees (per. com.) obtained a number of 1:1 ratios from backcross matings: also a wild melanic female from London (Totteridge, R.I.L., (frequency 63 per cent)) produced 10 *capucina* and 4 f. *typica*. Sib melanics were paired, from this brood (0/17/66) and gave 6 melanic and 0 f. *typica*. There is therefore no clear-cut evidence from broods of the inheritance of f. *capucina*, but the overall likelihood is that it is dominant. There is little evidence either from samples that the intermediates are the heterozygotes (Ford 1955a). More likely is it that they are of polygenic origin; both may contribute.

   There is also a jet-black form with a conspicuous crescent mark on the inner margin of the forewings. This form, which is much blacker than f. *capucina*, is at present spreading in industrial Yorkshire (J.B.), Essex (R.T. 1966), Oxfordshire (H.B.D.K. 1971), and elsewhere.

  f. *capucina* appears to be absent or rare in several separate areas:

(1) Southern and south-east England.

Kent	Folkestone	'no melanics' ($S = 19$, A.M.M. 1956)
	Ashford	'f. *typica* abundant, no melanics' (C.A.W.D.)
	East Malling	'no melanics' ($S = 12$, G.H.L.D. 1968)
Sussex	Arundel	'no melanics' (G.H.)
Isle of Wight		'one f. *capucina*' ($S = 613$, H.B.D.K.)
Hampshire (south)	Boldre	f. *typica* 100 per cent ($S = 100+$, R.W.W. 1967)
	Burley	f. *typica* 100 per cent ($S = 100+$, C.M.P. 1966–7)
Dorset	Lyme Regis	f. *typica* 100 per cent ($S = 10$, T.J.W. 1968)
(2) Wales	Bangor	f. *typica* 100 per cent ($S = 186$, M.J.M. 1960–67)

(3) North-west England

Westmorland	Beetham	no *capucina* ($S = 106$, J.B. 1968)
Lancashire (north)	Silverdale	f. *typica* 100 per cent ($S = 150+$, J.B. 1970)

(4) Scotland (north of Perthshire)

Inverness-shire	Newtonmore	f. *typica* 100 per cent. ($S = 1000+$, G.W.H. 1952–66)

(Note: 30 miles south, however in Perthshire, at rural Rannoch 17·6 per cent were melanic, $S = 17$, D.R.J.W. 1968)

Appendix B

(5) Ireland — No recent melanic record though the species is very common (Two doubtful records, E.S.A.B. 1966)

Samples showing percentage of melanics:

	intermediate %	melanics %	
Berkshire			
Kingsclere	39·3 (expected het. = 43)	9·5	($S = 356$, R.S. 1956–9)

(This sample taken over four consecutive years suggests that the intermediates could be the heterozygotes of f. *capucina*. This is not borne out by samples from elsewhere and is probably coincidental.)

	intermediate %	melanics %	
Reading	19	6	($S = 36$, C.J.C. 1966)
Buckinghamshire			
Bletchley	17·84	41·18	($S = 17$, J.E. 1968)
Chalfont	9·85	24·25	($S = 132$, E.A. 1958–67)
Cheshire			
Delamere		25	($S = 24$, W.E.A. 1956)
Derbyshire			
Chesterfield	53·12	39·06	($S = 64$, J.H.J. 1967)
Devonshire			
Bampton	0	17	($S = 12$, A.R. 1956)
Hertfordshire			
Tring	5	50	($S = 20$, L.G. 1967)
Middlesex, London Area			
Totteridge	28·79	62·89	($S = 132$, R.I.L., 1968–9)
Mill Hill		65·9	($S = 27$, B.G. 1956–7)
Northamptonshire			
Wellingborough		51·85	($S = 54$, P.J.G. 1951–4)
Oxfordshire			
Steeple Barton	12·99	15·62	($S = 177$, H.B.D.K. 1967–71)
Worcestershire			
Birmingham		37·5	($S = 8$, W.E.A. 1967–71)
Yorkshire			
Leeds	40·91	47·73	($S = 44$, S.L.S. 1967)
Bradford		100	($S = 25$, J.B. 1950–6)

Melanics have been recorded from the following places:
(Small or no definite sample size.)

	melanics %	
Hampshire		
Chandlers Ford	20	(B.G.)
Leicestershire		
Nuneaton	50	(D.P.M.)
Middlesex, London Area		
Epping	75	(D.P.M.)
Pinner	56	(W.E.M.)

	melanics %	
Northamptonshire		
Peterborough	50	(S.W.P.)
Somerset		
Bristol	25	(C.S.H.B.)
Weston-super-Mare	10	(C.S.H.B.)
Surrey		
Redhill	33	(R.S.)
Weybridge	25	(J.L.M.)
Witley	30	(J.L.M.)

Eumichitis adusta Esp.
  f. *aterrima* Const. Scotland.
  (f. *nera* Scawharda ? syn.)

Dryobota protea Schiff.
  f. *nigra*. Cockayne. Recurring form around London and elsewhere. I have taken it in Surrey. ? Industrial.

Antitype flavicincta Schiff.
  f. *infuscata* Porritt. Industrial melanic in Surrey, Yorkshire, and elsewhere.

Antitype chi L.
  Geographic and ancient, in Wales and northern England.
  Industrial in Yorkshire and Newcastle area.
  f. *nigrescens* Tutt.
  f. *suffusa* Robson. Both these forms occur as a polymorphism in the north of England.
  f. *nigra*. Industrial melanic, the blackest form in northern England.
  f. *olivacea* Stephens. This name seems to have provided a blanket cover for all melanic *chi*. The meagre evidence of its inheritance is that 'f. *olivacea*' is controlled by a single gene which is recessive to f. *typica* (see text).

Anchoscelis lunosa Haw.
  f. *intensa* Turner. Ancient melanic, occurs as a polymorphism throughout its range.

Cryphia perla Schiff
  f. *suffusa* Tutt. Recurring form near Folkestone, Kent, and elsewhere.

Cryphia muralis Forster
  ab. *nigra* Huggins. 'Rare at Dingle, Ireland, and recorded at Cork' (E.S.A.B.).
  f. *obscura* Tutt.
  f. *castanea* Ckyne. and Williams. Recurring form in southern Ireland, County Cork.

Apatele leporina L.
  ab. *melaleuca* Culot. One from Deptford, Kent.
  f. *melanocephala* Mansbridge. Recurring form in Manchester and Liverpool district. Industrial. Incomplete dominant.
  f. *nigra* Tutt. Refers to the blackest individuals, ? homozygote, dominant.

Apatele psi L.
  f. *suffusa* Tutt. Industrial melanic, in London and elsewhere.

Apatele alni L.
  f. *suffusa* Tutt. Industrial melanic, north Oxfordshire, Hertfordshire, Tring, and elsewhere.

f. *steinerti* Caspari. Industrial melanic, London district, also Yorkshire (Sheffield) and Lancashire. Cheshire (Delamere Forest) 'Completely black, 22 per cent' ($S = 54$, W.E.A. 1968). *Genetics. steinerti* is dominant to f. *typica* with the darkest individuals the homozygotes (H.B.D.K., R.I.L., B.G. 1971, unpublished)

f. *melaina* Schultze. Refers to the blackest individuals, probably homozygotes of *steinerti*; alternatively, another mutant.

Apatele strigosa Schiff.
  f. *nigrescens* Turner. Excessively rare, one British specimen only known to me from Cambridgeshire (R.–C.–K. Collection).

Apatele aceris L.
  f. *intermedia*. Tutt.
  f. *candelisequa* Esper. (syn. *infuscata* Haw.). Ancient melanic but now commonly industrial, in London and elsewhere. London, f. *typica* 51·4 per cent, intermediate 25·7 per cent, melanic 22·9 per cent ($S = 35$, R.I.L. 1957–8). Surrey, Ottershaw, 'Perhaps 10 per cent have forewings dark grey, rather than light grey, but there is no sharp distinction' ($S = 118$, R.F.B. 1946–59).

Apatele euphorbiae Schiff.
  ssp. *myricae* Guen. Ancient, geographic race in Scotland (similar to race montivaga Guen. of montane Europe, and to Irish *euphorbiae*).

Apatele menyanthidis Vieweg.
  f. *suffusa* Tutt. Industrial form occurring commonly in Yorkshire and northern England.

Apatele rumicis L.
  f. *salicis* Curtis. Industrial melanic form occurring in Cheshire (Delamere Forest) and elsewhere.
  f. *lugubris* Schultz. Industrial melanic form which is common in Yorkshire (Barnsley). It is phenotypically identical to the ancient melanics found in west Scotland (Isles of Skye and Lewis) and also to that in County Clare, Ireland.

Apatele megacephala Schiff.
  f. *nigra* Shaw. Industrial melanic in Yorkshire and elsewhere. I took three jet-black specimens at Chapeltown near Sheffield in 1956 with no non-melanic individuals.

Craniophora ligustri Schiff.
  f. *coronula* Haw. (? syn. ab. *nigra* Tutt and ab. *olivacea* Tutt). Geographic (northern) melanic, now industrial.

Samples (showing percentage of melanics):

		melanics %	
Yorkshire	Aysgarth	24	($S = 25$, H.B.D.K. 1956)
Cambridgeshire	Chippenham Fen	48 (varying from intermediate to dark)	($S = 110$, B.O.C.G. 1957)
Surrey	Witley	First 'sooty' specimen	(J.L.M. 1956)

| Oxfordshire | Steeple Barton | One only | (no. f. *typica*) (H.B.D.K. 1969) |

Simyra albovenosa Goeze.
    f. *murina* Aur. Ancient melanic polymorphism in Scandinavia.

Rusina umbratica Goeze (syn. *tenebrosa* Hueb.)
    f. *obscura* Tutt. Non-industrial melanic in north, ? industrial melanic elsewhere.

Scotland   melanics %
    Inverness-shire   ? 100   ($S = 350$, G.W.H. 1952–7)
    Rannoch   48·86   ($S = 88$, H.B.D.K. 1956)

England
    Cambridge   1·0   ($S = 11$, B.O.C.G. 1957)

    Melanics recorded from Hampshire (Rowland's Castle) (F.C.S.), Oxfordshire (Steeple Barton) (H.B.D.K.), and elsewhere. No melanics recorded in London N.20 (Totteridge) ($S = 80$).

Dypterygia scabriuscula L.
    ab. *suffusa* Crewdson. '1 in 45 taken in Delamere Forest, (1950). *Cheshire N. Wales nat. Hist.* **4**, 78).

Xylophasia monoglypha Hufn.
    f. *aethiops* Tutt.
    f. *obscura* Thierry-Mieg.
    f. *infuscata* white.

    Genetics unknown. I have failed to breed this species. Possibly two-year life-cycle in some districts. The intermediate forms bear a relationship as possible heterozygotes of f. *aethiops* in some samples, but not in others. f. *aethiops* is always the rarer.

    There is a range of forms from black to pale with the palest being f. *typica*, the most frequent. The blackest, f. *aethiops*, is usually distinct. I have found this cryptic species at rest by day in three very different situations:

(1) on tree trunks such as pine, also on rocks in Scotland;
(2) in shade under shelter such as outhouses;
(3) in detritus on the ground.

The melanics of this most interesting species are:
(*a*) ancient and non-industrial in origin (as in northern Britain);
(*b*) more recently, an industrial melanic in urban areas.

(a) *Non-industrial melanics*

	f. aethiops (black) %	inter- mediate melanics %	com- bined melanics %	Total	
SCOTLAND					
Aberdeenshire					
Strathdon	5·2	12·4	(17·6)	($S=1190$, H.B.D.K. 1968)	
	4·8	12·3	(17·2)	($S=861$, H.B.D.K. 1969)	
Inverness-shire					

    Newtonmore 'The percentage of melanic *monoglypha* varies greatly from year to year, ... usually 30–70 per cent as in 1956. In 1958,

Appendix B

it was unusually low, about 1 per cent jet black and 5 per cent intermediates.' (S=1000+, G.W.H. 1952-8)

Perthshire				
Rannoch	<0·9	<0·4	(<1·3)	(S=29, H.B.D.K. 1956)
Ross-shire	0·45	3·2	(3·7)	(S=217, R.S. 1958)
Hebrides				
Benbecula	44 (approximately)	44 (approximately)	88 approximately)	(S=100, A.K. 1960)
Shetland				
Hillswick	1·68	81·09	82·77	(S=476, H.B.D.K. 1961)
,,	1·08	35·25	36·33	(S=556, H.B.D.K. 1962)
Mangaster	0	20·29	20·29	(S=202, H.B.D.K. 1962)
Tingwall Valley	0·5	84·6	85·1	(S=1492, H.B.D.K. 1961)
Dunrossness				
S. Shetland	7·4	90·6	98·0	(S=149, H.B.D.K. 1960)
,,	1·0	89·0	90·0	(S=402, H.B.D.K. 1962)

ENGLAND

Berkshire				
Newbury	0·46	3·2	(3·7)	(S=217, R.S. 1958)
Sussex				
Horsham	0·1	?	—	(S=100+, R.M.L. 1947-59)
Worcestershire				
Knightwick	2·33	4·67	6·98	(S=43, R.H.C. 1947)
Cheltenham	0·79	2·8	3·59	(S=528, R.S.J. 1959)

(b) *Industrial melanics*

Yorkshire				
Bradford	—	—	3·74	(S=2275, J.B. 1956)
,,	—	—	12·8	(S=273, J.B. 1957)
,,	—	—	30·61	(S=624, J.B. 1958)
,,	—	—	12·29	(S=1172, J.B. 1959)
,,	—	—	15·89	(S=2064, J.B. 1960)
Chapeltown (Sheffield)	—	—	78·6	(S=56, H.B.D.K. 1956)
Chesterfield	5·0	—	—	(S=100+, J.H.J. 1957)
Leeds	—	—	73·6	(S=72, S.S., 1968-9)
Pontefract	27·74	—	—	(S=133, R.B.W. 1959)
London area				
Totteridge	1·08	1·51	2·59	(S=464, R.I.L. 1957)
,,	1·95	8·98	10·93	(S=256, R.I.L. 1958)

(c) *Other localities*

Berkshire				
Newbury	0·46	3·23	3·69	(S=217, R.S. 1958)

Hampshire				
Winchester	—	—	28	(S=89, D.W.H.F. 1957)
Kent				
Orpington	1·21	3·63	4·84	(S=83, R.G.C. 1957)
Northamptonshire				
Wellingborough	0·66	—	—	(S=301, P.J.G. 1958)
Surrey				
Ottershaw	0·03	—	—	(S=6758, R.F.B. 1946–59)
Yorkshire				
Aysgarth	9·52	19·05	28·57	(S=63, H.B.D.K. 1956)

Hydraecia (Apamea) oblonga Haw.
  f. *unicolor* Tutt. ? homozygote.
  f. *lunulina* Haworth ? heterozygote. Ancient melanics, ? now industrial.
Xylophasia (Apamea) remissa Hueb.
  f. *obscura* Haw. ? homozygote. Ancient melanic and now industrial.
  f. *submissa* Haw. ? heterozygote. Ancient geographic melanics and now industrial.
Xylophasia crenata Huf. (syn. *rurea*)
  f. *nigro-rubida* Tutt.
  f. *alopecurus* Esp. (? syn. of modified *nigro-rubida*)
  (1) Ancient geographic melanic with variable expression.
  (2) Industrial melanic in many urban areas.
  This species is unusual in that melanic polymorphism appears to exist throughout its range ((1788). *Europaischer Schmetterlinge* 4, 473, pl. 147 f3.).

(a) *Non-industrial melanics*

	melanics %	total
SCOTLAND		
Aberdeenshire		
Strathdon	41·18	(S = 34, H.B.D.K. 1968)
Inverness-shire		
Kingussie	50·0	(S = 100+, G.W.H. 1952–8)
ENGLAND		
Gloucestershire		
Cheltenham	58·82	(S = 17, R.S.J. 1959)
Hampshire		
Winchester	28·2	(S = 89, D.W.H.ff. 1957)
Sussex		
West Dean	60·0	(S = 10, H.B.D.K. 1972)

(b) *Industrial melanics*

Yorkshire		
Bradford	80·15	(S = 136, J.B. 1958)
,,	83·34	(S = 18, J.B. 1959)
,,	93·10	(S = 58, J.B. 1960)
Chapeltown	78·51	(S = 56, H.B.D.K. 1956)
Leeds	75·0	(S = 52, S.S. 1968)
,,	70·0	(S = 20, S.S. 1969)

(c) *Other localities*
Lincolnshire
  Louth   38·10   ($S = 21$, H.B.D.K. 1956)
Oxfordshire
  Steeple Barton   47·0   ($S = 17$, H.B.D.K. 1966)
       ,,   31·58   ($S = 19$, H.B.D.K. 1967)
       ,,   27·2   ($S = 18$, H.B.D.K. 1970)
Surrey
  Ottershaw   57·14   ($S = 94$, R.F.B. 1946–59)
Yorkshire
  Aysgarth   42·0   ($S = 100+$, H.B.D.K. 1956)

Xylophasia exulis Lef. (syn. *maillardi* Geyer)
  f. *assimilis* Doubleday. The usual form throughout the Scottish mainland. 94 per cent are similar to this form at Hillswick and Mangaster in west Shetland ($S = 107$, H.B.D.K. 1962).
  Only one per cent are of this form however on Unst ($S = 50+$, H.B.D.K. 1963).

Apamea sordens Huf. (syn. *basilinea* Schiff.)
  ab. nov. (Plate 5.17(a)). A new industrial melanic in Leeds, Yorkshire (S.S 1962).

Celaena secalis L.
  There are several melanic forms of this highly polymorphic species. Some are becoming commoner in industrial areas and beyond.
  f. *nigra*. Tutt. The blackest form is widespread and may constitute up to 70 per cent of some populations as in industrial Yorkshire. Because of the difficulty of recording the range of forms I have failed to obtain frequency records capable of analysis.

Procus sp.
  Several of these ground-resting species are at present increasing their melanic forms. Unfortunately, they are frequently difficult to differentiate. For example (*Procus strigilis* Clerck and *P. latrunculus* Schiff. not separated in the figures), in 1957, "217 black or nearly black, against 42 mottled; total for period 2516" (R.F.B., Ottershaw, Surrey).

Procus strigilis Clerck.
  f. *aethiops* Osthelder. Ancient, and now industrial, melanic.

(a) *Non-industrial melanics*

WALES	intermediate %	melanic %	
Bangor	62·8	23·1	($S = 78$, M.J.M. 1964)
Tregarth	54	36	($S = 11$, M.J.M. 1964)
ENGLAND			
Gloucestershire			
Cheltenham	54·3	34·3	($S = 245$, R.S.J. 1959)
Warwickshire			
Knightwick	22·0	56·1	($S = 214$, R.H.C. 1957)

(b) *Industrial melanics*

Yorkshire			
Bradford	12·1	86·5	($S = 505$, J.B. 1958–60)

Berkshire
    Leyton Park,
        Reading — 73 ($S = 15$, C.J.C. 1968)

Procus latrunculus Schiff.
  f. *unicolor* Tutt. Rural but more normally today an industrial melanic.
                                    melanic %
Berkshire
    Reading   100   ($S = 16$, C.J.C. 1968)
Buckinghamshire
    Chalfont St. Peter   'all melanic' ($S = 100+$, E.A. 1956)
Middlesex
    Mill Hill   'all melanic' ($S = 100+$, B.G. 1957)
Yorkshire
    Pontefract   84·5   ($S = 291$, R.B.W. 1959)

Procus versicolor Bork.
  f. *aethiops* Heydemann. Widespread but not common.
Miana literosa Haw.
  f. *aethalodes* Richardson. Industrial melanic only. Twenty to thirty years ago this species occurred in the Sheffield district as f. *typica* only. It has become extinct. Later it reappeared but all of the melanic form (Reid, per. com.) (see text). It has now appeared in several industrial areas.
Luperina testacea Schiff.
  f. *nigrescens* Tutt. Occasional dark forms occur throughout its range including Scotland (J.F. 1964). No evidence of Industrial Melanism.
                              intermediate % melanic %
Worcestershire
    Knightwick   5·3   42·1   ($S = 19$, R.H.C. 1957)

Meristis trigrammica Hufn. (syn. *trilinea* Hueb.)
  f. *obscura* Tutt. Ancient melanic always at low frequency, no evidence of Industrial Melanism. Melanic form recessive (see text). I cannot recognize an 'intermediate' form.

	inter-mediates %	melanic %	
Gloucestershire			
Cheltenham	25	6·25	($S = 16$, R.S.J. 1959)
London area			
Totteridge, N. 20	3·1	0	($S = 64$, R.I.L. 1957, 1960, 1961)
Oxfordshire			
Steeple Barton	—	5·7	($S = 123$, H.B.D.K. 1956, 1962, 1966)
Norfolk			
Fritton	—	0·7	($S = 301$, H.B.D.K. 1956)
Surrey			
Ottershaw	—	3·0	($S = 134$, R.F.B. 1957)

Appendix B

Sussex			
Horsham	—	2·0	($S = 51$, R.M.L. 1959–60)
West Dean	—	11·8	($S = 121$, H.B.D.K. 1972)
Storrington	8·3	8·3	($S = 24$, R.C.D. 1956)
Worcestershire			
Knightwick	3·1	6·2	($S = 32$, R.H.C. 1957)
Yorkshire			
Aysgarth	—	3·5	($S = 58$, H.B.D.K. 1956)

    f. *obscura* also recorded from Hampshire (Rowland's Castle) (F.C.S.), Somerset (Taunton) (E.G.N.), and probably throughout its range.

Celaena leucostigma Hueb.
    f. *fibrosa* Hübner. The darker and banded form throughout its range where it appears to be always polymorphic.
    ssp. *scotica* Cockayne. The smaller and darker race.

Cosmia affinis L.
    f. *nigrimaculata* Warren. ? Industrial melanic in Yorkshire.

Cosmia trapezina L.
    ab. *nigra* Tutt. Rare mutant.

Nonagria neurica Hueb.
    f. *fusca* Edleston. The darker form recorded by Edleston from Sussex.
    f. *nigra* Wightman. The darkest form from Sussex, (Cuckmere) ((1931). *Entomologist's Rec. J. Var.* **43**, 106).

Nonagria dissoluta Treit.
    This distinct melanic form was the original type.
    f. *arundineta* Schmidt refers to the normal pale form. The melanic type form is widespread, frequencies varying locally from 50 per cent ($S = 16$, B.G. 1951) Hampshire (Lyndhurst and Tichfield Haven (R. J.)) to 20 per cent at Arundel (S.H.) in Sussex. Pale f. *arundineta* only are recorded from Surrey (Redhill) (R.S.), Gloucestershire (Cheltenham) ($S = 16$, R.S. 1959), Hampshire (Totton) (B.G.) in small samples ($S < 10$). Melanics probably at a higher frequency in the eastern counties, but only 4 per cent at Norfolk (Hickling Broad) ($S = 24$, C.J.C. 1972).

Nonagria geminipuncta Hatch.
    f. *fusca-unipuncta* Tutt. The darkest form in a range from pale to dark brown. This is in contrast to *N. dissoluta* where there is usually a clearcut polymorphism. Norfolk (Hickling Broad) 'a gradation to black' ($S = 12$, C.J.C. 1972).

Nonagria typhae Thun.
    f. *fraterna* Bork. There are probably several melanic forms brown-black to jet-black. True f. *fraterna* is a non-industrial and ? industrial melanic.

	melanics %	
Hampshire		
Brockenhurst	15·0	($S = 20$, B.G. 1951)
Norfolk		
Barton Broad	<2	($S = 100+$, H.B.D.K. 1930)
Hickling Broad	<10	($S = 51$, C.J.C. 1972)

	melanics %	
Surrey		
Elstead	<15 (brown)	($S = 20+$, H.B.D.K. 1936)
Ottershaw	11·8	($S = 17$, R.F.B. 1946–59)

Nonagria cannae Ochs. (syn. *algae* Esp.)
    There is a gradation of forms in this species.
    f. *fusca* Bowles. The darkest (Kent).

Nonagria sparganii Esp.
    There is a gradation of forms.
    f. *rufa-unipuncta* Tutt. Dark reddish-brown when fresh.

The following genera contain species which have red or smoky morphs at varying frequencies throughout their range: *Rhizedra*, *Arenostola*, and certain Leucanid and Nonagrid species. All are associated with reeds, rushes, grasses or sedges (Phragmites, Typha, Graminae, and Carex). *Rhizedra lutosa* Hueb. f. *lechneri* Rebel f. *postradiata* Cockayne, *Arenostola fluxa* Hueb. f. *pulveros* Warren Seitz, *Arenostola pygmina* Haw. f. *fusca* Lempke, *Arenostola phragmitides* Hueb. f. *rufescens* Tutt, *Leucania straminea* Treit. ab. *ferrago* Cockayne, *Leucania pallens* L. f. *suffusa* Stephens, *Leucania favicolor* Bau. f. *obscura* Mathews ((1905). *Entomologist's Rec. J. Var.* **17**, 14).

Leucania albipuncta Schiff.
    f. *grisea* Tutt. Ancient melanic, single gene difference, from the brown f. *typica* segregated 50 per cent in several of my broods. Genetics unknown.
    ab. *suffusa* Tutt. Has superimposed dark scales.

Leucania conigera Schiff.
    ab. *suffusa* Tutt.
    ab. *obscura* Hoffman and Klass. In females.
    ab. *nov.* (R.–C.–K. Collection) Dark grey-brown male taken Steeple Barton, Oxfordshire (1963).

Leucania vitellina L.
    f. *saturatior* Dannehl. Dark reddish-brown form which is in part temperature controlled. Prolonged cool during imaginal development in pupa (='actiphase') of Devon origin produces such specimens. Possibly also genetic as in Canary Islands (see *Heliothis peltigera*) ((1929) *Mitt. Munch. ent. Ges.* **19**, 113).

Leucania unipuncta Haw.
    ab. *nigra-suffusa* Richardson. Recessive to f. *typica*, ((1958). *Ent. Gaz.* **9**, 128 with plate) from Scilly Isles.

The Orthosias have a range of forms, one in each species usually being melanic. They all occur in winter in the early months of the year, when polymorphisms are frequent. Probably non-industrial in origin, but melanic forms are occurring at increasing frequencies in and around industrial areas.

Orthosia cruda Schiff. (syn. *pulverulenta* Esp.)
    ab. *haggarti* Tutt. Occasional specimens in Surrey (R.L.M. Witley) and elsewhere in southern England. Industrial melanic.

Orthosia incerta Huf.
  f. *virgata-brunnea* Tutt. Dominant to pale forms.
  f. *fuscatus* Haworth. The darkest form.
Orthosia populi Stroem. (syn. *populeti* Fab.)
  f. *nigra* Tutt. Industrial melanic, London district (Ongar). Dominant.
Orthosia advena Schiff. (syn. *opima* Hueb.)
  f. *nigra* Lempke. Industrial melanic, found around London.
  f. *fuscus* Robson. From Lancashire (Wallasey).
Orthosia gracilis Schiff.
  f. *rufescens* Cockerell. Purplish-brown forms occur locally in certain places: Hampshire (New Forest), Perthshire (Scotland), and elsewhere, where this species is associated with Sweet Gale (*Myrica gale*), which has dark reddish coloured stems around which the moth rests. ? Still in north Kent.
Panolis griseovariegata Goeze (syn. *piniperda* Panzer)
  f. *purpureo-fusca* Preissecker. ? Continental. ((1922). *Verh. zool.-bot. Ges. Wien* **72**, 3).
  f. *grisea* Tutt. Non-industrial melanic. Widespread throughout its range. ((1892). *British Noctuae* **2**, 129).
Panemeria tenebrata Scop.
  ab. *nigrescens* Cockayne.
Sarrothripinus revayana Scop.
  f. *nigrescens* Shelden. Non-industrial and industrial melanic. Highly polymorphic species within which this melanic form is common particularly around London. Surrey (Witley), 'almost always quite black' (J.L.M. 1965).
Euclidimera mi Clerck.
  ab. *suffusa* Warren.
Phoberia lunaris Schiff.
  ab. nov. R.–C.–K. Collection, bred from wild female, Ham Street, Kent.
Catocala fraxini L.
  f. *moerens* Fuchs. Genetics, recessive (see text) widespread in Europe.
Catocala nupta L.
  ab. *brunnescens* Warren. At least 13 records around London (see text).
  ab. *nigra* Lempke. First recorded from north Kent. Recurring form at a low frequency around Totteridge, London N. 20 (1962–71 R.I.L. (see text)).
Colocasia coryli L.
  f. *melanotica* Haverkampf. Probably industrial melanic where bryophytes have disappeared from beech trunks. Melanic form dominant to f. *typica*. Confined to Chiltern escarpment.

	melanics %	
Buckinghamshire		
Chinnor	18	($S = 121$, A.F. 1966)
Chalfont St. Peter	15	($S = 34$, E.A. 1952–7)
,, ,, ,,	19	($S = 36$, E.A. 1971)
High Wycombe	(melanic form commoner than f. *typica*)	($S = 50+$, D.W.H.ff)

Also occurs near Reading (A.F. 1964–5). f. *typica* only, recorded from Surrey (R.S., J.L.M.), Hampshire (B.G.), and Inverness-shire (G.W.H.).

Cucullia chamomillae Schiff.
  f. *chrysanthemi* Hübner. Occurs occasionally in many southern counties of England (H.B.D.K., Steeple Barton, Oxfordshire).
Plusia gamma L
  ab. *nigricans* Spuler. A rare aberration.
Abrostola tripartita Hufn.
  f. *plumbea* Cockayne. Widespread industrial melanic. Genetics unknown to me but there is evidence that a polygenic as well as a unifactorial (and more extreme) form occurs.
  Ansorge states that, after an absence of 50 years from Chalfont St. Peter, Buckinghamshire, all his recent samples (50+) are darker than those he took earlier.
  Bretherton and Goater also record a 'gradation' in Surrey and London (Mill Hill) respectively.

	melanics %	
SCOTLAND		
Inverness	0	($S = 240+$, G.W.H. 1952–8)
ENGLAND		
Cheshire		
Heswell	54	($S = 13$, D.E.H. 1956)
Gloucestershire		
Cheltenham	7·9	($S = 63$, R.S.J. 1959)
Kent		
Orpington	100	($S = 40$, R.G.C. 1956)
London area		
Totteridge, N. 20	5	($S = 97$, R.I.L. 1956–8)
Worcestershire		
Knightwick	47	($S = 15$, R.H.C. 1957)
Yorkshire		
Aysgarth	10	($S = 39$, H.B.D.K. 1956)
Bradford	100	($S = 11$, J.B. 1952)
Chapeltown, Sheffield	100	($S = 3$, H.B.D.K. 1956)

Brephos parthenias L.
  ab. *obscura* Tutt.
  ab. *nigra* Tutt.
Cosymbia albipunctata Huf. (syn. *pendularia* L.)
  f. *subroseata* Woodforde. Common in Staffordshire. Genetics—recessive.
  ab. *decoraria* Newman. Recessive. ((1861). *Zoologist* **19**, 7798).
Cosymbia pendularia Clerck. (syn. *orbicularia* Hueb.)
  ab. *roseonigrata* Cockayne. Recessive.
Cosymbia trilinearia Bork (syn. *linearia* Hueb.)
  ab. *infuscata* Prout.
Cosymbia porata Fab.
  ab. *nigrosparsaria* Lempke.
Cosymbia pupillaria Hübn.
  f. *fasciata* Wagner } Dark red-brown forms are in part temperature con-
  f. *badiaria* Staudinger

trolled. The darkest reflect a prolonged pupal metamorphosis in cool during actiphase. Probably also genetically controlled in part (Kettlewell, unpublished).

Calothysanis amata L.
    ab. *nigra* Rebel. Exceedingly rare.

Scopula immutata L.
    ab. *pulverata* Cockayne.

Scopula marginepunctata Goeze.
    f. *aniculosata* Rambur. Geographic melanic—local races. In Scotland, Wigtownshire, and north Cornwall.
    f. *orphneata* Fuchs.

Sterrha eburnata Wocke.
    f. *obscura* Buckly. Geographic melanic. Occurs on high mountains. Dominant to f. *typica*.
    ab. *nigra* Müller. Dominant. (Bowater 1914, *a* and *b*; Ford 1937)

Sterrha virgularia Hüb. (syn. *seriata* Schrank).
    This species has been used extensively for laboratory genetical studies in Germany. These have disclosed at least 9 mutant forms several of which can be considered under the heading of melanism.
    f. *cubicularia* Peyer. Form with grey speckling. Incomplete dominance to absence of speckling (Alexander 1912). Variable heterozygote in outcrosses to continental stock.
    f. *atra* Baker. Incomplete dominant (Kühn and Engelhardt 1937, 1943; Wagner 1940). Phenotypic expression affected by larval environment particularly temperature.
    f. *nigra*. More extreme than f. *atra*, recessive.
    f. *cana*. Uniform grey, recessive, not at same locus as f. *atra* (Kühn 1944).

Sterrha biselata Huf.
    f. *griseata* Preissecker.

Sterrha aversata L.
    ab. *atrata* Fuchs.
    ab. *suffumata* Lambillion. Both very rare.

Sterrha subsericeata Haw.
    f. *obscura* Rebel. Geographic melanic, recurring form in north Cornwall, ? polymorphism.

Rhodometra sacraria L.
    f. *fumosa* Prout. I have bred several thousand specimens of this species under controlled temperature and humidity conditions. 'Cool' (40°–60°F) throughout the pupal period produces this form, but with a very high mortality rate. The dark crimson f. *rosea* Oberthur and somewhat paler f. *sanguinaria* Esper are also temperature-controlled, the form depending on the point in the pupal metamorphosis at which cold is applied (H.B.D.K. unpublished).

Larentia cervinalis Scop. (syn. *clavaria* Haworth)
    ab. *nov*. in R.–C.–K. Collection, from Cheshire.

Ortholitha limitata Scop. (syn. *chenopodiata*)
    f. *medodii* Thierry-Mieg. Northern melanic, specimens from Yorkshire (Barnard's Castle) and Scotland.

Ortholitha plumbaria Fab. (syn. *mucronata*)
   ab. *nigrescens* Cockerell. Rare aberration.
Ortholitha mucronata Scop. ssp. *scotica* Cockayne.
   ab. *luridaria* Bork. (syn. *nigrescens*). Northern melanic around 10 per cent at Aviemore, Perthshire. Harper states that it is very variable with extreme and intermediate melanics, one colony in Inverness-shire producing about 40 per cent melanic forms. Genetics—unifactorial; Cockayne queried whether paler individuals were heterozygotes (Cockayne (1941). *Entomologist's Rec. J. Var.* **53**, 26).
   ab. *multistrigaria* Heydemann.
Anaitis plagiata L.
   ab. *nigrescens* Hanneman. Rare form, two in R-C-K Collection, from Glasgow. (*Int. Ent. Z.* **26**, 415 (1932)).
   ab. *suffusa* Prout. ((1914). Seitz **4**, 177).
Trichopteryx polycommata Sch.
   ab. *caliginosa* Cockayne. Three specimens from Brighton, Sussex. ((1946). *Entomologist's Rec. J. Var.* **58**, 92.
Trichopteryx carpinata Bork.
   f. *nigra* Bretschneider. Industrial melanic widespread in Europe. Genetics—unifactorial ? dominant. (Bretschneider 1927a).
   f. *fasciata* Prout. J. Fradgley states that in Cardiganshire about half are of this form, which he considers to be a 'quasi-melanic' ((1901). *Entomologist's Rec. J. Var.* **13**, 336).
   f. *obscura* Lempke. ((1950). *Ent. Tijdskr.* **92**, 136).
Operophtera brumata L.
   f. *harrisoni* Prout. Industrial melanic occurring in Essex (Epping), Surrey, and elsewhere.
Oporinia dilutata Sch.
   f. *obscurata* Staudinger. Industrial melanic up to 100 per cent in Essex (Epping Forest). Genetics—probably multi-factorial.
   f. *regressa* Harrison. Dominant.
   f. *melana* Prout.
   f. *latifasciata* Prout. These three forms are all industrial melanics which have been taken in the vicinity of Durham.
Oporinia christyi Prout.
   f. *nigra* Harrison. Probably an industrial melanic in the north of England.
   f. *oblita* Allen. Industrial melanic from Durham ((1906). *Entomologist's Rec. J. Var.* **18**, 86).
Oporinia autumnata Bork.
   f. *intermedia* Clerck.
   f. *latifasciata* Vorbr.
   f. *melana* Prout. ? industrial melanic recorded from Kent ((1899). *Entomologist's Rec. J. Var.* **11**, 122).
Oporinia filigrammaria Herr.-Sch.
   f. *melana* Harrison.
   f. *distincta* Harrison. Northern melanic.

Philereme transversata Hufn. (syn. *rhamnata* Sch.)
  f. *hastedonensis* Lamb.
Lygnis testata L.
  f. *obscura* Bretschneider. Non-industrial melanic, widespread in Scotland, Perthshire and Inverness-shire. Intermediates occur, ? incomplete dominant. ? Industrial melanic in Yorkshire (Bretschneider, R. (1927). Eine neue melanistische Form. *Dt. ent. Z. Iris* **21**, 198–200).
Lygris populata L.
  f. *musauaria* Freyer. Non-industrial melanic, widespread in Scotland, intermediates occur. 'Melanism is directly proportional to altitude; at 2800 ft, at Dalwhinnie, all are intermediate or extreme melanics. On low ground melanics are very infrequent' (G.W.H., Inverness-shire). At Strathdon, Aberdeenshire, f. *typica* 76 per cent, melanic 12 per cent, intermediate 12 per cent ($S = 100+$, H.B.D.K.). Orkney mainland, f. *typica* 36·8 per cent, intermediate 47·4 per cent, melanic 15·8 per cent ($S = 19$, R.I.L. 1972). Genetics—incomplete dominant. (Walther, 1927).
  f. *fuscata* Prout. On moors and high ground.
Plemyria bicolorata Hufn. (syn. *rubiginata* Schiff.)
  f. *semifumosa* Cockayne. Northern melanic, recurring form in Scotland, Rannoch, Argyllshire. Fifty per cent at Forres, East Aberdeenshire. Genetics—dominant or incomplete dominant.
Thera firmata Hubn.
  f. *pupereobrunnea* Cockayne. Non-industrial melanic widespread in Scotland. ? Industrial melanic in Cheshire (Delamere).
Thera obeliscata Hubn.
  f. *obliterata* Buchanan-White. ? Industrial and non-industrial melanic. Berkshire (Reading) 'melanic forms have largely replaced normal ones' (C.J.C. 1966). Gloucestershire (Cheltenham) melanics 100 per cent ($S = 12$, R.S.J. 1959). Essex (Westcliff-on-Sea) 'one in 12 melanic in 1955, sample of 5, all melanic in 1956,' (H.C.H.). Surrey (Ottershaw) 1946–59, melanics 10–20 per cent ($S = 893$, R.F.B.).
  f. *nigrescens* Lempke. ? Industrial melanic recorded from Surrey and Cheshire.
Thera juniperata L.
  f. *infuscata* Schwingenschuss. ? Industrial melanic, recorded from Purley, Surrey.
Dysstroma truncata L.
  f. *nigerrimata* Fuchs. Industrial and non-industrial melanic occurring around Chester and also Salford in Lancashire (R.L.). ? heterozygote (ppNn) (Groth, M. (1935). *Flora Fauna Silkeborg* 73–100).
  f. *melaina* Müller. Industrial and non-industrial melanic, ? dominant homozygote (ppNN), darkest form is PPNN.
  ab. *nov*. Non-industrial melanic in Ireland (Kerry mountains).

Samples	intermediates %	melanics %	
Buckinghamshire			
Chalfont St. Peter	'All specimens taken here have been the dark forms'		($S = 50+$, E.A. 1956)

Gloucestershire			
Cheltenham	4	29	($S = 24$, R.S.J. 1959)
Lancashire			
Salford	10·7	88·1	($S = 84$, R.L.)
London area			
Dulwich, S.E.21	'all *truncata* black here'		(A.A. 1959)
Pinner		80	($S = 50+$, W.E.M.)
Surrey			
Ottershaw		52·6	($S = 19$, R.F.B. 1957)
Worcestershire			
Knightwick	21·7	39·1	($S = 23$, R.H.C. 1957)
Ireland			
Dublin	33·3	55·5	($S = 27$, E.S.A.B. 1957)
Scotland			
Inverness-shire	'all typical'	0	($S = 180$, G.W.H. 1952–8)

Dysstroma citrata L.
    f. *nigerrima* Schawerda. Industrial melanic recorded from Glasgow.
    f. *fusca* Prout. Non-industrial melanic from the Shetland Isles.
Xanthorhöe fluctuata L.
    f. *thules* Prout. Non-industrial melanic from Scotland and the Shetland Isles.
    f. *neopolisata* Mill. Industrial melanic from Liverpool and Cheshire, also elsewhere. ? dominant.
    ab. *obsolescens* Cockayne. Rare, from Aviemore, two only in R.–C.–K. Collection.
Xanthorhöe montanata Schiff.
    f. *shetlandica* Weir. Non-industrial, geographic melanic. Suffused darkish forms, occur regularly throughout the northern Isles.
    ab. *unicolor* Rebel. Very rare, one in R.–C.–K. Collection.
Calostigia salicata Hueb.
    f. *obscura* Schawerda. Non-industrial melanic from Braemar, Scotland.
Calostigia multistrigaria Haw.
    f. *nubilata* Tutt. Industrial melanic from Huddersfield, Sheffield, and elsewhere. Dominant. (Tutt (1896). *British Moths* 267.)
Calostigia didymata L.
    f. *nigra* Prout. Non-industrial melanic from Rannoch, Scotland.
    f. *rebeli* Wnuk. Non-industrial melanic from the Shetland Isles.
Entephria caesiata Sch.
    f. *glaciata* Germar. Non-industrial melanic.
    f. *nigricans* Prout. Non-industrial melanic, widespread in Scotland.
    f. *atrata* Lange. Non-industrial melanic from Shetland, industrial melanic from Paisley (Scotland) and elsewhere.
    ab. *paradoxa* Lange. Rare, one specimen only.
Entephria flavicinctata Hueb.
    f. *obscurata* Stdgr. Non-industrial melanic from Scotland.
Lampropteryx suffumata Schiff.
    f. *piceata* Stephens. Non-industrial melanic occurring locally in northern

England and in Scotland. ? Dominant. ? Industrial melanic in Yorkshire. Cumberland, Penrith, 'frequency varies from wood to wood, approximately 2–5 per cent.' Culgarth near Penrith, melanics 10 per cent (W.F.D.). Westmorland (Kendal) 'f. *piceata* non-existent in this district, f. *typica* common' (N.L.B.). Beetham, 'all typical' ($S = 58$, J.B. 1969–72). Yorkshire, ? industrial melanic at Bishopwood near Selby, also in the Gundal Valley near Pickering.

Euphyia bilineata L.
    ssp. *atlantica* Staudinger. Non-industrial melanic replaces f. *typica* in Shetland, also Outer Hebrides ((1892). *Dt. ent. Z. Iris,* **5**, 247).
    ssp. *hibernica* Tutt. In my experience confined alone to western coastline. Form *typica* further inland. ((1902). *Entomologist's Rec. J. Var.* **14**, 203.)
    f. *isolata* Kane. Non-industrial melanic taken up to 1967 at lighthouse on Tearach Isle, western Ireland, and recently a similar form has been taken on the Blasket Islands. ((1898). *Entomologist* **31**, 85).

Mesoleuca albicillata L.
    ab. *suffusa* Carrington. Very rare, two specimens in R.–C.–K. Collection.

Melanthia procellata Schiff.
    f. *extrema* Schawerda. Widespread on the Continent. ((1921). *Verh. zool.-bot. Ges. Wien* **71**; (1921). *Z. Öst. EntVer.* 62).
    ab. *nigrapicata* Cockayne. Very rare. (Cockayne, E.A. (1952). *Entomologist* **85**, 265–70).

Perizoma affinitata Steph.
    f. *unicolorata* Kane. Non-industrial melanic from Ireland and Arran. ((1897). *Entomologist* **30**, 236.)

Perizoma albulata Sch.
    ssp. *thules* Weir. Non-industrial melanic from Shetland. (1880). *Entomologist* **13**, 290.)

Hydriomena furcata Thun.
    f. *obscura* Peyer. Non-industrial melanic from Scotland (Aberdeen and Forres) also Cork, Ireland. ? Industrial melanic from Yorkshire and Lancashire.
    f. *nigra* Hackray. Industrial melanic from Sheffield and London, ? dominant.

Hydriomena impluviata Schiff. (syn. *coerulata* Fab.)
    f. *semifuscata* Prout. Non-industrial melanic in Arran, Scotland, and England. Industrial melanic, Newcastle district (A.G.L.) and elsewhere.
    f. *nigerrima* Harrison. Non-industrial melanic from Scotland and northern England. ((1911). *Entomologist* **44**, 413.)
    f. *gunillae* Nordstrom. Non-industrial melanic in Scotland, ? industrial melanic.

Hydriomena ruberata Freyer.
    f. *nigrocastanea* Cockayne. Industrial melanic from Durham.
    f. *suffusa* Loeberbauer. Non-industrial melanic from Orkney.
    f. *infuscata* Dannehl.

Earophila badiata Schiff.
    f. *planicolor* Lempke. Industrial melanic around London.
    f. *eckfordii* Smith. Industrial melanic from Cheshire. ((1947). *Proc. Chester Soc. nat. Sci. Lit. Art.* 72.)

ab. *alpestris* Nbgr. Northampton. ((1904). *Societas Ent.* **19**, 20.)
Venusia cambrica Curtis.
  f. *bradyi* Prout. Industrial melanic from Cheshire and Yorkshire. Dominant.
  f. *lofthousei* Prout. Industrial melanic from Yorkshire.
Hydrelia sylvata Schiff. (syn. *testaceata* Don.)
  f. *goodwini* Bankes. Non-industrial melanic recorded near Maidstone, Kent. ((1907). *Entomologist* **40**, 33.)
Eupithecia tenuiata Hubn.
  f. *johnsoni* Harrison. Non-industrial from north of England. ((1931). *Entomologist* **34**, 69.)
Eupithecia venosata Fab.
  ssp. *plumbea* Huggins. Found on Slea Head, Dingle Peninsula, and Blasket Islands, Ireland (E.S.A.B.).
  ssp. *fumosae* Gregson. Non-industrial melanic commonly found near *Silene maritima* on North Shetland.
Eupithecia albipunctata Haw. (syn. *tripunctaria* Herr. Sch.)
  f. *angelicata* Barrett. ? Industrial melanic in Midlands, ? dominant. ((1877). *Ent. Mon. Mag.* **13**, 278.)
Eupithecia vulgata Haw.
  f. *unicolor* Lempke. Industrial melanic from Lancashire (Watergate, Liverpool), Yorkshire (Keighley, Sheffield).
Eupithecia castigata Hub.
  f. *obscura* Dietze. Industrial in Midlands and around London.
  f. *obscurissima* Prout. Industrial melanic in Midlands, Lancashire, and Scotland.
Eupithecia nanata Hüb.
  f. *oliveri* Prout. Industrial melanic from Birmingham. ((1915). *Entomologist* **48**, 7.)
Eupithecia innotata Hufn.
  f. *unicolor* Prout. Industrial melanic from Durham. Spring broods (of *E. innotata*) generally larger and darker.
Eupithecia abbreviata Steph.
  f. *hirschkei* Bastelberger. Non-industrial melanic, more recently industrial Cheshire (Delamere), and Buckinghamshire (Brickhill).

melanics %

Surrey		
Ottershaw	15·4	($S = 65$, R.F.B. 1946–59)
Witley	'Small proportion of melanics and intermediates'	($S = 50+$, J.L.M. 1965)
Woking	'Most examples quite black	(C. de W. 1967)
Westmorland		
Ullswater	'Melanics over 80 per cent'	($S = 50+$, W.F.D.)

Melanics also found in Berkshire (Reading) and Hertfordshire.

  f. *nigra* Cockayne. Widespread recently in the Wye Valley and Forest of Dean, also occurs in Hertfordshire and Hampshire. ((1953). *Entomologist's Rec. J. Var.* **65**, 167.)

Eupithecia lariciata Freyer
   f. *nigra* Prout. Non-industrial and industrial melanic occurring in Isle of Man, Warwickshire (Sutton Coldfield), and Buckinghamshire (Chesham).

Chloroclystis rectangulata L.
   f. *nigrosericeata* Haworth. ? Dominant, ? industrial melanic from London and Sunderland.
   f. *anthrax* Dietze. Taken in north Kent, London (Dulwich), and Essex (Navestock).
   f. *cydoniata* Borkhausen. Taken in Lancashire, Surrey, and the London area (Epping and Streatham).

Recent rapid spread of melanic forms in industrial areas.

Samples of unclassified melanics:

		melanics %	
Berkshire			
Reading		'Melanics commoner than f. *typica*'	(C.J.C.)
Cambridgeshire			
Cambridge		'Melanics 100 per cent'	($S = 21$, B.O.C.G. 1957)
Hampshire			
Chandler's Ford		'f. *typica* 100 per cent'	($S = 50+$, B.G.)
Droxford		,, ,, ,, ,, ,,	($S = 16$, J.R.L. 1958)
Kent			
Orpington		100	($S = 50+$, L.W.S. 1958)
Shortlands		100	(M.G.M.)
London area			
Dulwich, S.E.21		'Five out of seven specimens melanic'	(A.A.)
Pinner		'Almost 100 per cent melanic'	($S = 50+$, W.E.M.)
Surrey			
Ottershaw		100	($S = 297$, R.F.B. 1946–59)
Witley		'Fairly common, not seen a green example'	(J.L.M.)

Anticollix sparsata Treit
   f. *obscura* Lempke. Common in many places around London e.g. Byfleet. ? Dominant.

Abraxas grossulariata L.
   At least eight melanic forms have been recorded. All except one (*nigrosparsata* Raynor which is an incomplete dominant) are recessive:
   f. *nigrosparsata* Raynor. Ten per cent in Yorkshire (Huddersfield).
   f. *nigra* Raynor.
   Poulton's sooty Oxford.
   f. *hazeleighensis* Raynor.
   f. *aberdoniensis* Raynor.
   f. *seminigra* Cockayne. Most of these melanic forms have been bred and with the exception of *nigrosparsata*, in Huddersfield, are rarely taken in the wild.

Abraxas sylvata Scop.
  f. *obscura* Tutt. Recorded from Oxfordshire, (Henley), Buckinghamshire, and Yorkshire (Sledmere). ? Recessive. ((1897). *Ent. Rec.* **9**, 307.)

Lomaspilis marginata L.
  f. *nigrounicolorata* Haverkampf. Continental. ((1904). *Annls Soc. ent. Belg.* **48**, 187.)

Ligdia adustata Schiff.
  ab. *plumbosa* Cockayne. One specimen from Kent (Bexley). ((1950). *Entomologist* **83**, 53.)

Cabera pusaria L.
  ab. *heyeraria* Herrich-Schaffer. Very rare, taken in London (Sydenham).
  ab. *melaina* Oberth. Taken at Tring, Hertfordshire. ((1896). *Etud. ent.* **20**, 70.)

Cabera exanthemata Scop.
  f. *plumbata* Hachray. ? Industrial melanic from Derbyshire.

Ellopia fasciaria Schiff.
  f. *manitiaria* Herrich–Schaffer. Industrial melanic from Delamere, Cheshire. Dominant.
  f. *grisearia* Fuchs. Non-industrial and industrial melanic from Perthshire, Cheshire (Delamere), Staffordshire, and Surrey.

Ennomos autumnaria Werneberg.
  ab. *schultzi* Siebert. Rare recessive (recorded from Dover, Kent), subviable (see text) (Bretschneider (1936). *Entomologist* **50**, 214, 260.)
  ab. *brunneata* Cockayne. Recessive recorded from Sussex (Cockayne 1952*b*).

Ennomos quercinaria Hufn.
  ab. *perfuscata* Hawkins. Rare recessive, ? industrial melanic from Regent's Park, London, also the Midlands. (Seitz, A. (1915). *Macrolepidoptera of the world* **4**, 323.)
  ab. *brunneata* Cockayne. Rare industrial melanic recorded from Newcastle. ((1952). *Entomologist's Rec. J. Var.* **64**, 240.

Deuteronomos fuscantaria Haw.
  f. *perfuscata* Rebel. Widespread, recorded from Essex (Enfield), Oxfordshire (Steeple Barton) (1972) and Yorkshire. ((1910). *Schmetterlingsbuch. Berg.* 388.)

Selenia tetralunaria. Hufn.
  ab. *notabilis* Thierry-Meig. Rare recessive. ((1910). *Annls. Soc. ent. Belg.* **54**, 386.)
  ab. *nigrescens* Cockayne and Kettlewell. Rare recessive recorded from Cheshire, (Delamere and Chester), also Surrey. ((1949). *Entomologist's Rec. J. Var.* **61**, 12.)

Selenia bilunaria Esp.
  ab. *eblanaria* Baynes. Incomplete dominant with variable heterozygotes (Cockayne (1952). *Entomologist's Rec. J. Var.* **64**, 5), occurring in Surrey.
  ab. *brunnearia* Mansbridge. Recessive, recorded from Cheshire (Delamere), Ireland (Cork), and Kent (Bexley). ((1911). *Entomologist's Rec. J. Var.* **23**, 228.)
  ab. *fumata* Smith. Recorded from Chester ((1950). *Proc. Chester Soc. nat. Sci. Lit. Art* **3**, 78, pl. 5, f. 3.)

ab. *nigrata* Smith. Rare recessive recorded from Cheshire ((1951). *Proc. Chester Soc. nat. Sci. Lit. Art* **4,** 78, pl. 5, f. 3.)

ab. *harrisoni* Wagner. This aberration was claimed to have been induced by J. W. H. Harrison and F. C. Garrett (1926) using salts of lead and manganese. A recessive recorded from Durham and Newcastle.

ab. *nigrofasciata* Smith. Rare aberration from Cheshire.

Gonodontis bidentata Clerck.

f. *nigra* Prout (? syn. f. *fusca* Lempke). Industrial melanic, from London and the north of England. Dominant (Onslow 1921*a* and *b*; Bowater 1914).

f. *bowateri* Cockayne. Industrial melanic from Cannock Chase near Birmingham. Dominant to *nigra* (Bowater (1964). *Entomologist's Rec. J. Var.* **12**, 334.).

f. *halperi* Wagner. Non-industrial from Scotland, Rannoch and Hebrides.

f. *medionigra* Cockayne. Sex-linked recessive from Manchester ((1937–8). *Proc. S. Lond. ent. nat. Hist. Soc.* **83**).

Melanics of this species are widespread throughout England, but are local. Imagines seem not to fly any distance. There are two forms of f. *typica*, one of which has a rufous band, and this complicates the phenotype when it occurs in the melanic forms. I have scored intermediates as melanics.

		melanics %	
Cheshire			
	Caldy	1·3	($S = 76$, C.A.C. 1962)
	Delamere	7·2	($S = 69$, W.E.A. 1956)
	,,	60·0	($S = 15$, W.E.A. 1968)
Lancashire			
	Liverpool	about 20	($S = 50+$, P.M.S. 1966)
	South Liverpool	31·6	($S = 19$, C.A.C. 1962)
	Manchester	55·4	($S = 56$, H.N.M. 1957)
London area			
	Dulwich, S.E.21	7·7	($S = 143$, A.A. 1958–9)
	Totteridge	27	($S = 11$, R.I.L. 1957)
	,,	0	($S = 13$, R.I.L. 1958)
	,,	30	($S = 22$, R.I.L. 1960)
Yorkshire			
	Bradford	90·6	($S = 64$, J.B. 1957–9)
	,,	52·9	($S = 17$, J.B. 1960)
	East Leeds	69·6	($S = 158$, S.S. 1967–9)

Melanic specimens have also been taken in Essex (R.T.), Leicestershire (J.D.), Cumberland, (Penrith), and Shropshire (R.W.). Melanic specimens *do not occur* in the following places: Hampshire (Chandler's Ford) (B.G.), Kent (Dover, (B.O.C.G.), Orpington (R.G.C.) and Shortlands (M.G.M.), Norfolk (Fritton) (H.B.D.K.), Surrey (Redhill) (R.S.), Sussex (Horsham) (R.M.L.), and Scotland (Inverness-shire) (G.H.).

Crocallis elinguaria L.

f. *fusca* Reutti. Non-industrial melanic, ? recently industrial on the continent. Recorded in Scotland, Isle of Harris, and Aberdeenshire (Strathdon) and Perthshire (H.B.D.K.) also the Isle of Wight. ((1898). *Lep. Badens* 120.)

Colotois pennaria L.
  f. *obscura* Aigner. Widespread. Seems to appear more frequently later in the season. ? Temperature effect.
Angerona prunaria L.
  f. *fuscaria* Prout. Non-industrial melanic (Williams, H. B. (1946–7). *Proc. S. Lond. ent. nat. Hist. Soc.* 125). Recorded from Huntingdonshire (Monk's Wood) and Kent (Eltham).
Cepphis advenaria Hueb.
  ab. *fulva* Guttmer. Recorded from Surrey. ((1904). *Int. Ent. Z. Iris.* **17**, 80.)
Pseudopanthera macularia L.
  ab. *fuscaria* Staudinger. Recorded from the New Forest.
  ab. *brunneata* Cockayne. Recorded from Sussex and the New Forest.
Semiothisa liturata Clerck.
  f. *nigrofulvata* Collins. Industrial melanic, widespread and spreading. Dominant. Recorded from Cheshire (Delamere) (A.M.M.), Lancashire (Formby) 50 per cent (A.H.), Staffordshire (Cannock) (25 per cent) ($S = 12$, J.H.S. 1958), Isle of Man (2 per cent) (A.H.), Somerset (Taunton) (E.G.N. 1958), Wiltshire (Warminster) '4 in many hundreds' (R.J.); not recorded in samples from Newtonmore, Inverness-shire ($S = 150+$, G.W.H.).
Itama wauaria L.
  ab. *nigraria* Hawarth. Rare aberration, recorded from Yorkshire (Barnard's Castle).
Isturgia limbaria Fab.
  ab. *fumata* Mathew. Rare recessive, one bred 1899.
Chiasmia clathrata L.
  f. *alboguttata* Fettig (syn. f. *nocturnata* Fuchs). Non-industrial melanic, frequent and widespread in south England, Wiltshire (Salisbury), Hampshire (Andover), also Sussex, Buckinghamshire, and Hertfordshire, many of these on chalk downs. This species flies both by day and by night. This surprising polymorphism should be investigated.
Erannis leucophaearia Schiff.
  f. *nigricaria* Hubner.
  f. *brunnescens* Lempke.
  f. *grisescens* Lempke. This form and the preceding one have been recorded from Buckinghamshire, Cheshire (Delamere), Kent, and Surrey.
  f. *funebraria* Thierry-Mieg. Recorded from Essex (Chingford).
  f. *fuscata* Haverkampf.
  f. *merularia* Weymer. Poland (Drozda 1970).
Erannis aurantiaria Esp.
  f. *fumipennaria* Hellweger. Industrial melanic occurring in Yorkshire (Leeds) and the south-west of the county.
Erannis progemmaria Hueb (syn. *marginaria* Bork.)
  f. *fuscata* Harrison. A black industrial melanic form from Yorkshire and Durham. Dominant.
  f. *uniformata* Fuchs. Industrial melanic (body paler) from Yorkshire (Barns-

ley, Huddersfield), Cheshire (Delamere), Lancashire (Liverpool), and Durham (Sunderland).

  f. *infumata* Fuchs. Industrial and non-industrial melanic from York, Kent, and Scotland (Glasgow and Perth).

Erannis defoliaria Clerck.

  f. *obscurata* Staudinger. Industrial melanic from Cheshire (Delamere), London area (Epping), Sussex, and Wales (Portmadoc). Widespread elsewhere.

  f. *suffusa* Cockerell. Industrial melanic from Yorkshire (Leeds), Hertfordshire (St. Albans), and London area (Epping).

  f. *nigra* Bandermann. Recorded from London area (Epping and Chingford), Lancashire (Burnley), Monmouthshire. (Hawkins, C.N. (1936–7). *Proc. S. Lond. ent. nat. Hist. Soc.* **82**).

  f. *transitoria* Lempke. Recorded from Cheshire (Delamere), London area (Epping), and Kent (Sevenoaks).

Phigalia pilosaria Schiff. (syn. *pedaria* Fab.)

  There are several melanic forms and in collections these appear to merge one into another. Lees (1971) has however shown that those he studied were controlled by different genes.

  f. *monacharia* Staudinger. Non-industrial and industrial melanic. The blackest form whose distribution is different in many ways from that of *Biston betularia* f. *carbonaria*, notably in that the melanic form has a high frequency in parts of Scotland where f. *carbonaria* is absent (see text). Dominant.

  f. 'intermediate' Lees. Is distinct entity whose gene is allelic to f. *monacharia*.

  f. *melanaria* Bretschneider. Occurs in females only; Y-chromosome inheritance.

  f. *obscurata* Schawerda. I am not sure whether this is a synonym of f. 'intermediate'. Occurs in Cheshire (Delamere), Durham (Sunderland and Teesdale) and Yorkshire (Sheffield and York).

  f. *extinctaria* Standfuss.

Phigalia pilosaria

Sample figures showing percentage of melanics from north to south
(Lees 1971),

	melanic %	Sample size	Year		melanic %	Sample size	Year
SCOTLAND				*Dumbartonshire*			
Easter Ross				Rowardennan	25·0	28	1968–9
Ardross	0·0	10	1968				
Inverness-shire				ENGLAND			
Fort Augustus	0·0	36	1968	*Cumberland*			
Newtonmore	0·1	1000+	1968	Penrith	26·1	46	1957–62
Perthshire				*Westmorland*			
Killiekrankie	23·5	493	1966–9	Milnthorpe	43·5	46	1968
Rannoch	19·2	52	1966–70	Beetham	57·1	21	1972
Blairgowrie	11·8	17	1968	*Lancashire*			
Aberfoyle	32·5	83	1969	Ince Blundell	76·0	46	1969

Appendix B

	melanic %	Sample size	Year
Cheshire			
Wilmslow	78·6	42	1968–9
Sandbach	73·2	41	1969
Bebington	75·9	29	1957
Yorkshire			
Bradford	80	5	1957
Staffordshire			
Keele	80·0	20	1969
Moddershall	68·5	54	1969
Ellenhall	40·4	89	1969
Shugborough	62·8	70	1969
Gailey	33·8	68	1969
Armitage	61·6	26	1969
Hopwas	63·1	111	1968–9
Walsall	75·9	58	1969
Warwickshire			
Middleton	67·7	93	1968
Coleshill	69·3	54	1967–8
Earlswood	34·4	67	1968
Newland	30·2	76	1968–9
Kenilworth	32·6	46	1968
Bretford	35·6	73	1969
Chesterton	11·4	53	1968
Alveston	9·8	72	1968–9
Snitterfield	26·8	67	1968
Alcester	20·0	30	1969
Weston	8·5	59	1968–9
Leicestershire			
Quorn	66·7	12	1968
Worcestershire			
Himley	60·0	20	1969
Lickey	35·2	156	1967–8
Spetchley	11·5	26	1969
Gloucestershire			
Wormington	6·3	48	1969
Chatcombe	19·0	100	1970
Wickwar	11·0	137	1970
Park End	31·5	73	1969
Oxfordshire			
Bruern	4·2	48	1969
Kiddington	6·9	73	1968–9
Steeple Barton	3·3	60	1968–9
Nuneham Courtenay	5·3	38	1968
Shilton	7·2	235	1970
Berkshire			
Wytham	4·3	326	1967–9
Radley	9·1	22	1968
Newbury	6·3	32	1967–8
Buckinghamshire			
Chalfont St. Peter	5·3	19	1968
Wiltshire			
Purton	1·6	126	1970
Herefordshire			
Hope under Dinmore	9·5	21	1970
Pontrilas	17·8	45	1970
Ledbury	15·0	20	1968
Cambridgeshire			
Madingley	11·1	36	1968–70
Suffolk			
Woolpit	11·2	18	1970
Tunstall	11·3	124	1970
Brandon	0·0	15	1970
Norfolk			
Harleston	19·6	51	1970
Swanton Novers	12·2	33	1970
Somerset			
Shipham	5·0	60	1968–9
Hampshire			
Winchester	7·1	14	1968
Minstead	8·9	45	1968–9
Surrey			
Ottershaw	6·1	247	1946–69
Woking	23·3	43	1968
Sussex			
Falmer	4·0	25	1965
WALES			
Caernarvonshire			
Tregarth	2·9	35	1963–6

	melanic %	Sample size	Year		melanic %	Sample size	Year
Merionethshire				*Monmouthshire*			
Maentwrog	11·9	142	1968–9	St. Arvans	23·9	406	1970
Carmarthenshire				Coldra	42·5	73	1970
Llandovery	19·0	63	1970	Bettwys			
Llanybri	14·2	28	1970	Newydd	36·8	87	1970
Llangain	21·8	55	1970	Ebbw Vale	68·8	48	1970
Cynwyl Elfed	15·8	63	1970	*Glamorgan*			
Llanbyther	18·6	102	1970	Llanishen	39·2	79	1970
Breconshire				Gelligaer	61·3	88	1970
Llangattock	25·5	169	1970	Llwyn Y Pia	50·9	55	1970
Senny Bridge	21·7	74	1970	St. Donats	15·6	32	1970
Talybont	42·6	221	1969–70	Margam	66·7	45	1970
Pembrokeshire				Pont Rhyd			
Canaston	14·3	21	1970	Y Fen	62·5	48	1969–70

I am grateful to Dr. D. R. Lees (1971) for permission to quote his work.

Apocheima hispidaria Schiff.
    f. *obscura* Kühne. Industrial melanic from the London area (Wimbledon), and Cheshire (Delamere).
    f. *fusca* Lempke. Recorded from Hampshire (New Forest), Gloucestershire (Forest of Dean), Cheshire (Delamere), Surrey. Industrial and non-industrial melanic.

Nyssia zonaria Schiff.
    ab. *obscura* Harrison.

Lycia hirtaria Clerck.
    f. *nigra* Cockayne. Recorded in London area. Recessive (see text).
    f. *fumaria* Haworth.
    f. *borealis* Toll. Poland (Drozda 1970).

Biston strataria Hufn.
    f. *melanaria* Koch (? syn. *indanaria* Koch). Recently common in Netherlands but only three specimens known to me in Britain. Two in New Forest (L.W.S.) and one from Kent (Ashford). ((1949). *Int. Ent. Z. Iris.* **59**, 137.)
    f. *nigricans* Oberth. Recorded from Cheshire, Kent, Oxfordshire, and elsewhere. ? Homozygote dominant lethal. Heterozygote 1:2 ratio.
    f. *robiniaria* Frings. Recorded from Kent, dominant. Poland (Drozda, 1970).

Biston betularia L.
    f. *carbonaria* Jordan (syn. *doubledayaria* Millière). Industrial melanic (see text). Dominant.
    f. *insularia* Thierry-Mieg. I regard this as a complex containing several forms at least two of which are allelic to *carbonaria*. Non-industrial and industrial melanic (see text).

Hemerophila abruptaria Thun.
    f. *fuscata* Tutt. Occurring in and around London (see text) and elsewhere,

Bedfordshire (Woburn), Buckinghamshire (Chalfont St. Peter, Bletchley), Cambridge, Kent (Folkestone) (less than 1 per cent, $S = 122$), Northamptonshire (Peterborough, Wellingborough). (Onslow (1921). *J. Genet.* **11**, 293–8.)

   f. *unicolor* Tutt. Recorded from industrial areas around London. Dominant. (Brett, G. A. (1936). *Proc. S. Lond. ent. nat. Hist. Soc.* (1935–6) 84–92; Brett, G. A. (1937). *J. Genet.* **34**, 307–323.)

Cleora cinctaria Schiff.
   ab. *fuscaria* Cockayne. Hampshire (Bournemouth district and New Forest).
   ab. *nigraria* Rebel. Hampshire (New Forest). Recessive.
   ab. *caminariata* Fuchs. Poland (Drozda 1970) and Britain.

Cleora rhomboidaria Schiff. (syn. *gemmaria* Brahm)
   f. *perfumaria* Newman. Widespread, industrial and non-industrial melanic. Dominant to *typica* (Ford 1937). This form rapidly populated the London area at the beginning of the century. Also recorded from Kent (Folkestone), Lancashire (Prestwich), Yorkshire (Bradford).
   f. *rebeli* Aigner (syn. *nigra* Adkin). Industrial melanic, widespread. Dominant (Williams 1933).

	melanics %	
Cheshire		
Bebington	58·8	($S = 34$, A.C. 1957)
Heswall	29	($S = 7$, D.H. 1956)
Hampshire		
Warblington	11	($S = 27$, A.H.S. 1963)
Kent		
Maidstone	17	($S = 28$, A.M.M. 1956)
Somerset		
Weston-super-Mare	2	($S = 50$, C.S.H.B. 1956)

   Also recorded from Norfolk (constantly recurring), Northamptonshire (Peterborough (S.W.P.), Wellingborough (P.J.G.)), Surrey (Redhill, Witley (R.L.M.)), Eire (Dublin (E.S.A.B.)).

Cleora ribeata Schiff. (syn. *abietaria* Clerck)
   All appear to be melanic in Sussex (Arundel (S.H.)).
   f. *sericearia* Curtis. Industrial and non-industrial melanic, dominant. Cheshire (Delamere), Hampshire (New Forest), Surrey.
   f. *nigra* Cockayne. Industrial and non-industrial melanic, dominant to *sericearia* and *typica* (Onslow 1920b). Recorded from Boxhill, Surrey.

Cleora repandata L.
   There are several similar black forms which are likely to be genetically different, but which are probably allelic. Non-industrial examples are in Caledonian Pine Forests (Rannoch), in Eire (Killarney (E.S.A.B.), Kerry, Dingle (H.C.H.)), Northern Ireland (Belfast) (G.A.C.). Industrial melanics occur in Surrey, the London area, Yorkshire, and Lancashire, and elsewhere in the north of England.
   f. *nigricata* Fuchs. Non-industrial melanic, dominant. Recorded from Rannoch Wood and elsewhere in Scotland (see text).

f. *nigra* Tutt. Industrial melanic, dominant.

melanics %

Kent
   Aylesford   'about 20'   (G.A.N.D. 1958)

London area
   Totteridge   32·8   ($S = 61$, R.I.L. 1957–8)

Surrey
   Redhill   'about 33'   (R.S. 1956)

Yorkshire
   Bradford   100   ($S = 100+$, J.B. 1958–60)

Also from Cheshire (Delamere), Lancashire (Formby, Knowsley).

f. *ochronigra* Mansbridge. From Lancashire and Yorkshire.

f. *nigropallida* Mansbridge. Industrial melanic from Lancashire.

Cleora glabraria Schiff. (syn. *jubata* Thunberg)

   f. *obscura* Fuchs. Non-industrial melanic from Hampshire (New Forest).

Cleora lichenaria Hufn.

   f. *obscuraria* Schneider. Non-industrial melanic from Hampshire (New Forest) and north Devon.

   f. *perfumaria* Dannehl. Non-industrial melanic.

Boarmia roboraria Schiff.

   f. *infuscata* Staudinger. Industrial melanic, widespread. Dominant. Recorded from Berkshire (Windsor) (melanics = 90 per cent, $S = 50+$, 1970–2, A.H.H.H.), Essex, Hampshire (Warblington) (A.H.S.), Kent (C. de W.), Surrey (Ottershaw) (R.F.B.).

   f. *melaina* Schulze. Industrial melanic, dominant (see text). Recorded from Warwickshire (Coventry) and Surrey.

   ab. *varia* Cockayne. Rare, dominant, two specimens in R.–C.–K. Collection from Warwickshire, Coventry.

Boarmia punctinalis Scop. (syn. *consortaria* Fab.)

   f. *consobrinaria* Borkhausen. Dominant, recorded from Kent, London, and Surrey (Harrison 1932*b*; Onslow 1920*a*).

   f. *humperti* Humpert. ? Industrial melanic widespread in southern England. (Hasebroek 1934).

   f. *nigra* Warnecke. Frequent in north Europe.

Ectropis crepuscularia Hueb.

   (There is considerable confusion between the melanic forms of *E. crepuscularia* and *E. bistortata*.)

   f. *delamerensis* White. Industrial melanic recorded from Cheshire, Delamere (Harrison and Garrett 1926).

   f. *varia* Cockayne. Industrial melanic, dominant to *typica*, recessive to *delamerensis*. Occurs in Durham (Teesdale) and Yorkshire (Barnsley). ((1948). *Entomologist's Rec. J. Var.* **60**, 10.)

   f. *nigra* Thierry-Meig. Dominant, recorded from Glamorgan, Gloucestershire, and Hertfordshire (Tring).

Ectropis bistortata Goeze.

   f. *passetii* Thierry-Meig. ? Industrial melanic recorded from Gloucestershire (Forest of Dean), Glamorgan (Swansea), Hampshire (Mudeford) (F.M.B.C.),

Hertfordshire (Bushey) (B.G.), Kent (Aylesford) (G.A.N.D.), Oxford (melanics = 20 per cent, $S = 15$, D.J.L.A.), Surrey (Godalming (H.B.W.), Ottershaw (R.F.B.)).

    f. *obscurata* Heinrich. Recessive. Recorded from Hertfordshire (Tring) (Harrison and Garrett 1926).

    f. *schillei* Klein. Recorded from Cheshire (Delamere), Gloucestershire, and Glamorgan (Swansea).

Ectropis consonaria Hueb.

    f. *nigra* Bankes. ? Industrial melanic. Dominant. Recorded from Stroud district Gloucestershire, melanics = 32·8 per cent ($S = 61$, H.B.D.K. and A.R. 1956) (see text).

    Kent (Maidstone) (Onslow 1919*b*).

    f. *waiensis* Richardson. ? Non-industrial recorded from Gloucestershire (Forest of Dean) and Monmouthshire (Wye Valley).

Ectropis luridata Bork. (syn. *extersaria* Hueb.)

    f. *cornelseni* Hoffman. ? Industrial melanic, dominant, occurring in Kent, but more frequently on the continent (Hasebroek 1934; Drozda 1970).

Ectropis punctulata Schiff. (syn. *punctularia* Hueb.)

    f. *intermedia* Lempke. ((1952). *Tijdschr. Ent.* **95**, 253.)

    ab. *nigra* Rebel. (Drozda 1970).

    A halved somatic mosaic has been recorded from Kent of which one side is melanic.

Gnophos obscurata Schiff.

    f. *obscuriorata* Prout. Geographic melanic. Phenotypically similar melanic forms found on heather and peat soils in Hampshire (New Forest), Surrey, and elsewhere.

    f. *saturata* Prout. Geographic melanic found in same areas as f. *obscuriorata* (? synonym).

    f. *anastomosis* Staudinger. Non-industrial melanic found in Scotland.

Gnophos obfuscaria Schiff. (syn. *myrtillata* Thun.)

    Melanic race from Co. Clare, Ireland.

Pachycnemia hippocastanaria Hueb.

    f. *nigrescens* Lempke. Non-industrial melanic from Hampshire (Bournemouth and New Forest).

Ematurga atomaria L.

    f. *unicoloraria* Staudinger. ? Industrial melanic from Lancashire, Bury and Burnley.

Bupalus piniarius L.

    f. *funebris* Cockayne. (? syn. *albomacula* Dziur. and *flavomaculata* Hanneman) ? Industrial melanic, dominant. Widespread in Surrey, Woking 'all specimens here of this form' (C. de W.).

    f. *fuscantaria* Knilikowski. Industrial and ? non-industrial melanic recorded from Cheshire (Chester and Delamere) and Surrey (Oxshott).

    f. *albomacula* Dziurzynski. Recorded from Cheshire (Delamere).

    The following melanic forms occur on the Continent:

    f. *nigricarius* Backhaus.

    f. *albopuncta* Dziurzynski.

f. *magnusaria* Gumpenberg.
f. *nigricans* Dziurzynski.
Dyscia fagaria Thun.
   f. *signata* Cockayne. Recorded from Surrey (Oxshott).
   f. *fusca* Lempke. Recorded from Cheshire (Delamere).
Crocota gilvaria Schiff.
   ssp. *burrenensis* Cockayne. Geographic race from Ireland.
   ab. *nigricans* Jourd. Recorded from south-east Kent.
Perconia strigillaria Hueb.
   f. *atra* Cockayne. Non-industrial melanic occurring in Hampshire (New Forest, Brockenhurst).
   f. *fumosa* Fjeldberg.
   f. *fuscata* Hannemann. Recorded from Surrey (Oxshott).
Apoda avellana L. (syn. *limacodes* Hufn.)
   f. *asella* Esper. Occurs in Hampshire (Liphook).
   f. *suffusa* Seitz. Recorded in Kent (Ashford).
Heterogenea asella Schiff.
   f. *nigra* Tutt. Males only. Recorded in Hampshire (New Forest).
Zygaena trifolii Esper.
   ab. *nigricans* Oberth. Recessive. Recorded as a rare but recurring form in Sussex (Grosvenor 1927).
   ab. *obscura* Tutt. Recorded in Sussex.
Zygaena filipendulae L.
   ab. *chrysanthemi* Borkhausen. Recessive recorded from Lancashire (St. Annes-on-Sea). (See text p. 213).
   ab. *brunnescens* Cockayne. Rare recessive recorded from Kent (Orpington) and Lancashire (Fleetwood). (Cockayne (1940). *Entomologist's Rec. J. Var.* **52**, 91.)
Hepialus humuli L.
   ssp. *thulensis* Newman. Shetlands race (see text).
Hepialus lupulinus L.
   ab. *nigrescens* Cockayne. Rare aberration, two specimens from north Kent in R.-C.-K. Collection.

Résumé. 259 British species have 449 melanic forms of which 175 are probably industrial.

APPENDIX C

Phenotype frequencies of *Biston betularia* and its two melanics, f. *carbonaria*, and f. *insularia* from centres in Britain, 1952–70.
(Samples of 50 and under; percentages in brackets because of small sampling.)

Locality	Date	Observer	Per cent f. *typica*	Per cent f. *carbonaria*	Per cent f. *insularia*	Total
BEDFORDSHIRE						
Leighton Buzzard	1952–56	J.F.R.	15·43	82·28	2·29	175
Woburn	1952–56	S.H.K.	(4·76)	(95·23)	0	21
BERKSHIRE						
Newbury	1952–56	R.S.	79·37	11·11	9·52	63
	1957–64	R.S.	69·26	14·34	16·49	244
	1969	R.S.	70·90	13·40	15·70	223
	1970	R.S.	63·30	12·70	24·00	71
	1969–70	R.S.	67·10	13·05	19·85	294
Abingdon	1957–64	R.J.S.	(33·33)	(50·00)	(16·67)	6
Reading	1963	A.F.	43·10	44·83	12·07	58
	1964	A.F.	(38·77)	(46·94)	(14·28)	49
	1965	A.F.	(25·00)	(50·00)	(25·00)	4
Windsor Park	1971–72	A.H.H.H.	10·00	80·00	10·00	100+
BERWICKSHIRE						
Duns	1952	A.G.L.	96·10	3·90	0	77
	1957–64	A.G.L.	(100·00)	(0)	(0)	7
	1965	A.G.L.	(100·00)	(0)	(0)	1

Appendix C   363

BUCKINGHAMSHIRE					
Chalfont	1957–64	E.A.	17·78	78·10	4·12	315
	1967	E.A.	9·30	83·10	7·60	148
	1969	E.A.	10·89	85·15	3·96	101
	1967 & 69	E.A.	10·10	84·12	5·78	249
Granborough	1968	J.E.	(26·47)	(64·70)	(8·83)	34
CAMBRIDGESHIRE					
Cambridge	1952–56	H.B.D.K.	4·54	92·95	3·41	88
		B.O.C.G.				
	1957–64	B.O.C.G.	4·35	94·78	0·87	115
CHESHIRE					
Chester	1952–56	S.G.S.	5·65	93·55	0·81	124
Delamere	1952–56	W.E.A.	6·68	90·86	2·46	569
	1961–63	C.G.M.	8·15	87·87	3·99	577
Heswall	1952–56	D.E.H.	(6·82)	(93·18)	0	44
Wirral	1952–56	A.C.	1·56	98·44	0	192
	1967	A.C.	(11·90)	(85·70)	(2·40)	42
Ashton	1967	W.E.A.	6·60	93·40	0	75
	1968	W.E.A.	2·67	94·00	3·33	150
	1967–68	W.E.A.	4·63	93·70	1·67	225
Birkenhead	1959–63	W.R.	4·35	94·57	1·09	92
Caldy	1959–64	C.A.C.	7·17	91·70	1·13	3988
	1965	P.M.S.	8·37	90·37	1·26	478
	1966–70			91·12		2872
Meols	1961–64	P.M.S.	11·86	87·12	1·02	590
	1965	P.M.S.	10·52	86·84	2·64	266
	1966–70	P.M.S.		87·10		1054
CUMBERLAND					
Penrith	1952–56	W.F.D.	60·71	26·79	12·50	56
	1957–64	W.F.D.	77·78	11·11	11·11	72

Appendix C

Locality	Date	Observer	Per cent f. typica	Per cent f. carbonaria	Per cent f. insularia	Total
DERBYSHIRE						
Chesterfield	1952–56	J.H.J.	0·59	99·41	0	170
	1967	J.H.J.	3·70	95·06	1·24	81
	1968	J.H.J.	1·42	98·11	0·47	212
	1967–68	J.H.J.	2·56	96·58	0·85	293
DEVON AND CORNWALL						
Ashburton	1952–56	S.T.S.				
Bude	1952–56	A.H.				
Plymouth	1952–56	F.W.J.	100·0	0	0	500+
Tavistock	1952–56	D.J.W.				
Tiverton	1952–56	F.H.L.				
Torquay	1952–56	F.H.L.	100·0	0	0	100
Axminster	1957–64	T.J.W.	93·32	1·12	0·56	179
	1964–66	T.J.W.	100·00	0	0	70
Bampton	1957–64	A.R.	98·8	0	1·20	88
Exmouth	1957–64	R.S.	(100·0)	0	0	20
Lyme Regis	1959–64	T.J.W.	98·62	0·35	1·03	290
	1966	T.J.W.	98·15	0	1·85	54
DORSET						
Broadmayne	1952–56	V.W.P.	97·02	0	2·98	168
Deanend Wood (near Sixpenny Handley)	1952–56	H.B.D.K.	93·64	0·87	5·49	346
ESSEX						
Bradwell-on-Sea	1952–56	A.J.D.	6·69	86·25	6·57	818
	1967	A.J.D.	8·85	78·76	12·39	113
Westcliffe-on-Sea	1952–56	H.C.H.	5·95	85·55	8·50	353
	1957–64	H.C.H.	0	(85·71)	(14·29)	7
Brentwood	1963	K.M.G.	(2·0)	(94·0)	(4·0)	49
	1965	K.M.G.	5·41	91·35	3·24	185
	1967	K.M.G.	4·00	92·00	4·00	150
	1968	K.M.G.	5·62	87·64	6·74	178
	1969	K.M.G.	7·25	89·64	3·11	193
	1970	K.M.G.	6·34	88·03	5·63	142
	1965–70	K.M.G.	5·73	89·73	4·54	848

Appendix C   365

GLOUCESTERSHIRE					
Hardwicke	1952–56	R.D.	38·89	23·61	37·50	72
Nailsworth	1952–56	A.R.	48·65	15·68	35·67	185
Cheltenham	1957–64	R.S.J.	34·18	12·66	53·16	79
HAMPSHIRE					
Borden	1952–56	D.W.	65·43	20·29	14·28	3095
	1966	D.W.	60·91	34·28	4·82	353
Bournemouth	1952–56	F.M.B.C.	(60·00)	(30·00)	(10·00)	40
Chandler's Ford	1952–56	B.G.	82·10	10·49	7·41	324
Fleet	1952–56	A.W.R.	47·31	46·03	6·66	315
Fordingbridge	1952–56	P.J.B.	(86·84)	(5·26)	(7·89)	38
Winchester	1952–56	R.S.M.W.	(84·85)	(15·15)	0	33
	1957–64	W.H.D. & J.W.F.	78·00	12·00	10·00	200
Droxford	1957–64	J.R.L.	(77·78)	(8·33)	(13·89)	36
Hook	1957–64	J.W.F.	55·81	27·91	16·28	86
	1966	J.W.F.	62·16	29·73	4·05	74
Lyndhurst	1957–64	J.R.G.T.	89·74	5·13	5·13	78
Minstead	1957–64	J.R.G.T. & L.W.S.	90·54	7·14	2·32	518
	1965	L.W.S.	89·39	4·54	6·06	132
	1966	L.W.S.	88·70	6·96	4·34	115
	1967	L.W.S.	83·94	5·11	10·95	137
	1968	L.W.S.	90·65	0·94	8·41	107
	1969	L.W.S.	82·93	8·13	8·94	123
	1970	L.W.S.	88·00	9·33	2·67	75
Rowlands Castle	1965–70	L.W.S.	87·27	5·84	6·89	689
Southsea, Portsmouth	1957–64	F.C.S.	84·14	11·03	4·83	145
	1957–64	J.R.L.	(88·89)	0	(11·11)	9
Boldre	1967	R.W.W.	87·37	6·32	6·32	95
Southampton	1966	P.H.	69·86	16·44	13·70	73
			(39·29)	(21·43)	(39·29)	28
HEREFORDSHIRE					
Ross-on-Wye	1952–56	J.E.K.				

Appendix C

Locality	Date	Observer	Per cent f. typica	Per cent f. carbonaria	Per cent f. insularia	Total
HERTFORDSHIRE						
Bishop's Stortford	1952–56	C.C.	4·76	89·18	6·06	231
	1957–64	C.C.	8·76	87·55	3·69	217
Rothamsted	1952–56	C.B.W.	11·08	83·69	5·23	325
Tring	1952–56	L.G.	16·89	74·30	9·81	214
	1957–64	L.G.	22·42	72·12	5·46	165
IRELAND (Ulster)						
Belfast district	1952–56	W.S.W.	93·10	6·90	0	58
	1957–64	W.S.W.	94·12	5·88	0	51
	1965	D.G.	(84·44)	(15·56)	0	45
	1966	D.G.	(82·35)	(5·89)	(11·76)	17
	1967	D.G.	(71·42)	(14·29)	(14·29)	14
	1965–67	D.G.	79·40	11·91	8·69	76
IRELAND (Eire)						
Dublin district	1952–56	E.S.A.B.	(96·30)	(3·70)	0	27
	1957–64	E.S.A.B.	(86·05)	(13·95)	0	43
	1967	E.S.A.B.	(100)	0	0	1
ISLE OF MAN*						
Santon	1952–56	A.H.	50·70	13·11	36·20	69
ISLE OF WIGHT						
Freshwater	1957–64	R.K.J.	85·61	11·03	2·36	127
	1967	R.K.J.	74·58	22·03	3·39	59
	1968	R.K.J.	77·65	15·29	7·06	85

* *1972.* A 1971 survey of the Island organised by Professor C. A. Clarke, suggests that these earlier records may be in error. His latest sample ($n = 68$) from Santon', not only shows that f. *carbonaria* has dropped from 13·1 per cent to 1·5 per cent but that f. *insularia* is at a low frequency in all localities. I have seen the 1971 insects and can confirm this, but must add in mitigation of the earlier records that f. *typica* is unusually darkly mottled on the Isle of Man, so may therefore have been misclassified. As for the frequency difference in f. *carbonaria*, I think that the recorder (A.H.) must have included samples from along the north-east coast where the frequency of this form in 1971 did not drop below 12 per cent. The samples show in fact, a north-east/south-west cline throughout the island (1971, $n = 219$).

KENT					
Bromley	1952–56	M.G.M.	9.26	88.89	1.85	54
Folkestone	1952–56	A.M.M.	46.43	42.86	10.71	140
	1957–64	A.M.M.	(50.00)	(38.89)	(11.11)	36
Maidstone	1952–56	J.R.G.	21.43	73.21	5.36	224
	1957–64	J.R.G.	17.24	82.76	0	58
	1967	G.H.L.D.	(25.81)	(38.71)	(35.48)	31
	1968	G.H.L.D.	12.82	65.38	21.79	78
Aylesford	1957–64	G.A.N.D.	14.22	79.41	6.37	204
Ham Street	1957–64	H.B.D.K.	88.14	3.39	8.47	59
Orpington	1957–64	L.W.S.	16.06	76.60	7.34	218
Tunbridge Wells	1957–64	J.K.B.	(52.63)	(47.37)	0	19
LANCASHIRE					
Formby	1952–56	N.G.L.	5.32	94.68	0	94
Manchester	1952–56	H.N.M.	0	98.00	2.00	350
	1957–64	H.N.M.	0	99.02	0.97	410
	1966–69	L.M.C., R.R.A. & J.A.B.	2.57	97.43	0	972
Southport	1952–56	K.L.G.	10.00	90.00	0	120
Liverpool	1962–63	C.A.C.	2.70	97.30	0	148
LEICESTERSHIRE					
Market Harborough	1952–56	H.A.B.	11.83	84.95	3.32	93
Leicester	1968	J.D.	(15.56)	(84.44)	0	36
LINCOLNSHIRE					
Grimsby	1952–56	G.A.T.J.	7.41	88.89	3.75	81
	1957–64	G.A.T.J.	2.20	96.70	1.10	91
Louth	1952–56	H.B.D.K.	6.33	91.13	2.53	158
LONDON AREA					
Whetstone	1952–56	R.I.L.	5.81	90.21	3.98	327
	1957–64	R.I.L.	4.87	92.97	2.16	185

368   *Appendix C*

Locality	Date	Observer	Per cent f. typica	Per cent f. carbonaria	Per cent f. insularia	Total
Mill Hill	1955	B.G.	8·86	83·85	7·29	192
	1956	B.G.	(4·35)	(95·65)	0	23
	1957–64	B.G.	11·22	84·69	4·19	98
	1965	B.G.	12·93	80·17	6·89	116
Barnes	1970	G.H.	(9·75)	(82·93)	(7·32)	41
Totteridge	1967	R.I.L.	3·75	90·07	6·18	453
	1969	R.I.L.	5·56	88·13	6·31	539
	1970	R.I.L.	6·31	87·86	5·83	206
	1967–70	R.I.L.	5·21	88·68	6·11	1198
MIDDLESEX						
Pinner	1952–56	W.E.M.	9·14	84·95	5·91	558
	1970	W.E.M.	(15·63)	(78·12)	(6·25)	32
NORFOLK						
Cromer	1952–56	J.B.	(20·93)	(67·44)	(11·63)	43
Fritton	1952–56	H.B.D.K.	14·60	77·37	8·03	137
Stalham	1957–64	R.S.	(100)	0	0	7
NORTHAMPTONSHIRE						
Wellingborough	1952–56	P.J.G.	14·24	79·88	5·90	288
	1957–64	P.J.G.	11·54	82·05	6·41	78
Ashton	1957–64	C.L.	(15·15)	(72·73)	(12·12)	33
Oundle	1965	I.F.T.	12·00	80·00	8·00	50
	1966	I.F.T.	(23·68)	(76·32)	0	38
OXFORDSHIRE						
Oxford district	1952–56	H.B.D.K., P.M.S.	41·00	34·03	24·97	717
Steeple Barton	1952–56	H.B.D.K.	49·13	34·26	16·61	289
	1957–64	H.B.D.K.	45·93	31·94	22·13	479
	1966	H.B.D.K.	36·54	34·62	28·85	52
	1967	H.B.D.K.	43·14	41·15	15·69	51
	1968	H.B.D.K.	40·50	31·65	27·85	79
	1969	H.B.D.K.	41·77	32·91	25·32	79
	1970	H.B.D.K.	35·87	40·22	23·91	92
	1966–70	H.B.D.K.	39·56	36·11	24·33	353
	1971	H.B.D.K.	36·47	38·82	24·71	85
	1972	H.B.D.K.	34·88	42·67	22·45	129

Appendix C 369

Location	Years	Observer			
Elsfield	1957–64	C.L.	(40·00)	(36·00)	(24·00)	25
	1968	M.H.	36·95	41·31	21·74	92
Wytham	1957–64	G.V., C.J.C. & S.L.S.	48·33	30·00	21·67	120
Boar's Hill	1966	D.J.L.A.	33·61	31·97	34·43	244
Chinnor Hill	1962	A.F.	(50·0)	(50·0)	0	42
	1963	A.F.	(40·91)	(54·54)	(4·54)	22
Chipping Norton	1968	B.W.	42·36	21·23	36·11	144
SCOTLAND						
Edinburgh	1951–61	E.C.P-C.	(15·4)	(84·6)	0	26
Glasgow	1952–56	H.B.D.K.	(10·33)	(89·66)	0	29
Kinloch Rannoch	1952–56	R.L.	100·0	0	0	285
	1961	E.C.P-C.	(93·3)	0	(6·6)	15
Newtonmore	1952–56	G.H.	100·0	0	0	100
Port Appin, Argyllshire	1956–61	E.C.P-C.	100·0	0	0	83
Winchburgh, West Lothian	1964–69	E.C.P-C.	(52·20)	(47·80)	0	23
SHROPSHIRE						
Broseley	1969	R.U.W.	43·84	52·17	4·35	69
	1970	R.U.W.	(30·44)	(60·87)	(8·69)	23
SOMERSET						
Portishead	1952–56	J.A.B.	(50·00)	(20·00)	(30·00)	10
Taunton	1957–64	E.G.N.	(94·74)	(2·63)	(2·63)	38
STAFFORDSHIRE						
Cannock Chase	1952–56	R.P.D.	(7·50)	(90·00)	(2·50)	40
SUFFOLK						
Lowestoft	1952–56	J.B.	17·86	75·00	7·14	56
Sudbury	1967	L.W.S.	0	(66·67)	(33·33)	9

370   *Appendix C*

Locality	Date	Observer	Per cent f. typica	Per cent f. carbonaria	Per cent f. insularia	Total
SURREY						
Cobham	1952–56	J.B.P.	17·98	76·86	5·17	484
Cranleigh	1952–56	H.B.D.K.	(50·00)	(39·30)	(10·70)	28
Ottershaw	1952–56	R.F.B.	13·73	79·93	6·34	1435
	1957–64	R.F.B.	11·64	82·59	5·77	919
Woking	1952–56	C. de W. & E.T.	16·84	76·66	6·88	1615
	1957–64	C. de W.	13·47	82·64	3·89	334
	1967	C. de W.	9·20	86·80	4·00	150
Bramley	1964	R.F.B.	45·52	41·64	13·01	123
	1965	R.F.B.	36·45	50·47	13·08	107
	1966	R.F.B.	37·93	52·88	9·19	87
	1967	R.F.B.	40·79	43·42	15·79	76
	1968	R.F.B.	34·65	62·20	3·15	127
	1969	R.F.B.	42·60	47·30	10·10	258
	1970	R.F.B.	38·60	51·00	10·40	146
Witley	1965–70	J.L.M.	38·50	51·21	10·29	801
	1965	J.L.M.	56·57	23·90	19·53	251
	1966	J.L.M.	59·16	22·90	17·94	262
	1967	J.L.M.	58·50	24·83	16·67	294
	1968	J.L.M.	57·09	26·64	16·26	289
	1969	J.L.M.	57·35	26·51	16·14	347
	1965–69	J.L.M.	57·73	24·96	17·31	1443
SUSSEX						
Billingshurst	1952–56	P.D.	86·82	11·63	1·55	129
Brighton	1952–56	R.W.D.	(41·18)	(52·94)	(5·88)	17
	1957–64	R.W.D.	(47·06)	(47·06)	(5·88)	17
Eastbourne	1952–56	R.E.E.	40·57	26·42	33·02	106
East Grinstead	1952–56	M.G.	(66·70)	(53·30)	0	21
Hastings	1952–56	C.F.A.	74·15	15·67	10·18	383
Petworth	1952–56	P.D.	79·66	11·86	8·47	59
Horsham	1957–64	P.H.L., R.M.L.	68·27	28·85	2·88	104
West Dean	1972	H.B.D.K.	(70·00)	0	(30·00)	10

Appendix C 371

WALES					
Caernarvonshire					
Llandudno	1952–56	J.A.T.	100·0	0	100
Bangor	1957–64	M.J.M.	(96·15)	(3·85)	26
Tregarth	1957–64	M.J.M.	(90·0)	(10·0)	40
	1964–68	M.J.M.	100·0	0	58
	1969	M.J.M.	(92·5)	(7·5)	12
	1970	M.J.M.	(100)	0	22
Merionethshire					
Dolgelley	1952–56	T.T.	95·37	2·78	108
Cardiganshire					
Llandysul	1966	J.W.F.	64·78	30·98	71
WARWICKSHIRE					
Birmingham	1952–56	H.B.D.K.	8·94	87·09	1611
	1957–64	W.B.	16·00	80·00	75
Tysoe	1952–56	T.T.	(22·22)	(66·67)	18
WESTMORLAND					
Kendal	1952–56	N.L.B.	(40·43)	(48·94)	47
Beetham	1970	J.B.	40·23	48·28	174
WILTSHIRE					
Marlborough College	1952–56	⎰C.R.P.	79·78	8·91	460
	1957–64	⎱R.W.C.V., S.L.S., D.M.C, A.G.H.	76·97	11·80	534
Warminster	1965	P.W.	68·42	17·11	76
Codford St. Mary	1952–56	R.J.	83·33	9·26	54
Marlborough	1957–64	T.W.W.B, M.H.	(73·08)	(11·54)	26
	1965	A.S.	72·63	6·67	285
	1967	A.S.	(66·6)	0	21
Westwoods	1965	P.W.	(52·38)	(4·76)	42
			(66·66)	(4·76)	21

Appendix C

Locality	Date	Observer	Per cent f. typica	Per cent f. carbonaria	Per cent f. insularia	
WORCESTERSHIRE						
Malvern	1952–56	R.K.J.	46·30	27·80	25·90	54
	1957–64	R.K.J.	38·27	36·27	25·49	102
Wyre Forest	1952–56	H.B.D.K.	31·34	64·18	4·48	67
YORKSHIRE						
Aysgarth	1952–56	H.B.D.K.	4·00	94·00	2·00	100
Bradford	1952–56	J.B.	1·97	95·67	2·36	508
	1957–64	J.B.	3·52	96·48	0	142
Grassington	1952–56	G. de C.F.	0	100·0	0	60
Sheffield	1952–56	H.B.D.K.	0·50	99·0	0·50	409
Pontefract	1957–64	R.B.W.	2·17	95·65	2·18	92
Leeds	1967	S.S.	0	(97·92)	(2·08)	48
	1968	S.S.	0	100·0	0	58
	1969	S.S.	0	(100·0)	0	27
	1967–69	S.S.	0	99·31	0·69	133

List of recorders

Agassiz, D. J. L., Ansorge, Sir Eric, Astbury, C. F., Ashworth, W. E., Askew, R. R., Ashton, A., Aston, A.

Baker, J. A., Barnwell, S., Battern, A. G. M., Baynes, E. S. A., Bird, J. K., Birkett, Dr. N. L., Bishop, Dr. J. A., Blathwayt, C. S. H., Bowater, Col. W. W., Brain, T. W. W., Bretherton, R. F., Briggs, J., Buckler, H. A., Burton, P. J.

Cadbury, Dr. C. J., Carr, Rev. F. M. B., Chappel, D. M., Chatelain, R. G., Clarke, Prof. C. A., Clarke, R. H., Cole, G. A., Collinson, W. E., Conn, Dr. D. L. T., Cook, Dr. L. M., Craufurd, C., Creaser, A.

Davey, S. R., David, P., Davidson, W. F., Davies, W. F., Davis, Dr. G. A. N., Demuth, R. P., Dewick, A. J., de Worms, Baron C., Dicker, G. H. L., Douglas, J., Dowdeswell, Dr. W. H., Duffield, C. A. W., Dyson, R. C.

Edwards, T. G., Ellerton, Capt. J., Ellis, E., Ellison, R. E., Evans, L. J.

Ffennell, D. W. H., Fradgley, J. W., Frazer, de Courcy, G., Fitter, A., Ford, Prof. E. B., Ford, R. L. E.

Gardiner, B. O. C., Gent, P. J., Gibbins, M. J., Goater, B., Goodson, L., Gotto, D., Greenwood, K. L., Grimwood, K. M., Groves, Miss J. R., Gurdon, J. B.

Haggett, G. W., Haggett, S., Harbottle, Rev. A. H., Hardstaff, M., Hardy, D. E., Harper, Dr. L. W., Harper, Commander G. W., Harper, Dr. P., Hedges, A., Howard, Dr. G. Huggins, H. C., Hughes, M.

Jackson, Capt. R., Jackson, R. S., Jeffrey, F. W., Jeffs, G. A. T., Johnson, J. H., Jones, R. K.

Kennard, A., Kershaw, S. H., Kettlewell, Dr. H. B. D., Kettlewell, Mrs. H. M., Knight, J. E., Knill-Jones, S. A.

Langmaid, J. R., Lane, C., Lawson, P. H., Leaver, C. J., Leech, N. J., Lees, Dr. D. R., Lees, F. H., Leverton, R., Long, A. G., Long, R. M., Lorimer, R. I., Lovell, R., Lyon, F. H.

Mackworth-Praed, C., Messenger, J. L., Michaelis, H. N., Minnion, W. E., More, D., Morgan, Mrs. M. J., Morley, A. M., Morris, M. G., Murray, Rev. D. P.

Neale, Dr. E. G., Noble, F. A., Nolan, P.

Pelham-Clinton, E. C., Philpott, V. W., Pooles, L., Pooles, S. W., Purefoy, J. B.

Reid, J. F., Reid, W., Richards, Dr. A. W., Richardson, A., Ridge, A., Rivers, C. F., Robinson, H.

Saundby, Air Marshal Sir Robert, Scholes, S., Scott, C., Seacowe, R. J., Sheppard, Mrs. A., Sheppard, Prof. P. M., Siggs, L. W., Smith, R., Smith, S. G., Sperring, A. H., Stanley, F. C., Stidson, S. T., Styles, J. H., Sutton, Dr. S. L., Swain, H. W., Symes, H.

Temple, V., Thomas, I. F., Thomas, J. A., Thorpe, J., Tinbergen, Prof. N., Todd, G., Tomlinson, R., Tremewan, W. G., Trought, T., Trundell, E. E., Turner, Dr. J. R. G.

Uffen, R. W. J.

Varley, Prof. H.

Walker, R. B., Wallace, Dr. D. R. J., Wallace, T. J., Watson, R. W., Whiting, R., Whitney, Dr. R. V., Williams, C. B., Williams, H. B., Williams, P., Withers, B., Wood, E. F., Wright, D., Wright, Capt. W. S.

Also members of Marlborough College, Oundle School, Rannoch School, and Winchester College.

The Hardy–Weinberg table of gene and phenotype frequencies

REC. HOM.	REC. GENE	DOM. GENE	HET.	DOM. HOM.	DOM. PHEN.
0·00	0·000	100·000	0·000	100·000	100·00
1·00	10·000	90·000	18·000	81·000	99·00
2·00	14·142	85·858	24·284	73·716	98·00
3·00	17·321	82·679	28·641	68·359	97·00
4·00	20·000	80·000	32·000	64·000	96·00
5·00	22·361	77·639	34·721	60·279	95·00
6·00	24·495	75·505	36·990	57·010	94·00
7·00	26·458	73·542	38·915	54·085	93·00
8·00	28·284	71·716	40·569	51·431	92·00
9·00	30·000	70·000	42·000	49·000	91·00
10·00	31·623	68·377	43·246	46·754	90·00
11·00	33·166	66·834	44·332	44·668	89·00
12·00	34·641	65·359	45·282	42·718	88·00
13·00	36·056	63·944	46·111	40·889	87·00
14·00	37·417	62·583	46·833	39·167	86·00
15·00	38·730	61·270	47·460	37·540	85·00
16·00	40·000	60·000	48·000	36·000	84·00
17·00	41·231	58·769	48·462	34·538	83·00
18·00	42·426	57·574	48·853	33·147	82·00
19·00	43·589	56·411	49·178	31·822	81·00
20·00	44·721	55·279	49·443	30·557	80·00
21·00	45·826	54·174	49·652	29·348	79·00
22·00	46·904	53·096	49·808	28·192	78·00
23·00	47·958	52·042	49·917	27·083	77·00
24·00	48·990	51·010	49·980	26·020	76·00
25·00	50·000	50·000	50·000	25·000	75·00
26·00	50·990	49·010	49·980	24·020	74·00
27·00	51·962	48·038	49·923	23·077	73·00
28·00	52·915	47·085	49·830	22·170	72·00
29·00	53·852	46·148	49·703	21·297	71·00
30·00	54·772	45·228	49·545	20·455	70·00
31·00	55·678	44·322	49·355	19·645	69·00
32·00	56·569	43·431	49·137	18·863	68·00
33·00	57·446	42·554	48·891	18·109	67·00
34·00	58·310	41·690	48·619	17·381	66·00
35·00	59·161	40·839	48·322	16·678	65·00
36·00	60·000	40·000	48·000	16·000	64·00
37·00	60·828	39·172	47·655	15·345	63·00
38·00	61·644	38·356	47·288	14·712	62·00
39·00	62·450	37·550	46·900	14·100	61·00
40·00	63·246	36·754	46·491	13·509	60·00
41·00	64·031	35·969	46·062	12·938	59·00

The Hardy–Weinberg table of gene and phenotype frequencies

REC. HOM.	REC. GENE	DOM. GENE	HET.	DOM. HOM.	DOM. PHEN.
42·00	64·807	35·193	45·615	12·385	58·00
43·00	65·574	34·426	45·149	11·851	57·00
44·00	66·332	33·668	44·665	11·335	56·00
45·00	67·082	32·918	44·164	10·836	55·00
46·00	67·823	32·177	43·647	10·353	54·00
47·00	68·557	31·443	43·113	9·887	53·00
48·00	69·282	30·718	42·564	9·436	52·00
49·00	70·000	30·000	42·000	9·000	51·00
50·00	70·711	29·289	41·421	8·579	50·00
51·00	71·414	28·586	40·829	8·171	49·00
52·00	72·111	27·889	40·222	7·778	48·00
53·00	72·801	27·199	39·602	7·398	47·00
54·00	73·485	26·515	38·969	7·031	46·00
55·00	74·162	25·838	38·324	6·676	45·00
56·00	74·833	25·167	37·666	6·334	44·00
57·00	75·498	24·502	36·997	6·003	43·00
58·00	76·158	23·842	36·315	5·685	42·00
59·00	76·811	23·189	35·623	5·377	41·00
60·00	77·460	22·540	34·919	5·081	40·00
61·00	78·102	21·898	34·205	4·795	39·00
62·00	78·740	21·260	33·480	4·520	38·00
63·00	79·373	20·627	32·745	4·255	37·00
64·00	80·000	20·000	32·000	4·000	36·00
65·00	80·623	19·377	31·245	3·755	35·00
66·00	81·240	18·760	30·481	3·519	34·00
67·00	81·854	18·146	29·707	3·293	33·00
68·00	82·462	17·538	28·924	3·076	32·00
69·00	83·066	16·934	28·132	2·868	31·00
70·00	83·666	16·334	27·332	2·668	30·00
71·00	84·261	15·739	26·523	2·477	29·00
72·00	84·853	15·147	25·706	2·294	28·00
73·00	85·440	14·560	24·880	2·120	27·00
74·00	86·023	13·977	24·047	1·953	26·00
75·00	86·603	13·397	23·205	1·795	25·00
76·00	87·178	12·822	22·356	1·644	24·00
77·00	87·750	12·250	21·499	1·501	23·00
78·00	88·318	11·682	20·635	1·365	22·00
79·00	88·882	11·118	19·764	1·236	21·00
80·00	89·443	10·557	18·885	1·115	20·00
81·00	90·000	10·000	18·000	1·000	19·00
82·00	90·554	9·446	17·108	0·892	18·00
83·00	91·104	8·896	16·209	0·791	17·00
84·00	91·652	8·348	15·303	0·697	16·00
85·00	92·195	7·805	14·391	0·609	15·00
86·00	92·736	7·264	13·472	0·528	14·00
87·00	93·274	6·726	12·548	0·452	13·00
88·00	93·808	6·192	11·617	0·383	12·00
89·00	94·340	5·660	10·680	0·320	11·00
90·00	94·868	5·132	9·737	0·263	10·00
91·00	95·394	4·606	8·788	0·212	9·00

The Hardy–Weinberg table of gene and phenotype frequencies

REC. HOM.	REC. GENE	DOM. GENE	HET.	DOM. HOM.	DOM. PHEN.
92·00	95·917	4·083	7·833	0·167	8·00
93·00	96·437	3·563	6·873	0·127	7·00
94·00	96·954	3·046	5·907	0·093	6·00
95·00	97·468	2·532	4·936	0·064	5·00
96·00	97·980	2·020	3·959	0·041	4·00
97·00	98·489	1·511	2·977	0·023	3·00
98·00	98·995	1·005	1·990	0·010	2·00
99·00	99·499	0·501	0·997	0·003	1·00

Bibliography

ADAMS, P. A. and HEATH, J. E. (1964). Temperature regulation in the Sphinx moth *Celerio lineata*. *Nature, Lond.* **201**, 20–2.
ADAMS, P. A., and HEATH, J. E. (1968). *Nature, Lond.* **205**, 309–10.
ADKIN, R. (1915–16). *Proc. S. Lond. ent. nat. Hist. Soc.* 1915–16, 134–5.
ADKIN, R. (1916). Notes on breeding from a melanic race of *Boarmia gemmaria*. *Proc. ent. Soc. Lond.* 1915, 122–3.
ADKIN, R. (1925–26). Melanism in the Lepidoptera. *Proc. S. Lond. ent. nat. Hist. Soc.* 1925–26, 7–21.
ADKIN, R. (1927a). Mongrel races of *Diacrisia mendica*. *Proc. ent. Soc. Lond.* **2**, 15–16.
ADKIN, R. (1927b). Mongrel races of *Diacrisia mendica*. *Proc. ent. Soc. Lond.* **2**, 66.
AE, S. A. (1961). A study of interspecific hybrids in black swallowtails in Japan. *J. Lepid. Soc.* **15**, 175–90.
AE, S. A. (1963). A further study of interspecific hybrids in black Swallowtails in Japan. *J. Lepid. Soc.* **17**, 163–9.
AITKENHEAD, P. and BAKER, C. R. B. (1964). The larvae of the British Hepialidae. *Entomologist*, **97**, 25–38.
ALEXANDER, W. B. (1912). Further experiments on the cross breeding of the two races of the moth *Acidalia virgularia*. *Proc. R. Soc. B.*, **85**, 45–52.
ALLEN, P. M. B. (1955). Review of E. B. Ford's 'Moths'. *Entomologist's Rec. J. Var.* **67**, 104.
ARKLE, J. (1889). Natural pairing of *Amphidasys betularia* and var. *doubledayaria*. *Entomologist*, **22**, 236.
ARNOLD, G. A. and CROCKER, J. (1967). *British Spider Study Group Bulletin* **35**.
ASHWELL, D. A. (1953–4). Experiments with *Abraxas grossulariata*. *Proc. S. Lond. ent. nat. Hist. Soc.* (1953–4), 129–42.
ASKEW, R. R., COOK, L. M., and BISHOP, J. A. (1971). Atmospheric pollution and melanic moths in Manchester and its environs. *J. appl. Ecol.* **8**, 247–56.
Atmospheric Pollution Bulletin (Fuel Research Station, Greenwich, London, England).
BACOT, A. W. (1901). Larvae of *Lasiocampa quercus* and its varieties ... and of cross pairings between these races. *Entomologist's Rec. J. Var.* **8**, 114–17.
BACOT, A. W. (1905a). On *Triphaena comes*. *Proc. ent. Soc. London*. 1905, 67–71.
BACOT, A. W. (1906). Heredity experiments with *Triphaena comes*. *Entomologist's Rec. J. Var.* **17**, 340–1.
BARBER, H. N. (1954). Genetic polymorphism in the rabbit in Tasmania. *Nature, Lond.* **173**, 4417, 1227–9.
BARRETT, C. G. (1892). Scent of *Hepialus humuli*. *Entomologist's mon. Mag.* **28**, 217.
BARRETT, C. G. (1895–1902). *The Lepidoptera of the British Isle* (11 vols.). Reeve & Co., London.

BATESON, W. (1892). On variation in the colour of cocoons, pupae, and larvae: further experiments. *Trans. ent. Soc. Lond.* 1892, 205–14.
BATESON, W. (1898). On progress in the study of variation. *Sci. Prog. London.*, **7**, 53–68.
BATESON, W., ONSLOW, H., DIXEY, F. A., and POULTON, E. B. (1922). Experiments in inheritance of colour in Lepidoptera. *Rep. Br. Ass. Advmt. Sci.* 1922, 318.
BEADNELL, C. M. (1927). Acquired melanism in moths transmitted by inheritance. *Br. med. J.* (1927), 362.
BEAVER, H. (1954). *Report of the Committee on air pollution.* H.M.S.O., London.
BEIRNE, B. P. (1943). The relationships and origins of the Lepidoptera of the Outer Hebrides, Shetlands, Faroes, and Iceland. *Proc. R. Ir. Acad.* **49B**, 91–101.
BEIRNE, B. P. (1945). Lepidoptera of Shetland. *Entomologist's Rec. J. Var.* **57**, 37–40.
BEIRNE, B. P. (1955). Natural fluctuations in abundance of British Lepidoptera. *Entomologist's Gaz.* **6**, 21–52.
BELL, W. (1909). Parallel variation in larvae and imagines of *Lasiocampa quercus*. *Entomologist's Rec. J. Var.* **21**, 45.
BERRY, R. J. and DAVIS, P. E. (1970). Polymorphism and behaviour in the Arctic skua, *Stercorarius parasiticus* L. *Proc. R. Soc.* B **175**, 255–67.
BETREM, J. G. (1929). Resultaten van kweeken van *Dasychira pudibunda* L. *Ent. Ber.*, **7**, 411–12.
BISHOP, J. A. and HARPER, P. S. (1970). Melanism in the moth *Gonodontis bidentata*. A cline within the Merseyside conurbation. *Heredity, Lond.* **25**, 449–56.
BISSETT, G. W., FRAZER, J. F. D., ROTHSCHILD, M., and SCHACHTER, M. (1960). *Proc. R. Soc.* B **152**, 255.
BLAIR, W. F. (1941). Annotated list of mammals of the Tularosa Basin. *Am. Midl. Nat.* **26** (1), 218–29.
BLAIR, W. F. (1943). Ecological distribution of Mammals in the Tularosa Basin, New Mexico. *Contr. Lab. vertebr. Biol. Univ. Mich.* No. 20, 20–4, 1 map.
BLEASDALE, J. K. A. (1952). Atmospheric pollution and plant growth. *Nature Lond.* **169**, 376.
BLEST, A. D. (1963a). Relations between moths and predators. *Nature, Lond.* **197**, 1046–7.
BLEST, A. D. (1963b). Longevity, palatibility, and natural selection in five species of New World Saturniid moths. *Nature, Lond.* **197**, 1183–6.
BLEST, A. D. (1964). *Zoologica, N.Y.* **49**, 3.
BOISDUVAL, J. B. A. (1834). *L'Icones historique des Lepidoptères nouveaux ou peu connus* (p. 17), Paris.
BOWATER, W. W. (1913). Heredity of melanism in the Lepidoptera. *Rep. Br. Ass. Advmt Sci.* 1913, 514–15.
BOWATER, W. W. (1914). Heredity of melanism in Lepidoptera. *J. Genet.* **3**, 299–315.
BOWATER, W. W. (1915). Notes on breeding *Odontopera bidentata*. *Ent. Rec. J. Var.* **27**, 109–15.

Bowater, W. W. (1918). Inbreeding *Amphidasys betularia*. *Entomologist's Rec. J. Var.* **30**, 41–2.
Bowden, S. R. (1962). Übertragung von *Pieris napi*, Genen auf *Pieris bryoniae* durch wiederholte Ruckkreuzung (Lep. Pieridae). *Z. ArbGem., ost. Ent.* **14**, 12–18.
Bowden, S. R. (1966). A variant 'brown-face' in *Pieris bryoniae* (Lep. Pieridae). *Entomologist* **99**, 281–3.
Bretherton, R. F. (1970). *Entomologist's Gaz.* **21**, 255–61.
Bretschneider, R. (1936). Über das Herausmendeln von rezessivem Melanismus durch Inzucht. *Ent. Z., Frankf. a. M.* **50**, 207–60.
Bretschneider, R. (1939). Neus uber Melanismus der Schmetterlinge in Gau Sachen. *Ent. Z., Frankf. a. M.* **53**, 59.
Brett, G. A. (1935–6). Some breeding experiments with *Hemerophila abruptaria* Thnbg. *Proc. S. Lond. Nat. Hist. Ent. Soc.* 1935–6, 84–92.
Brett, G. A. (1937). Some breeding experiments on the Geometrid moth *Hemerophila abruptaria* Thnbg. and two of its melanic varieties. *J. Genet.* **34**, 307–23.
British Medical Journal (1970). **4**, 256. Smokeless fuels: shortages.
Brodo, I. M. (1960). Lichen growth and cities: a study on Long Island, New York. *Bryologist*, **69** (4), 427–49.
Brower, J. van Z. (1958a). Experimental studies of mimicry in some North American butterflies. I. *Evolution, Lancaster, Pa.* **12**, 32.
Brower, J. van. Z. (1958b). Experimental studies of mimicry in some North American butterflies. II. *Battus philenor, Papilio troilus, P. polyxenes,* and *P. glaucus. Evolution, Lancaster, Pa.* **12**, 123–36.
Brower, J. van. Z. (1960). Experimental studies of mimicry. IV. The reactions of starlings to different proportions of models and mimics. *Am. Nat.* **94**, 271–82.
Brower, J. van. Z. and Brower, L. P. (1961). Palatability of North American model and mimic butterflies to caged mice. *J. Lepid. Soc.* **15**, 23–4.
Brower, L. P. (1959a). Speciation in butterflies of the *Papilio glaucus* group. I. Morphological relationship and hybridization. *Evolution, Lancaster Pa.* **13**, 40–63.
Brower, L. P. (1959b). Speciation in butterflies of the *Papilio glaucus group*. II. Ecological relationship and interspecific sexual behaviour. *Evolution, Lancaster Pa.* **13**, 212–28.
Brower, L. P., Alcock, J., and Brower, J. van Z. (1971). Avian feeding behaviour and the selective advantage of incipient mimicry, in *Ecological genetics and evolution*, ed. Creed. Blackwell Scientific Publications, Oxford.
Brower, L. P., Brower, J. V. Z., and Cranston, P. P. (1965). Courtship behaviour of the Queen Butterfly, *Danaus gilippus berenice* Cramer. *Zoologica N.Y.*, **50**, 1–39.
Brower, L. P., Pough, F. H., and Meck, H. R. (1970). Theoretical investigation of automicry. I. Single trial learning. *Proc. natn. Acad. Sci. U.S.A.* **66**, 4, 1059–66.
Brown, R. H. (1951). Redstart feeding young on hairy caterpillars. *Br. Birds* **44**, 37.

BUCKLER, W. (1887). *The larvae of British butterflies and moths*, vol. 2 (ed. H. T. Stainton). Adlard, London.
BURNS, J. M. (1966). Preferential mating versus mimicry: disruptive selection and sex-limited dimorphism in *Papilio glaucus. Science, N.Y.* **153**, 551.
BURNS, J. M. (1967). Selective forces in *Papilio glaucus. Science, N.Y.* **156**, 534.
BUXTON, P. A. (1923). *Animal life in deserts*. Edward Arnold & Co., London.
BYTINSKI-SALZ, H. (1939). New and little known forms of *Hepialus humuli* mostly from Britain. *Entomologist's Rec. J. Var.* **51**, 81–4.
CADBURY, C. J. (1969). *Melanism in moths with special reference to selective predation by birds*. D. Phil. Thesis, Oxford University.
CAIN, A. J., and SHEPPARD, P. M. (1950). Selection in the polymorphic land snail *Cepaea nemoralis. Heredity, Lond.* **4**, 275–94.
CAIN, A. J. (1954). Natural Selection in *Cepaea. Genetics, Princeton* **39**, 89–116.
CAROLSFIELD, and KRAUSE, A. G. (1959). The mating of Hepialidae. *Entomologist's Rec. J. Var.* **71**, 33–4.
CARPENTER, G. D. H., and FORD, E. B. (1933). *Mimicry*. Methuen, London.
CARR, F. M. B. (1902). Varieties of *Amphidasys betularia. Entomologist* **35**, 218.
CASPARI, E. (1948). Genetic and environmental conditions affecting a behaviour trait in *Ephestia kühniella. Anat. Rec.* **101**, 690.
CASPARI, E. (1952). Pleiotropic gene action. *Evolution, Lancaster, Pa.* **6**, 1–18.
CASSAL, R. (1904). *Amphidasys betularia* var. *doubledayaria* in the Isle of Man. *Entomologist's Rec. J. Var.* **16**, 49.
CHAPMAN, J. W. (1888). *Entomologist's mon. Mag.* **25**, 40.
CHARLESWORTH, J. K. (1955). The late glacial history of the Highlands and Islands of Scotland. *Trans. R. Soc. Edinb.* **62**, 769–928.
CHEVALIER, H. (1969). Taxonomie et biologie des grand *Arion* de France (Pulmonata: Arionidae). *Malacologia* **9**(1), 73–8.
CHRISTY, W. M. (1897). *Bombyx quercus* taking probably only one year to complete its metamorphosis in Caithness. *Entomologist's Rec. J. Var.* **9**, 62–3.
CLARKE, B. (1962). Balanced polymorphism and the diversity of sympatric species. In *Taxonomy and Geography* (ed. D. Nichols). Systematics Association, Oxford.
CLARKE, C. A. (1971). Female of the cross *Papilio glaucus* × *P. rutulus*, treated with ecdysone. *Proc. R. ent. Soc. Lond.* C **36**, No. 2.
CLARKE, C. A., and SHEPPARD, P. M. (1955). The breeding in captivity of the hybrid *Papilio rutulus* female × *Papilio glaucus* male. *Lepid. News* **9**, 46–8.
CLARKE, C. A., and SHEPPARD, P. M. (1956a). The genetics of some mimetic forms of *Papilio dardanus* Brown and *Papilio glaucus* L. *Proc. R. ent. Soc. Lond.* C **21**, 40–52.
CLARKE, C. A., and SHEPPARD, P. M. (1956b). Hand-pairing of butterflies. *Lepid. News* **10**, 47–53.
CLARKE, C. A., and SHEPPARD, P. M. (1957). The breeding in captivity of the hybrid *Papilio glaucus* female × *Papilio eurymedon* male. *Lepid. News* **11**, 201–5.
CLARKE, C. A., and SHEPPARD, P. M. (1959a). The genetics of some mimetic forms of *Papilio dardanus* Brown, and *Papilio glaucus* L. *J. Genet.* **56**, 236–60.

CLARKE, C. A., and SHEPPARD, P. M. (1959b). The genetics of *Papilio dardanus* Brown. I. Race *cenea* from S. Africa. *Genetics, Princeton.* **44**, 1347–8.

CLARKE, C. A., and SHEPPARD, P. M. (1960a). The evolution of dominance under disruptive selection. *Heredity, Lond.* **14**, 73–87, 163–73.

CLARKE, C. A., and SHEPPARD, P. M. (1960b). The genetics of *Papilio dardanus* Brown. II. Races *dardanus, polytrophus, meseres*, and *tibullus. Genetics, Princeton.* **45**, 439–57.

CLARKE, C. A., and SHEPPARD, P. M. (1960c). Super-genes and mimicry. *Heredity, Lond.* **14**, 175–85.

CLARKE, C. A., and SHEPPARD, P. M. (1962a). The genetics of the mimetic butterfly *Papilio glaucus*. *Ecology* **43**, 159–61.

CLARKE, C. A., and SHEPPARD, P. M. (1962b). Disruptive selection and its effects on a metrical character in the butterfly *Papilio dardanus*. *Evolution, Lancaster, Pa.* **16**, 214–26.

CLARKE, C. A., and SHEPPARD, P. M. (1963). Frequencies of the melanic forms of the moth *Biston betularia* L. on Deeside and in adjacent areas. *Nature, Lond.* **198**, 1219.

CLARKE, C. A., and SHEPPARD, P. M. (1964). Genetic control of the melanic form *insularia* of the moth *Biston betularia* L. *Nature, Lond.* **202**, 215–16.

CLARKE, C. A., and SHEPPARD, P. M. (1966). A local survey of the distribution of industrial melanic forms in the moth *Biston betularia* and estimates of the selective values of these in an industrial environment. *Proc. R. Soc.* B **165**, 424–39.

CLARKE, C. A., WYNNE EDWARDS, J., HADDOCK, D. R. W., HOWES-EVAN, A. W., McCONNELL, R. B., and SHEPPARD, P. M. (1956). ABO blood groups and secretor character in duodenal ulcer. *Br. med. J.* **2**, 725.

COCKAYNE, E. A. (1919). Inheritance of colour in *Diaphora mendica* Clerck and var. *rustica* Hübn. *Entomologist's Rec. J. Var.* **31**, 101–4.

COCKAYNE, E. A. (1925). A note on the genetics of *Grammaria trigrammica*. *Entomologist's Rec. J. Var.* **37**, 142.

COCKAYNE, E. A. (1926a). The induction of melanism in the Lepidoptera and its subsequent inheritance. *Entomologist's Rec. J. Var.* **38**, 65.

COCKAYNE, E. A. (1926b). Experimental melanic changes. *Entomologist's Rec. J. Var.* **38**, 44–5.

COCKAYNE, E. A. (1926c). New aberrations of *Hyloicus pinastri* L., and their mode of inheritance. *Entomologist's Rec. J. Var.* **38**, 65–8.

COCKAYNE, E. A. (1930). *Entomologist's Rec. J. Var.* **42**, 139.

COCKAYNE, E. A. (1932). Larval variation. *Proc. R. ent. Soc. Lond.* **7**, 51–2.

COCKAYNE, E. A. (1937). The genetics and status of *Xylomania conspicillaris* L., ab. *intermedia* Tutt, and ab. *melaleuca* View. *Entomologist's Rec. J. Var.* **49**, 81.

COCKAYNE, E. A. (1938). Aberrations of *Hemerophila abruptaria* (Lep. Geometridae). *Entomologist's Rec. J. Var.* **50**, 110.

COCKAYNE, E. A. (1940a). *Dasychira pudibunda* L. ab. *bicolor*, and ab. *concolor* Stdg. *Entomologist's Rec. J. Var.* **52**, 86.

COCKAYNE, E. A. (1940b). The buff forms of *Biston betularia* L. ab. *lomasaria* Cottam and ab. *decolarata*, their history and genetics. *Entomologist's Rec. J. Var.* **52**, 93–6.

COCKAYNE, E. A. (1941). *Phigalia pedaria* L. ab. *melanaina* Bretschneider, an example of Y chromosome inheritance. *Entomologist's Rec. J. Var.* **53**, 95.
COCKAYNE, E. A. (1945). Some of the contributions of entomology to genetics. *Proc. R. ent. Soc. Lond.* C **9**, 47-55.
COCKAYNE, E. A. (1947-8). *Arctia caja* L.: its variation and genetics. *Proc. S. Lond. ent. Nat. Hist. Soc.* 1947-8, 155-91.
COCKAYNE, E. A. (1949a). Aberrations of *Hemerophila abruptaria*. *Entomologist* **82**, 149-52.
COCKAYNE, E. A. (1949b). L. W. Newman. *Entomologist's Rec. J. Var.* **61**, 80-1.
COCKAYNE, E. A. (1951a). *Entomologist's Rec. J. Var.* **63**, 162 (p. 5, f. 4).
COCKAYNE, E. A. (1951b). Aberrations of British macro-lepidoptera. *Entomologist* **84**, 241-5.
COCKAYNE, E. A. (1952a). The problem of *Lasiocampa quercus* Linnaeus ab. *olivaceo-fasciata* Cockerell, ab. *olivacea* Tutt, and melanic larvae. *Entomologist's Rec. J. Var.* **64**, 306-9.
COCKAYNE, E. A. (1952b). Aberrations of British Geometridae. *Entomologist's Rec. J. Var.* **64**, 237-42.
COCKAYNE, E. A., and KETTLEWELL, H. B. D. (1949). *Silenia tetralunaria* Hufn. ab. *nigrescens*, ab. nov., with an account of its genetics. *Entomologist's Rec. J. Var.* **61**, 9-12.
COCKAYNE, E. A., and KETTLEWELL, H. B. D. (1953). The Rothschild-Cockayne-Kettlewell-Collections of British Lepidoptera. *Entomologist's Rec. J. Var.* **65**, 303-4.
COCKAYNE, E. A., and NEWMAN, L. W. (1931). *Papilio machaon* ab. *nigra* Reutti. *Proc. R. ent. Soc. Lond.* **6**, 95-6.
COCKERELL, T. D. A. (1889). On the variation of insects. *Entomologist* **22**, 1-6.
COLLINS, J. L. (1927). A low temperature type of albinism in Barley. *J. Hered.* **18**, 331-4.
COLLINSON, W. E. (1958). The black eggar (Annual paper read to Yorkshire Naturalists' Union, 1958).
COOK, L. M. (1960). *A study of some aspects of the ecology of the scarlet tiger moth Panaxia dominula in Britain*. D. Phil. Thesis, Oxford University.
COOK, L. M., ASKEW, R. R., and BISHOP, J. A. (1970). Increasing frequency of the typical form of the Peppered Moth in Manchester. *Nature, Lond.* **227**, 1155.
COOK, L. M., and KETTLEWELL, H. B. D. (1960). Radioactive labelling of lepidopterous larvae: a method of estimating late larval and pupal mortality in the wild. *Nature, Lond.* **187**, 301-2.
COOKE, H. (1877). *Entomologist* **10**, 92-6, 151-3.
COPPINGER, R. P. (1970). Effect of experience and novelty on avian feeding behaviour with reference to evolution of warning coloration in butterflies. II. Reactions of naïve birds to novel insects. *Am. Nat.* **104**, 938, 323-35.
COTT, H. B. (1940). *Adaptive Colouration in Animals*. Methuen, London.
COTT, H. B. (1946). The edibility of birds. *Proc. zool. Soc. Lond.* **116**, 371-524.
CREED, E. R. (1963). *Polygenic situations and polymorphism especially in* Maniola, *ladybird species and* Crangonyx. D. Phil. thesis, Oxford University.

CREED, E. R. (1966). Geographic variations in the Two-Spot Ladybird in England and Wales. *Heredity, Lond.* **21**, 57–72.

CREED, E. R. (1971). Melanism in the Two-spot Ladybird *Adalia bipunctata*, in Great Britain, in *Ecological genetics and evolution*, ed. Creed. Blackwell Scientific Publications, Oxford.

CREED, E. R., DOWDESWELL, W. H., FORD, E. B., and MCWHIRTER, K. G. (1959). Evolutionary studies on *Maniola jurtina*: the English mainland, 1956–7. *Heredity, Lond.* **13**, 363–91.

CROSBY, J. L. (1963). Evolution and nature of dominance. *J. theor. Biol.* **5**, 35–51.

CROTCH, W. D. (1865). *Hepialus humuli* var. *thulensis*. *Entomologist* **2**, 176–7.

CUÉNOT, L. (1925). Serie 'Encyclopedie scientifique' 135, 327. Gaston Dom, Paris.

CUNO, W. (1932). Einiges über die Zucht von *Arctia caja* L. *Ent. Z., Frankf. a M.* **46**, 73.

CURIO, E. (1965a). Die Schutzenpassungen dreier Raupen einer Schwärmers (Lep. Sphingidae) auf Galapagos. *Zool. Jb. (Syst.)* **92**, 487–522.

CURIO, E. (1965b). Ein Falter mit 'Falchem Kopf'. *Natur Mus., Frankf.* **95**, 43–6.

DAKIN, W. J. (1927). Melanism in the Lepidoptera and its evolutionary significance. *Nature, Lond.* **119**, 318.

DARLINGTON, P. J. (1943). Carabidae of mountains and islands: data on the evolution of isolated faunas and on atrophy of wings. *Ecol. Monogr.* **13**, 37–61.

DARWIN, C. R. (1884). *Journal of Researches*. London.

DARWIN, C. R. (1887). Letter to J. D. Hooker, 7 March 1855, in *Life and Letters of Charles Darwin* (ed. Darwin, F.). Murray, London.

DE BEER, G. R. (1958). *Endeavour* **17** (66), 61.

DE BEER (1964). *Atlas of Evolution*. Nelson, London.

DECKER, A. and McGINNIS, J. (1947). *Proc. Soc. exp. Biol. Med.* **66**, 224–8.

DE RUITER, L. (1953). Some experiments on the camouflage of stick caterpillars. *Behaviour* (3), 222–32.

DETHIER, V. G. (1939). *Jl N.Y. ent. Soc.* **47**, 131.

DETWYLER, T. R. (1971). *Man's impact on environment*. McGraw-Hill Book Co., U.S.A.

DE WORMS, C. (1958). Lepidoptera in Finland, June 1958. *Entomologist* **91**, 242–6.

DE WORMS, C. (1961). A review of the Lepidoptera of the London area for 1960 and 1961. *Lond. Nat.* **41**, 60–5.

DICE, L. R. (1929). *Occ. Pap. Mus. Zool. Univ. Mich.* **203**, 1–4.

DICE, L. R. (1930). Mammal distribution in the Alanogordo region, New Mexico. *Occ. Pap. Mus. Zool. Univ. Mich.* **213**, 1–32.

DICE, L. R. (1936–42). *Contr. Lab. vertebr. Biol. Univ. Mich.* Nos. 10, 12, 13, 15, 18, 19.

DICE, L. R. (1944). *Contr. Lab. vertebr. Biol. Univ. Mich.* No. 30.

DICE, L. R. (1947). Effectiveness of selection by owls on Deermice (*Peromyscus maniculatus*) which contrast in colour with their background. *Contr. Lab. vertebr. Biol. Univ. Mich.* No. 34, 1–20.

DICE, L. R., and BLOSSOM, P. M. (1937). Studies of mammalian ecology in

south-western North America, with special attention to the colours of desert mammals. *Publs. Carnegie Instn.* **485**, 1–129.
DIGBY, P. S. B. (1955). Factors affecting the temperature excess of insects in sunshine. *J. exp. Biol.* **32**, 279–98.
DIXEY, F. A. (1894). Mr. Merrifield's experiments in temperature variation as bearing on theories of heredity. *Trans. ent. Soc. Lond.* 1894, 439–446.
DOBZHANSKY, TH. (1959). Changes in inversion frequencies in Californian populations of *Drosophila pseudoobscura* since 1941. *Proc. 15th Int. Congr. Zool.* 169–70.
DOBZHANSKY, TH. (1962). *Mankind evolving.* Yale University Press, New Haven and London.
DOBZHANSKY, TH. (1964). *Heredity and the nature of man.* Allen & Unwin, London.
DOBZHANSKY, TH., ANDERSON, W. W., PAVLOVSKY, O., SPASSKY, B., and WILLS C. J. (1964). Genetics of natural populations. XXXV. A progress report on genetic changes in populations of *Drosophila pseudoobscura* in the American south-west. *Evolution, Lancaster, Pa.* **18**, 164–76.
DOBZHANSKY, TH., and PAVLOVSKY, O. (1957). An experimental study of interaction between genetic drift and natural selection. *Evolution, Lancaster, Pa.* **11**, 311–19.
DONCASTER, L. (1903). Mendel's law of heredity in insects. *Entomologist's Rec. J. Var.* **15**, 142–4.
DONCASTER, L. (1960a). Mendel's laws of heredity. *Entomologist's Rec. J. Var.* **18**, 19–20.
DONCASTER, L. (1906b). Collective enquiry as to progressive melanism in insects. *Entomologist's Rec. J. Var.* **18**, 165–254.
DONCASTER, L., and RAYNOR, G. H. (1906). *Proc. zool. Soc. London.* 1906, Part I, 125–33.
DOWDESWELL, W. H. (1959). *Practical animal ecology.* Methuen, London.
DOWDESWELL, W. H., FISHER, R. A., and FORD, E. B. (1940). The quantitative study of populations in the Lepidoptera. I. *Polyommatus icarus* Rott. *Ann. Eugen.* **10**, 123–36.
DOWDESWELL, W. H., FISHER, R. A., and FORD, E. B. (1949). The quantitative study of populations in the Lepidoptera. II. *Maniola jurtina. Heredity, Lond.* **3**, 67–84.
DOWNES, J. A. (1964). Arctic insects and their environment. *Can. Ent.* **96**, 279–307.
DROZDA, A. (1970). *Studia i Materialy Entomologiczne*, pp. 7–75. Bytom, Poland (Rocznik Museum Gornoslaskiego w Bytomui).
EDLESTON, R. S. (1860). Remarks on *Bombyx quercus* and the variety *B. callunae* of Palmer. *Zoologist* **18**, 6815–16.
EDLESTON, R. S. (1864–5). *Amphydasis betularia. Entomologist* **2**, 150.
EHLICH, P. R., and RAVEN, P. H. (1965). Butterflies and plants: a study in evolution. *Evolution, Lancaster, Pa.* **18**, 586–608.
ELTRINGHAM, H. (1925). On the abdominal brushes of certain male noctuid moths. *Trans. R. ent. Soc. Lond.* 1925, 1–5.
ESPER, E. J. C. (1788). *Der europäischen Schmetterlinge* (Eulenphalenen). **4**, 387.

EVANS, L. J. (1953). Notes on breeding *Cycnia mendica* Clerck. *Entomologist's Rec. J. Var.* **65**, 4–6.
FAURE, J. C. (1943a). The phases of the lesser army worm. *Fmg S. Afr.* **18**, 69–78.
FAURE, J. C. (1943b). Phase variation in the army worm *Lephygma exempta* Walk. *Scient. Bull. Dep. Agric. S. Afr.* No. 234.
FEDERLEY, H. (1910). *Dicranura vinula* L. und ihre Nordischen Rassen. *Acta Soc. Fauna Flora fenn.* **33**(1), 1–20.
FEDERLEY, H. (1915). Chromosomenstudien an Mischlingen. II. Die Spermatogenese des Bastards *Dicranura erminea* female × *D. vinula* male. *Öfvers finska Vetensk. Soc. Förh. A*, 57(30), 26 pp.
FEDERLEY, H. (1916). Die Vererpung de Raupendimorphismus von *Chaerocampa elpenor* L. *Öfvers. finska Vetensk.Soc. Förh.* A, **58** (17), 13pp.
FEDERLEY, H. (1920). Die Bedeutung der Polymeren faktoren fur die Zeichung der Lepidopteren. *Hereditas* **1**, 221–69.
FEDERLEY, H. (1923). Über polymere Faktoren bei Lepidopteren. *Z. indukt. Abstamm.-u. VerebLehre* **30**, 284–6.
FEDERLEY, H. (1937) Fusion zweier Chromosomen als Folge einer Kreuzung. *Acta. Soc. Fauna Flora a fenn.* **60**, 685–95.
FEDERLEY. H. (1939). Geni e cromosomi. *Scientia genet.* **1**, 186–205.
FEDERLEY, H. (1940). Gene und Chromosomen. *Boll. Soc. ital. Biol. sper.* **15**, 81–97.
FEDERLEY, H. (1943). Zytogenetische Untersuchungen an Mischlingen der Gattung *Dicranura* (Lepidoptera). *Hereditas* **29**, 205–54.
FEDERLEY, H. (1945). Polyploidie und Non-Disjunstion in der Gametogenese einiger Lepidopteren. *Soc. Sci. Fennica Com. Biol.* **9**(17), 1–9.
FEDERLEY, H. (1953). Kreuzungsversuche mit Lepidopteren. *Z. Lepid.* **3**, 1–32.
FENTON, A. F. (1964). Atmospheric pollution of Belfast and its relationship to the lichen flora. *Ir. Nat. J.* **14**, 237–45.
FISCHER, E. (1923). Erblichkeitsverhältnisse bei *Argynnis paphia valesina* Esp. *Schweizer ent. Anz.* **2**, 3–5.
FISCHER, E. (1924). Über die Zweibrutigkeit der *P. bryoniae* O. *Mitt. Münch. ent. Ges.* **14**, 8–10.
FISHER, J. (1952). *The Fulmar.* Collins, London.
FISHER, R. A. (1927). *Trans. ent. Soc. Lond.* **75**, 269–78.
FISHER, R. A. (1928). The possible modification of the response of the wild type to recurrent mutations. *Am. Nat.* **62**, 115–26.
FISHER, R. A. (1930a). *The genetical theory of natural selection.* Clarendon Press, Oxford.
FISHER, R. A. (1930b). Distribution of gene ratios for rare mutations. *Proc. R. Soc. Edinb.* **50**, 204–19.
FISHER, R. A. (1931). *Biol. Rev.* **6**, 345–68.
FISHER, R. A. (1933). On the evidence against the chemical induction of melanism in Lepidoptera. *Proc. R. Soc.* B **112**, 407–16.
FISHER, R. A. (1935). *Phil. Trans. R. Soc.* B **225**, 197–226.
FISHER, R. A. (1937). The wave of advance of advantageous genes. *Ann. Eugen.* **7**, 360.

FISHER, R. A. (1938). *Proc. R. Soc.* B **125**, 25–48.
FISHER, R. A. (1950). Gene frequencies in a cline determined by selection and diffusion. *Biometrics* **6**, 353–61.
FISHER, R. A., and FORD, E. B. (1926). The variability of species. *Nature, Lond.* **118**, 515–16.
FISHER, R. A., and FORD, E. B. (1947). The spread of a gene in natural conditions in a colony of the moth *Panaxia dominula*. *Heredity, Lond.* **1**, 143–74.
FISHER, R. A., and FORD, E. B. (1950). The 'Sewall Wright effect'. *Heredity, Lond.* **4**, 117–9.
FISHER, R. A., and YATES, F. (1964). *Statistical tables for biological agricultural and medical research*. Oliver and Boyd, Edinburgh.
FITZPATRICK, T. B., BRUNET, P., and KUKITA, A. (1958). *The biology of hair growth* (ed. W. Montague). Academic Press.
FLETCHER, T. B. (1943). A note on the scent brushes in the Hepialidae. *Entomologist's Rec. J. Var.* **55**, 40–1.
FORD, E. B. (1937). Problems of heredity in the Lepidoptera. *Biol. Rev.* **12**, 461–503.
FORD, E. B. (1938). The genetic basis of adaptation, In *Evolution* (ed. G. R. de Beer), Oxford.
FORD, E. B. (1940a). Polymorphism and taxonomy. In *New Systematics* (ed. J. S. Huxley). Clarendon Press, Oxford.
FORD, E. B. (1940b). Genetic research in the Lepidoptera. *Ann. Eugen.* **10**, 227–52.
FORD, E. B. (1945). Polymorphism. *Biol. Rev.* **20**, 73–88.
FORD, E. B. (1949). Industrial melanism. *Proc. 8th int. Congr. Genet.* 571.
FORD, E. B. (1951). The experimental study of evolution. *Rep. Brisbane Meet. of Austr. and N.Z. Ass. Advent Sci.* **28**, 143–54.
FORD, E. B. (1953). The genetics of polymorphism in the Lepidoptera. *Adv. Genet.* **5**, 43–87.
FORD, E. B. (1953b). Polymorphism and taxonomy. *Rep. 7th Sci. Congr. R. Soc. N.Z.*, 245–53.
FORD, E. B. (1954a). Problems in the evolution of geographical races. In *Evolution as a Process* (ed. J. S. Huxley, A. C. Hardy, and E. B. Ford). Allen & Unwin, London.
FORD, E. B. (1954b). Evolution in polymorphic forms. *Proc. 9th int. Congr. Genet.* **1**, 463–8.
FORD, E. B. (1955a). *Moths*. Collins, London.
FORD, E. B. (1955b). Rapid evolution and the conditions which make it possible. *Cold Spring Harb. Symp. quant. Biol.* **20**, 230–8.
FORD, E. B. (1955c). Polymorphism and taxonomy. *Heredity, Lond.* **9**, 255–64.
FORD, E. B. (1957). *Butterflies*. Collins, London.
FORD, E. B. (1958). Darwinism and the study of evolution in natural populations. *J. Linn. Soc. Zool.* **44**, 41–8.
FORD, E. B. (1960a). *Mendelism and evolution*. Methuen, London.
FORD, E. B. (1960b). Evolution in progress. In *Evolution after Darwin*, Vol. I, *The Evolution of Life*. (ed. S. Tax). University of Chicago Press.

FORD, E. B. (1961). The theory of genetic polymorphism. *Symp. R. ent. Soc. Lond.* **1**, 11–19.

FORD, E. B. (1963). Mimicry. *Proc. 16th int. Congr. Zool.* **4**, 184–6.

FORD, E. B. (1965). *Genetic polymorphism.* Faber & Faber, London.

FORD, E. B. (1966). Genetic polymorphism. *Proc. R. Soc.* B **164**, 350–61.

FORD, E. B. (1971). *Ecological Genetics*, 3rd edition. Chapman & Hall, London.

FORD, E. B. and HUXLEY, J. S. (1927). Mendelian genes and rates of development in *Gammarus chevreuxi*. *Br. J. exp. Biol.* **8**, 112–34.

FORD, E. B. and SHEPPARD, P. M. (1969). The medionigra polymorphism of *Panaxia dominula*. *Heredity, Lond.* **24**, 561–9.

FORDHAM, W. H. (1956). Advance of grey squirrels, *Sciurus carolinensis*, and incidence of melanism in north Hertfordshire. *Proc. zool. Soc. Lond.* **126**, 170–2.

FOSTER, M. (1966). Mammalian pigment genetics. *Adv. Genet.* **13**, 311–39.

FOX, H. M., and VEVERS, G. (1960). *The Nature of animal colours.* Sedgwick & Jackson, London.

FRASER, F. C. (1947). *Entomologist's mon. Mag.* **83**, 218.

FRAZER, J. F., and ROTHSCHILD, M. (1960). Chemical defence mechanisms. *Proc. 11th int. Congr. Ent.* **3**, 249.

FRINGS, C. (1905). *Lasiocampa quercus* L. ab. nov. *paradoxa* Frgs. *Sociestas ent.* **20**, 89–90.

FRYER, J. C. F. (1913). *Phil. Trans. R. Soc.* B **204**, 227–54.

FRYER, J. C. F. (1928). Polymorphism in the moth *Acalla comariana* Zeller. *J. Genet.* **20**, 157–78.

FRYER, J. C. F. (1931). Further notes on the Tortricid moth *Acalla comariana* Zeller. *J. Genet.* **24**, 195–202.

GARDNER, E. J. (1968). *Principles of genetics.* Wiley, London.

GARRETT, F. C. (1927). On the inheritance of acquired characters. *Durham Univ. J.* **25**, 100–2.

GARRETT, F. C. and HARRISON, J. W. H. (1926). On the induction of melanism in Lepidoptera. *Entomologist's mon. Mag.* **62**, 210–11.

GEIGER, R. (1958). *The Climate near the ground* (translated by M. N. Stewart). Harvard University Press, Cambridge, Mass. U.S.A. (*Das Klima der Lodewahen.* Luftschict. 3rd ed. Braunschweig, Vieweg.)

GEROULD, J. H. (1924). Seasonal changes in melanic pigmentation in butterflies of the genus *Colias*. *Anat. Rec.* **29**, 94.

GERSHENSON, S. (1945). *Genetics, Princeton* **30**, 207.

GERSCHLER, M. W. (1915). Melanismus bei Lepidopteren als mutation und individuelle Variation. *Z. indukt. Abstamm. -u. VererbLehre* **13**, 58–87.

GODWIN, H. (1956). *The History of the British Flora.* Cambridge University Press.

GOLDSCHMIDT, R. B. (1921). Erblichkeitsstudien an Schmetterlinge. III. Der Melanismus der Nonne *Lymantria monacha* L. *Z. indukt. Abstamm. -u. VererbLehre* **25**, 89–163.

GOLDSCHMIDT, R. B. (1924). Erblichkeitsstudien an Schmetterlinger. IV. Weitere Untersuchungen über die Vererbung des Melanismus. *Z. induk. Abstamm.-u. VererbLehre* **34**, 229–43.

GOLDSCHMIDT, R. B. (1934). *Lymantria*. *Bibliogr. genet.* **11**, 1–186.

GOLDSCHMIDT, R. B. (1938). *Physiological Genetics*. McGraw-Hill, New York.
GOLDSCHMIDT, R. B. (1940). *The Material basis of evolution*. Yale University Press, New Haven.
GOLDSCHMIDT, R. B. (1945). Mimetic polymorphism, a controversial chapter of Darwinism. *Q. Rev. Biol.* **20**, 147–64, 205–30.
GOLDSCHMIDT, R. B. (1948). A note on industrial melanism in relation to some recent work with *Drosophila*. *Am. Nat.* **81**, 474–6.
GOLDSCHMIDT, R. B. and FISCHER, E. (1922). *Argynnis paphia—valesina*, ein fall geschlechtskontrollierter Vererbung. *Genetica* **4**, 247–78.
GÖNNER, P. (1928). *Argynnis paphia* L. mut. *valesina* Esp. *Ent. Z., Frankf. a. M.* **42**, 229–31.
GOULD, L. T. (1892). Experiments in 1890 and 1891 on the colour relation between lepidopterous larvae and their surroundings. *Trans. ent. Soc. Lond.* 1892, 215–46.
GRIBBLE, F. C. (1958). Census of black-headed gulls in England and Wales, 1958. *Bird Study* **9**, 56–71.
GRIFFITH, A. F. (1929). Notes on Shetland birds in Yell and Unst in June 1928. *Brit. Ornith. Club Bull.* **49**, 99.
GROSVENOR, T. H. L. (1926–7). Annual address. *Proc. S. Lond. ent. nat. Hist. Soc.* 1926–7, 88–97.
GUILER, E. R. (1953). Distribution of the British possum in Tasmania. *Nature, Lond.* **172**, 1091.
HAGGETT, G. M. (1963). Researches in colour variation of the moth *Leucania vitellina* Hübner. *Proc. S. Lond. ent. nat. Hist. Soc.* 1963, Part II, 78–92.
HAGGETT, G. M., and WILLIAMS, H. B. (1950). Genetics and the collector. *Entomologist's Gaz*, **1**, 142–9.
HALDANE, J. B. S. (1922). Sex ratio and unisexual sterility in hybrid animals. *J. Genet.* **12**, 101–9.
HALDANE, J. B. S. (1924). A mathematical theory of natural and artificial selection. *Trans. Cam. Phil. Soc.* **23**, 26.
HALDANE, J. B. S. (1927). The comparative genetics of colour in rodents and Carnivora. *Biol. Rev.* **2**, 199–212.
HALDANE, J. B. S. (1930). A note on Fisher's theory of the origin of dominance, and on a correlation between dominance and linkage. *Am. Nat.* **64**, 87–90.
HALDANE, J. B. S. (1935). *J. Genet.* **31**, 317–26.
HALDANE, J. B. S. (1940). The estimation of recessive gene frequencies by inbreeding. *Proc. Indian Acad. Sci.* B **15**, 109–14.
HALDANE, J. B. S. (1948). The theory of a cline. *J. Genet.* **48**, 277–84.
HALDANE, J. B. S. (1953). *Animal populations and their regulation*. New Biol. (Penguin Books) **15**, 9–24.
HALDANE, J. B. S. (1955). On the biochemistry of heterosis and the stabilisation of polymorphism. *Proc. R. Soc.* B **144**, 217–220.
HALDANE, J. B. S. (1956). The theory of selection for melanism in the Lepidoptera. *Proc. R. Soc.* B **145**, 303–8.
HALDANE, J. B. S. (1957). The cost of natural selection. *J. Genet.* **55**, 511–24.

HALDANE, J. B. S. (1959). *Darwin's biological works* (pp. 127–47). Cambridge University Press.
HAMILTON, Lady C. (1885). *Louis Pasteur, his life and labours.* Longmans & Co., London.
HAMILTON, F. D. (1962). Census of Black-headed Gulls in Scotland 1958. *Bird Study* **9**, 72–80.
HAMLING, T. H. (1903–). Notes on breeding *Gonodontis bidentata* ab. *nigra. Trans. C. Lond. ent. Soc.* (1903), 40–3.
HAMLING, T. H. (1905). *Hemerophila abruptaria*—heredity statistics. *Trans. C. Lond. ent. Soc.* 1905, 5.
HARDY, E. (1937). *Country Life* **81**, 676.
HARE, E. J. (1957). Unst revisited. *Entomologist's Rec. J. Var.* **69**, 80–3.
HARPER, B. W. (1958). Northern lights—a visit to Unst. *Entomologist's Rec. J. Var.* **70**, 286–8.
HARRIS, E. (1904). Brood of *Hemerophila abruptaria. Proc. ent. Soc. Lond.* (1904), 72.
HARRIS, E. (1905a). Breeding experiment with *Synopsia abruptaria. Trans. C. Lond. ent. Soc.* 1905, 13–14.
HARRIS, E. (1905b). On *Hemerophila abruptaria. Proc. ent. Soc. Lond.* (1905), 63–4.
HARRISON, A. (1908). On *Aplecta nebulosa. Entomologist* **41**, 314.
HARRISON, A. (1909). On *Aplecta nebulosa. Proc. ent. Soc. Lond.* (1908). 61–2.
HARRISON, A. and BACOT, A. (1905). *Pachys betularia* heredity statistics. *Trans. C. Lond. ent. Soc.* 1905, 5.
HARRISON, A. and MAIN, H. (1909–10). On *Aplecta nebulosa. Proc. S. Lond. ent. nat. Hist. Soc.* 1909–10, 84.
HARRISON, A. and MAIN, H. (1911a). Varieties of *Aplecta nebulosa. Proc. ent. Soc. Lond.* 1911, 21.
HARRISON, A. and MAIN H. (1911b). On *Aplecta nebulosa. Entomologist's Rec. J. Var.* **23**, 230–1.
HARRISON, J. M. and HARRISON, J. C. (1962). Albinism and melanism in birds. *Bull. Br. Orn. Club* **82**, 101–9.
HARRISON, J. W. H. (1919). A preliminary study of the effects of administering ethyl alcohol to the lepidopterous insect *Selenia bilunaria* with particular reference to the offspring. *J. Genet.* **9**, 39–52.
HARRISON, J. W. H. (1920a). Genetical studies in moths of Geometrid genus *Oporabia* (Oporinia) with a special consideration of melanism in Lepidoptera. *J. Genet.* **9**, 195–280.
HARRISON, J. W. H. (1920b). The inheritance of melanism in the genus *Tephrosia* (*Ectropis*) with some consideration of the inconstancy of unit characters under crossing. *J. Genet.* **10**, 61–85.
HARRISON, J. W. H. (1923). The inheritance of wing colour and pattern in the Lepidoptera genus *Tephrosia* (*Ectropis*) with an account of the origin of a new allelomorph. I. Experiments involving melanic *T. crepuscularia. J. Genet.* **13**, 333–52.
HARRISON, J. W. H. (1926a). The inheritance of wing colour and pattern in the lepidopterous genus *Tephrosia* (*Ectropis*). II. Experiments involving melanic *T. bistortata* and typical *T. crepuscularia. J. Genet.* **17**, 1–19.

HARRISON, J. W. H. (1926b). Miscellaneous observations on the induction, incidence, and inheritance of melanism in the Lepidoptera. *Entomologist* **59**, 121–3.

HARRISON, J. W. H. (1927a). Some thoughts on melanism and melanchroism in the Lepidoptera. *Vasculum* **13** (3), 103–5.

HARRISON, J. W. H. (1927b). The inheritance of melanism in hybrids between continental *Tephrosia crepuscularia* and British *T. bistortata*. *Genetica* **9**, 467–80.

HARRISON, J. W. H. (1927c). The induction of melanism in the Lepidoptera and its evolutionary significance. *Nature, Lond.* **119**, 127–9.

HARRISON, J. W. H. (1927d). Melanism in the Lepidoptera and its evolutionary significance. *Nature, Lond.* **119**, 318.

HARRISON, J. W. H. (1928a). Notes on melanism. *Proc. ent. Soc. Lond.* **3**, 25–28.

HARRISON, J. W. H. (1928b). A further induction of melanism in the lepidopterous insect, *Selenia bilunaria* Esp. and its inheritance. *Proc. R. Soc.* B **102**, 338–47.

HARRISON, J. W. H. (1928c). Induced changes in the pigmentation of the pupae of the butterfly *Pieris napi*, and their inheritance. *Proc. R. Soc.* B **102**, 347–53.

HARRISON, J. W. H. (1931). The colours of the larvae of the peppered moth. *Vasculum* **17**, 153.

HARRISON, J. W. H. (1932a). The recent development of melanism in the larvae of certain species of Lepidoptera with an account of its inheritance in *Selenia bilunaria* Esp. *Proc. R. Soc.* B **111**, 188–200.

HARRISON, J. W. H. (1932b). The inheritance of melanism in crosses between continental *Tephrosia crepuscularia* and its melanic *T. bistortata*. *Genetica* **14**, 151–9.

HARRISON, J. W. H. (1935). The experimental induction of melanism and other effects in the Geometrid moth *Selenia bilunaria*. *Proc. R. Soc.* B **117**, 78–92.

HARRISON, J. W. H. (1956a). Parallel variation in the larvae and imago of *Oporinia dilutata* Bkh. *Entomologist* **89**, 70.

HARRISON, J. W. H. (1956b). Melanism in the Lepidoptera. *Entomologist's Rec. J. Var.* **68**, 172–81.

HARRISON, J. W. H. and GARRETT, F. C. (1926). The induction of melanism in the Lepidoptera and its subsequent inheritance. *Proc. R. Soc.* B **99**, 241–63.

HARRISON, J. W. H. and MAIN, H. (1908–9). *Proc. S. Lond. ent. nat. Hist. Soc.* 1908–9, 84.

HARRISON, J. W. H. and MAIN, H. (1909–10). *Proc. S. Lond. ent. nat. Hist. Soc.* 1909–10, 164.

HARSHBERGER, J. W. (1970). *The Vegetation of the New Jersey Pine-Barrens.* Dover Publications Inc., New York.

HASEBROËK, K. (1909). Über *Cymatophora or* F. ab. *albingensis* Warn. und die entwicklungsgeschichtliche Bedeutung ihres Melanismus. *Ent. Rdsch.* **26**, 51–3.

HASEBROËK, K. (1911). Wie haben wir Hamburger unsere melanistische *Cymatophora or* F. ab. *albigensis* Warn. nach den Mendelschen Regeln in Kreuzungszucht zu nehmen? *Int. ent. Z.* **5**, 9–11.

HASEBROËK, K. (1914). Über die Entstehung des neuzeitlichen Melanismus der

Schmetterlinge und die Bedeutung der Hamburger Formen für dessen Ergrundung. *Zool. Jb. Syst.* **37**, 567–600.

HASEBROËK, K. (1925). Untersuchungen zum problem des neuzeitlichen Melanismus der Schmetterlinge. *Fermentforschung* **8**, 199–226.

HASEBROËK, K. (1925–26). Die prinzipielle Losung des Problems des Grosstadt und Industriemelanismus der Schmetterlinge. *Int. ent. Z.* **19**, 78–9.

HASEBROËK, K. (1928–29). Atmosphare und Luftstromüngen in ihren Beziehungen zum Industrie und Grosstadtmelanismus. *Int. ent. Z.* **22**, 313–18, 321–35.

HASEBROËK, K. (1929). Über den Industrie und Grosstadt-melanismus der Schmetterlinge. *Z. induct. Abstamm.-u. VererbLehre* **50**, 201–18.

HASEBROËK, K. (1934). Industrie und Grosstadt als Ursache des neuzeitlichen Melanismus der Schmetterlinge in England und Deutschland. *Zool. Jb. Allg. Zool.* **53**, 411–60.

HAWKINS, C. N. (1940). A new British variety of *Lycia hirtaria* Clerck. *Entomologist* **73**, 28.

HAWKSWORTH, D. L. and ROSE, F. (1970). A qualitative scale for estimating sulphur dioxide air pollution in England and Wales using epiphytic lichens. *Nature, Lond.* **227**, 145.

HERSHKOVITZ, P. (1968). Metachromism or the principle of evolutionary change in mammalian tegumentary colours. *Evolution, Lancaster, Pa.* **22**, 556–75.

HEWSON, R. (1953). 'Black' larvae of *Lasiocampa quercus* in Yorkshire. *Entomologist's Rec. J. Var.* **65**, 1–2.

HEYDEMANN, F. (1927–8). Der Gebirgs- und Küstenmelanismus und -Nigrismus. Zugleich ein Beitrag zur Frage des Industriemelanismus. *Int. ent. Z.* **21**, 247–52, 271–6, 283–8, 291–6, 303–5, 315–18, 327–37.

HOFFMEYER, S. (1948). *De Danske Spindera*. Universitetsforlaget i, Aarhus, Denmark.

HOFFMEYER, S. (1949). *De Danske Ugler*. Universitetsforlaget i, Aarhus, Denmark.

HOLLOM, P. A. D. (1940). Report on the 1938 survey of Black-headed Gull colonies. *Br. Birds* **33**, 202–44.

HOPKINS, F. G. (1892). Pigments of Lepidoptera. *Nature, Lond.* **45**, 581.

HOVANITZ, W. (1943). The distribution of gene-frequencies in wild populations of *Colias*. *Genetics, Princeton*, **29**, 1–30.

HOVANITZ, W. (1946). Comparative dispersion of female colour types of *Colias*. *Genetics, Princeton* **31**, 218.

HOVANITZ, W. (1948). Differences in the field activity of two female colour phases of *Colias* butterflies at various times of the day. *Centr. Lab. vertebr. Biol. Univ. Mich.* **41**, 1–37.

HOVANITZ, W. (1951). The biology of *Colias* butterflies. III. Variation of adult flight in the Arctic and sub-Arctic. *Wasmann J. Biol.* **9**, 1–9.

HOWARD, G. (1967). Lepidoptera in Lapland 1966. *Entomologist* **100**, 1.

HOWARD, G. (1969). Lepidoptera in Lapland. *Entomologist's Rec. J. Var.* **81**, 75–7.

HRUBÝ, K. (1958). Caterpillar colour inheritance in *Biston betularia* L. *Proc. 10th Int. Congr. Genet.* **2**, 125–6.

Hudson, G. V. (1928). *The Butterflies and Moths of New Zealand*, Wellington.
Huggins, H. C. (1956). Some aspects of melanism. *Entomologist* **89**, 185-7.
Hughes, A. W. (1931-2). On *Lasiocampa quercus*. *Proc. S. Lond. ent. nat. Hist. Soc.* 1931-2, 47.
Hughes, A. W. (1932). Induced melanism in the Lepidoptera. *Proc. R. Soc.* B **110**, 378-402.
Hutt, F. B. (1964). *Animal genetics*. Ronald Press, New York.
Huxley, J. S. (1942). *Evolution: the modern synthesis*. Allen & Unwin, London.
Huxley, J. S. (1955). Morphism and evolution: Bateson Lecture. *Heredity, Lond.* **9**, 1-52.
Isely, F. B. (1938). Survival value of acridian protective colouration. *Ecology* **19**, 370-89.
Jackson, C. H. N. (1940). *Ann. Eugen.* **10**, 332-69.
James, J. W. (1965). Simultaneous selection for dominant and recessive mutants. *Heredity, Lond.* **20**, 142-4.
Jarvis, F. V. L. (1941-2). *Proc. S. Lond. ent. nat. Hist. Soc.* 1941-2, Plate 1.
Jeannel, R. (1925). L'apterism chez les insectes insulaires. *C. r. hebd. Séanc. Acad. Sci., Paris.* **180**, 1222-4.
Johnson, J. H. (1956a). Viability of melanic and typical forms of *Gonodontis bidentata* Clerck compared. *Entomologist* **89**, 117-21.
Johnson, J. H. (1956b). *Ectropis bistortata* Goeze: melanics and 'mosaics'. *Entomologist's Rec. J. Var.* **68**, 92-5.
Johnson, J. H. (1961). Further observations on the egg-laying capacity of melanic and typical forms of *Gonodontis bidentata* Clerck. *Entomologist* **94**, 173-8.
Johnson, J. H. (1964). Polymorphism in N.E. Derbyshire. *Entomologist's Rec. J. Var.* **76**, 282-6.
Johnston, R. F. and Selander, R. M. (1964). House sparrows: rapid evolution of races in North America. *Science, N.Y.* **144**, 548-50.
Jones, D. A., Parsons, J., and Rothschild, M. (1962). Release of hydrocyanic acid from crushed tissue of all stages in the life-cycle of species of the Zygaeninae (Lepidoptera). *Nature, Lond.* **193**, 4810, 52.
Jones, E. W. (1952). Some observations on the lichen flora of tree boles, with special reference to the effects of smoke. *Revue bryol. lichen.* **21**, 96-115.
Kalmus, H. (1941). Egg-laying of ducks as an enforced relaxation oscillation. *Nature, Lond.* **148**, 626-7.
Kane, W. de V. (1896). Observations on the development of melanism in *Camptogramma bilineata*. *Ir. Nat.* **v**, 75-80.
Kennicott, R. (1857). Quadrupeds of Illinois injurious and beneficial to the farmer. *Exec. Doc. 65, 34th Congr. 3rd Session Rep. Comm. Patents, 1856*, 52-110.
Kettlewell, H. B. D. (1931). Notes on *Heliothis peltigera*. *Entomologist's Rec. J. Var.* **42**, 62-4.
Kettlewell, H. B. D. (1942-3). A survey of the insect *Panaxia (Callimorpha) dominula* L. *Proc. S. Lond. ent. nat. Hist. Soc.* 1942-3, 1-49.
Kettlewell, H. B. D. (1943-4). Temperature experiments on the pupae of

Heliothis peltigera Schiff. and *Panaxia dominula* L. *Proc. S. Lond. ent. nat. Hist. Soc.* 1943-4, 69-81.

KETTLEWELL, H. B. D. (1945). Exhibition of *Panaxia dominula* aberrations. *Proc. R. ent. Soc. Lond.* C **9**, 29.

KETTLEWELL, H. B. D. (1946a). Further observations on *Panaxia dominula*. *Entomologist* **79**, 31.

KETTLEWELL, H. B. D. (1946b). Female assembling scents with reference to an important paper on the subject. *Entomologist* **79**, 8-14.

KETTLEWELL, H. B. D. (1946c). Further observations on the season 1945 with special reference to *Pontia daplidice*. *Entomologist* **79**, 111.

KETTLEWELL, H. B. D. (1947). On genetic control of migration. *Proc. R. ent. Soc.* C **12**, 43.

KETTLEWELL, H. B. D. (1952a). Use of radioactive tracer in the study of insect populations (Lepidoptera). *Nature, Lond.* **170**, 584-86.

KETTLEWELL, H. B. D. (1952b). A possible genetic explanation and understanding of migration of continuous brooded insects. *Nature, Lond.* **169**, 832-3.

KETTLEWELL, H. B. D. (1955a). How industrialisation can alter species. *Discovery, Lond.* **16**, 12, 507-11.

KETTLEWELL, H. B. D. (1955b). Selection experiments on Industrial Melanism in the Lepidoptera. *Heredity, Lond.* **9** (3), 323-42.

KETTLEWELL, H. B. D. (1955c). Brood size and dispersal in *Biston betularia* L. *Entomologist* **88**, 50.

KETTLEWELL, H. B. D. (1955d). Recognition of appropriate backgrounds by the pale and black phases of Lepidoptera. *Nature, Lond.* **175**, 943-4.

KETTLEWELL, H. B. D. (1955e). Labelling Locusts with radioactive isotopes. *Nature, Lond.* **175**, 821.

KETTLEWELL, H. B. D. (1955f). A further case of similar assembling scents. *Entomologist* **88**, 19.

KETTLEWELL, H. B. D. (1956a). A southern migration of *Vanessa atalanta* L., the Red Admiral. *Entomologist* **89**, 1112, 18-19.

KETTLEWELL, H. B. D. (1956b). Melanism and an answer to J. W. H. Harrison. *Entomologist's Rec. J. Var.* **68**, 286-92.

KETTLEWELL, H. B. D. (1956c). Further selection experiments on Industrial Melanism in the Lepidoptera. *Heredity, Lond.* **10** (3), 287-301.

KETTLEWELL, H. B. D. (1956d). A résumé of investigations on the evolution of melanism in the Lepidoptera. *Proc. R. Soc.* B **145**, 297-303.

KETTLEWELL, H. B. D. (1956e). Evolution in action. *Times Science Review*, No. 19.

KETTLEWELL, H. B. D. (1956f). An answer to one thought 'on reading Dr. Ford's book *Moths*'. *Entomologist's Rec. J. Var.* **68**, 286-92.

KETTLEWELL, H. B. D. (1956g). *Xanthorhoë montanata* and *X. spadicearia* assembling to females of *Biston betularia*. *Entomologist* **89**, 130.

KETTLEWELL, H. B. D. (1957a). The contribution of Industrial Melanism in the Lepidoptera to our knowledge of evolution. *Advmt. Sci., Lond.* **52**, 245.

KETTLEWELL, H. B. D. (1957b). Industrial Melanism in moths and its contribution to our knowledge of evolution. *Proc. R. Instn. Gr. Br.* **36**, 164, 1-13.

KETTLEWELL, H. B. D. (1957c). Problems in Industrial Melanism. *Entomologist* **90**, 98-105.

Kettlewell, H. B. D. (1958a). A survey of the frequencies of *Biston betularia* L. (Lep.) and its melanic forms in Britain. *Heredity, Lond.* **12**, 51–72.
Kettlewell, H. B. D. (1958b). Evolution and the environment. *New Scient.* **4**, 297–9.
Kettlewell, H. B. D. (1958c). Industrial Melanism in the Lepidoptera and its contribution to our knowledge of evolution. *Proc. 10th int. Congr. Ent.* **2**, 831–41.
Kettlewell, H. B. D. (1958d). The importance of the micro-environment to evolutionary trends in the Lepidoptera. *Entomologist* **91**, 214–24.
Kettlewell, H. B. D. (1959a). Darwin's missing evidence. *Scient. Am.* **200**, 48–53.
Kettlewell, H. B. D. (1959b). New aspects of the genetic control of Industrial Melanism in the Lepidoptera. *Nature, Lond.* **183**, 918–21.
Kettlewell, H. B. D. (1959c). Brazilian insect adaptations. *Endeavour* **72**, 200–10.
Kettlewell, H. B. D. (1959d). Industrial Melanism in the Lepidoptera and its contribution to our knowledge of evolution. *The Society of American Zoologists* (lecture records), 1959, 25–40.
Kettlewell, H. B. D. (1960a). Radioactive migrants—a request. *Entomologist* **93**, 75.
Kettlewell, H. B. D. (1960b). Migrateurs radioactifs. *Entomologiste* **41**, 1–2.
Kettlewell, H. B. D. (1960c). A report on the investigation into radioactive migrants 1960. *Entomologist* **94**, (Pl. 3 and 4) 49–52.
Kettlewell, H. B. D. (1961a). The phenomenon of Industrial Melanism in the Lepidoptera. *A. Rev. Ent.* **6**, 245–62.
Kettlewell, H. B. D. (1961b). The radiation theory of female assembling in the Lepidoptera. *Entomologist* **94**, 59–65.
Kettlewell, H. B. D. (1961c). Geographic Melanism in the Lepidoptera of Shetland. *Heredity, Lond.* **16**, 393–402.
Kettlewell, H. B. D. (1961d). Selection experiments on melanism in *Amathes glareosa* Esp. (Lepidoptera). *Heredity, Lond.* **16**, 415–34.
Kettlewell, H. B. D. (1963a). Recent advances in our knowledge of melanism in the Lepidoptera. *Proc. 16th int. Congr. Zool.* **2**, 198–9.
Kettlewell, H. B. D. (1963b). Lepidoptera as scientific tools. *J. Lepid. Soc.* **17**, 173–7.
Kettlewell, H. B. D. (1963c). The genetical and environmental factors which affect colour and pattern in the Lepidoptera with special reference to a migratory species. *Entomologist* **96**, 127–30.
Kettlewell, H. B. D. (1963d). The life-history of *Utetheisa pulchella* L. and its possible adaptive significance. *Entomologist* **96**, 102–7.
Kettlewell, H. B. D. (1964a). The inherited and environmental contributions to the patterning of *Utetheisa pulchella*. *Entomologist* **97**, 169–72.
Kettlewell, H. B. D. (1964b). Adaptation mechanisms in the genetic armoury of Lepidoptera—ancient and modern. *Zenith, Oxford* **1**, 11–14.
Kettlewell, H. B. D. (1964c). Natural history and air pollution. *New Scient.* **22**, 34.
Kettlewell, H. B. D. (1965a). Insect adaptations in Brazil. *Animals* **5**, No. 18.

KETTLEWELL, H. B. D. (1965b). Insect adaptations in Africa. *Animals* **5**, No. 19.
KETTLEWELL, H. B. D. (1965c). Insect adaptations in Britain. *Animals* **5**, No. 20.
KETTLEWELL, H. B. D. (1965d). A 12-year survey of the frequencies of *Biston betularia* L. and its melanic forms in Great Britain. *Entomologist's Rec. J. Var.* **77**, 195–218.
KETTLEWELL, H. B. D. (1965e). Insect survival and selection for pattern. *Science, N.Y.* **148**, 1290–5.
KETTLEWELL, H. B. D. (1965f). Hibernation and pupation habits of *Cossus cossus* L. (Lep.). Cossidae. *Entomologist's Rec. J. Var.* **77**, 1–3.
KETTLEWELL, H. B. D. (1969). Evolution today. In *Purnell's Encyclopaedia of Animal Life*, No. 22.
KETTLEWELL, H. B. D. and BERRY, R. J. (1961). The study of a cline. *Heredity, Lond.* **16**, 403–14.
KETTLEWELL, H. B. D. and BERRY, R. J. (1969). Gene flow in a cline. *Amathes glareosa* Esp. and its melanic f. *edda* Stdg. (Lep.) in Shetland. *Heredity, Lond.* **24**, 1–14.
KETTLEWELL, H. B. D., BERRY, R. J., CADBURY, C. J., and PHILLIPS, G. C. (1969). Differences in behaviour, dominance, and survival within a cline. *Heredity, Lond.* **24**, 15–25.
KETTLEWELL, H. B. D. and CADBURY, C. J. (1963). Investigations on the origins of Non-industrial Melanism. *Entomologist's Rec. J. Var.* **75**, 149 60.
KETTLEWELL, H. B. D., CADBURY, C. J., and LEES, D. R. (1971). Recessive melanism in the moth *Lasiocampa quercus* L. In *Ecological Genetics and Evolution*, Blackwell, Oxford.
KETTLEWELL, H. B. D. and COOK, L. M. (1960). Radioactive labelling of lepidopterous larvae: a method of estimating late larval and pupal mortality in the wild. *Nature, Lond.* **187**, 301–2.
KETTLEWELL, H. B. D. and HEARD, M. J. (1961). Accidental radioactive labelling of a migratory moth. *Nature, Lond.* **189**, 676–7.
KIKKAWA, H. (1953). Biochemical genetics of *Bombyx mori*. *Adv. Genet.* **5**, 107–40.
KLAAT, B. (1928). Eine melanistische mutation beim Schwammspinner. *Zool. Anz.* **78**, 257–60.
KLOET, G. S. and HINCKS, W. D. (1945). *A check list of British insects*. Buncle, Arbroath.
KOMAI, T. (1956). Genetics of ladybeetles. *Adv. Genet.* **8**, 155–88.
KÜHN, A. (1944). *Biol. Zbl.* **64**, 154–7.
KÜHN, A. (1963). Die Zeitliche Folge der Bildung der Ommochrome in der Imaginelaugen von *Ephestia* und *Ptychopoda*. *Z. Naturf.* **186**, 252–4.
KÜHN. A. and ENGELHARDT, M. von (1937). Über eine melanistische Mutation von *Ptychopoda seriata* Schrk. *Biol. Zbl.*, **57**, 329–47.
KÜHN, A. and ENGLEHARDT, M. von, (1943). Über zwei melanistische Mutationen (*At und ni*) von *Ptychopoda seriata* Schrk. *Biol. Zbl.* **63**, 251–67.
KÜHN, A. and HENKE, K. (1932). Genetische und Entwicklungsphysiologische Untersuchungen an der Mehlmoth *Ephestia kuhniella* Zeller. VIII–XII. *Abh. Ges. Wiss.*, Göttingen, Math–Phys. Kl. VI (Biol.) **15**, 1–121.
LANCUM, F. H. (1920). *Entomologist* **53**, 236.

LANDSBOROUGH THOMSON (1964). *A new dictionary of birds*. Nelson, London and Edinburgh.
LANE, C. and ROTHSCHILD, M. (1965). A case of Müllerian mimicry of sound. *Proc. R. ent. Soc. Lond.* A **40**, 10–12, 156–8.
LEES, D. R. (1968). Genetic control of the melanic form *insularia* of the peppered moth *Biston betularia* L. *Nature, Lond.* **220**, 1249–50.
LEES, D. R. (1970). The *medionigra* polymorphism of *Panaxia dominula* in 1969. *Heredity, Lond.* **25**, 470–5.
LEES, D. R. (1971). Distribution of melanism in the Pale Brindled Beauty moth, *Phigalia pedaria*, in Britain. In *Ecological Genetics and Evolution*, (ed. E. R. Creed) pp. 152–74. Blackwell, Oxford.
LEIGH, G. F. (1904). *Trans. ent. Soc. Lond.* 1904, 677–94.
LEMCHE, H. (1931). *Amphidasys betularia* (L.) and its melanic varieties. *J. Genet.* **24**, 235–41.
LEMPKE, B. J. (1938). Catalogue der Nederlandsche Macrolepidoptera. *Tijdschr. Ent.* **9**, 63–4.
LEMPKE, B. J. (1946). A hereditary form of *Spilosoma lubricipeda* L. (*menthastri* Esp.). *Entomologist's Rec. J. Var.* **58**, 96–8.
LEMPKE, B. J. (1947). The variation of *Lymantria monacha* L. *Entomologist's Rec. J. Var.* **59**, 81–7.
LEMPKE, B. J. (1948). A hereditary form of *Spilosoma lutea* Hufn. *Entomologist's Rec. J. Var.* **60**, 4–5.
LEMPKE, B. J. (1951). A contribution to the genetics of *Lasiocampa quercus* L. *Entomologist's Rec. J. Var.* **63**, 200–3.
LEMPKE, B. J. (1959). *Tijdschr. Ent.* **102** (supplem.), 57–134 (with plates 3–14).
LEMPKE, B. J. (1960). Catalogus der Nederlanse Macrolepidoptera. *Tijdschr. Ent.* **103**, 145–215.
LERNER, A. B. (1955). Melanin pigmentation. *Am. J. Med.* **19**, 470–84.
LEVITICUS (c. 500 B.C.). Chapter 11, verse 13.
LEWIS, T. H. (1949). Dark colouration in the reptiles of the Tularosa Malpais, New Mexico. *Copeia* **3**, 181–4.
LONG, D. B. (1953). Effects of population density on larvae of Lepidoptera. *Trans. R. ent. Soc. Lond.* **104**, 543–85.
LORKOVIC, Z. (1962). The genetics and reproductive isolating mechanism of the *Pieris napi—bryoniae* group. *J. Lepid. Soc.* **16**, 5–19, 105–27.
LOWE, P. R. (1912). Observations on the genus *Coereba* together with an annotated list of the species. *Ibis* **9**, 6, 489.
LUSIS, J. J. (1961). On the biological meaning of the colour polymorphisms of the ladybeetle *Adalia bipunctata*. *Latv. Ent.* **4**, 3–9.
LYCKLAMA A NYEHOLT, H. J. (1932). Melanisme bij Lepidoptera. *Tijdschr. Ent.* **75** (supplem.), 29–34.
MCDIARMID, A. (1948). The occurrence of tuberculosis in the wild wood-pigeon. *J. comp. Path. Ther.* **58**, 128–33.
MACKIE, D. W. (1960). *Ostearius melanopygius* (O.P.C.). *Bulletin of the British Spider Study Group* No. **8**, 3–4.
MACKIE, D. W. (1964). A melanic form of *Salticus scenicus* Clerck. *Bulletin of the British Spider Study Group* No. **24**, 4.

MACKIE, D. W. (1965). An enquiry into the habits of *Drapetisca socialis*. *Bulletin of the British Spider Study Group* No. **27**, 4–6.
MCPHAIL, J. D. (1969). Predation and evolution of a stickleback (*Gasterosteus*). *J. Fish. Res. Bd., Can.* **26**, 3183–208.
MADDISON, T. (1893). On breeding *Polia chi* var-*olivacea*, *Entomologist's Rec. J. Var.* **4**, 3.
MAGNUS, D. B. E. (1958). Experimental analysis . . . of the Fritillary butterfly *Argynnis paphia* L. (Nymphalidae). *Proc. 10th int. Congr. Ent.* **2**, 405–8.
MAIN, H. and HARRISON, A. (1905). On *Colias edusa* and *Amphidasys betularia*. *Proc. ent. Soc. Lond.* 1905, 6.
MALAN, D. E. (1918). Ergebnisse anatomischer a Untersuchungen an Standfuss'schen Lepidopteren Bastards. *Mitt. Ent. Zurich*, **4**, 201–60.
MANSBRIDGE, W. (1893a). Melanism in Yorkshire Lepidoptera. *Entomologist's Rec. J. Var.* **4**, 110–11.
MANSBRIDGE, W. (1893b). Notes on melanism in Yorkshire Lepidoptera. *Proc. S. Lond. ent. nat. Hist. Soc.* **1893**, 97–9.
MANSBRIDGE, W. (1909). On *Aplecta nebulosa*. *Entomologist* **42**, 127.
MANSBRIDGE, W. (1909–10). On *Aplecta nebulosa*. *Proc. S. Lond. ent. nat. Hist. Soc.* 1909–10, 64–5.
MANSBRIDGE, W. (1917). *Aplecta nebulosa* Hufn. var. *plumbosa* var. nov. *Entomologist* **50**, 49.
MANSBRIDGE, W. (1918). Notes on breeding *Boarmia repandata*. *Entomologist's Rec. J. Var.* **30**, 120.
MANSBRIDGE, W. (1927–8). Examples of melanism in Lepidoptera. *Proc. ent. Soc. Lond.* **2**, 20–2.
MATHER, K. (1955). Polymorphism as an outcome of disruptive selection. *Evolution, Lancaster, Pa.* **9**, 52–61.
MATHER, K. (1964). *Statistical analysis in biology*. Methuen, London.
MATTHEE, J. J. (1945). Biochemical differences between solitary and gregarious phases of locust and Noctuids. *Bull. ent. Res.* **36**, 343–71.
MATHEE, J. J. (1947). Phase variation in the lawn caterpillar (*Spodoptera abyssinian* Guen.) *J. ent. Soc. Sth Afr.* **10**, 16–23.
MATTHEWS, L. H. (1952). *British mammals*. Collins, London.
MAYR, E. (1942). *Systematics and the origin of species*. Columbia University Press, New York.
MAYR, E. (1954). Change of genetic environment and evolution. In *Evolution as a Process* (ed. J. Huxley, A. C. Hardy, and E. B. Ford), pp. 157–80. Allen & Unwin, London.
MAYR, E. (1955). Is the Great White Heron a good species? *Auk* **73**, 71–7.
MAYR, E. (1963). *Animal species and evolution*. Harvard University Press, Cambridge, Mass. U.S.A.
MENDEL, G. (1866). Versuche über Pflanzenhybriden. *Verh. naturf. Ver. Brünn* 1866.
MEIJERE, J. C. H. de (1910). *Z. indukt. Abstamm. -u. VererbLehr.* **3**, 161–81.
MERA, A. W. (1925). Increase in melanism in the last half-century. *Lond. Nat.* 1925, 3–9.
MERRIFIELD, F. (1888). *Trans. ent. Soc. Lond.* 1888, 123–36.

MERRIFIELD, F. (1889). Incidental observations in pedigree moth breeding. *Trans. ent. Soc. Lond.* 1889, 79–97.

MERRIFIELD, F. (1890). Systematic temperature experiments on some Lepidoptera, in all their stages. *Trans. ent. Soc. Lond.* 1890, 131–59.

MERRIFIELD, F. (1892). The effects of artificial temperature in the colouring of several species of Lepidoptera . . . *Trans. ent. Soc. Lond.* 1892, 33–44.

MERRIFELD, F. (1894). Temperature experiments in 1893 on several species of *Vanessa* and other Lepidoptera. *Trans. ent. Soc. Lond.* 1894, 425–38.

MICHAEL, P. (1949). An impromptu experiment with *Hepialus humuli*. *Entomologist* **82**, 175.

MILLER, E. (1913). An account of the breeding of *Amphidasys betularia* and ab. *doubledayaria*. *Entomologist's Rec. J. Var.* **25**, 109–11.

MINNION, W. E. (1957). *Ennomos autumnaria* Wernb. ab. *brunneata* Cockayne. *Entomologist's Rec. J. Var.* **69**, 77–8.

MOREAU, R. E. (1930). On the age of some races of birds. *Ibis* (series 12) **6**, 229–39.

MORLEY, B. (1905). Notes on the melanism of *Larentia multistrigaria* in the neighbourhood of Skelmanthorpe (Huddersfield). *Entomologist's Rec. J. Var.* XVII, 170–1.

MORLEY, C. (1903–14). *Ichneumons of Great Britain* (5 vols). Keys, Plymouth & Brown, London.

MORLEY, C. and RAIT SMITH, W. (1933). The hymenopterous parasites of the British Lepidoptera. *Trans. R. ent. Soc. Lond.* **81**, 133–83.

MOSEBACH-PUKOWSKI, E. (1937). *Z. Morph. Ökol. Tiere* **33**, 358.

MURGATROYD, W. (1957). The accident at Windscale. *Nature, Lond.* **180**, 1093–4.

MURRAY, D. (1941). Secondary sexual characters in British moths. *Entomologist's Rec. J. Var.* **53**, 73–5.

MURRAY, D. (1943). Hair pencils and scent brushes. *Entomologist's Rec. J. Var.* **55**, 19–21.

MURRAY, K. F. (1928). *Entomologist's Rec. J. Var.* **40**, Pl. 3.

NEWMAN, E. (1855). *Zoologist* **13**, 4994.

NEWMAN, E. (1865a). Life-history of *Bombyx callunae*. *Entomologist* **2**, 137–9.

NEWMAN, E. (1865b). Differentiation of the two allied species, *Bombyx callunae* and *B. quercus*. *Entomologist* **2**, 140–1.

NEWMAN, E. (1865c). Singular geographical race of *Hepialus humuli*. *Entomologist* **2**, 162–3.

NEWMAN, L. H. (1943). A short history of *Endromis versicolora* L. ab. *lapponica* Bau. *Entomologist* **76**, 217–20.

NEWMAN, L. W. (1908). Yellow *Arctia dominula*. *Entomologist* **41**, 255.

NEWMAN, L. W. (1910). Melanism of *Ematurga atomaria*. *Proc. ent. Soc. Lond.* 1910, 36.

NEWMAN, L. W. (1916). On *Aplecta nebulosa*. *Entomologist's Rec. J. Var.* **58**, 22.

NICHOLSON, C. (1937). Larvae of *Callimorpha dominula* eaten by cuckoos. *Entomologist's Rec. J. Var.* **49**, 61–2.

NIEPELT, W. (1911). Zur Biologie von *Lasiocampa quercus* ab. *olivaceofasciata* Cockayne. *Int. ent. Z.* **5**, 185.

NORDSTROM, F., WAHLGREN, E., TULLGREN, A., and LJUNGDAHAL, D. (1941). *Svenska Fjärilar*, Stockholm.
O'DONALD, P. (1959). Possibility of assortative mating in Arctic Skuas. *Nature, Lond.* **183**, 1210–1.
O'DONALD, P. and DAVID, P. E. (1959a). *Genetics, N.Y.* **9**.
O'DONALD, P. and DAVIS, P. E. (1959b). The genetics of the colour phases of Arctic Skuas. *Heredity, Lond.* **13**, 481–6.
OERTEL, H. (1910). Merkwürdige Färbung einer Raupe von *Chaerocampa elpenor* L. *Int. ent. Z.* **4**, 48–9.
ONSLOW, H. (1916). On the development of the black markings on the wings of *Pieris brassicae*. *Biochem. J.* **10**, 26–30.
ONSLOW, H. (1919a). *J. Genet.* **7**, 209–58.
ONSLOW, H. (1919b). Inheritance of wing colour in Lepidoptera. II. Melanism in *Tephrosia consonaria* (var. *nigra* Bankes). *J. Genet.* **9**, 53–60.
ONSLOW, H. (1920a). The inheritance of wing colour in Lepidoptera. III. Melanism in *Boarmia consortaria* (var. *consobrinaria* Bkh.). *J. Genet.* **9**, 339–46.
ONSLOW, H. (1920b). Inheritance of wing colour in Lepidoptera. IV. Melanism in *Boarmia abietaria*. *J. Genet.* **10**, 135–40.
ONSLOW, H. (1921a). Inheritance of wing colour in Lepidoptera. V. Melanism in *Abraxas grossulariata* L. (var. *varleyata* Porritt). *J. Genet.* **11**, 123–139.
ONSLOW, H. (1921b). The inheritance of wing colour in Lepidoptera. VI. *Diaphoria mendica* Clerck and var. *rustica* Hübn. *J. Genet.* **11**, 277–92.
ONSLOW, H. (1921c). The inheritance of wing colour in Lepidoptera. VII. Melanism in *Hemerophila abruptaria* (var. *fuscata* Tutt). *J. Genet.* **11**, 293–8.
OUDEMANS, J. T. (1927). *Dasychira pudibunda* L. ab. *concolor* Stdg. *Ent. Ber., Amst.* **7**, 171–4.
OWEN, D. F. (1961). Industrial melanism in North American moths. *Am. Nat.* **95**, 227–33.
OWEN, D. F. (1962). The evolution of melanism in six species of North American Geometrid Moths. *Ann. ent. Soc. Am.* **55**, 695–703.
OWEN, D. F. and ADAMS, M. S. (1963). The evolution of melanism in a population of *Catocala ilia* (Noctuidae). *J. Lepid. Soc.* **17**, 159.
PARIS, R. (1963). In *Chemical plant taxonomy* (ed. T. Swain), pp. 337–58. Academic Press, London.
PARRY, D. A. (1951). Factors determining the temperature of terrestrial arthropods in sunlight. *J. Exp. Biol.* **28**, 445–62.
PARSLOW, J. L. F. (1967). Changes in status among breeding birds in Britain and Ireland (part 3). *Br. Birds* **60**, 177–202.
PARSONS, J. (1963). *J. Physiol., Lond.* **169**, 80.
PEARSALL, W. H. (1950). *Mountains and moorlands*. Collins, London.
PEARSON, J. (1937). Tasmanian Brush Opossum: its distribution and colour variation. *Pap. Proc. R. Soc. Tas.* 21–9.
PERKINS, J. F. (1959–60). Handbooks for the identification of British Insects. Hymenoptera. Ichneumonidae. Vol. 7, parts 2a, i and ii. R. ent. Soc. Lond.
PERRINS, C. M. (1959). Melanistic adult Great Tit and brood of young. *Br. Birds* **52**, 131.

PETERSEN, B. (1952). The relations between *Pieris napi* L. and *Pieris bryoniae* Ochs. *Trans. 9th int. Congr. Ent.* **1**, 83–7.
PETERSEN, B. (1955). Geographische variation von *Pieris* (*napi*) *bryoniae* durch Bastardierung mit *Pieris napi. Zool. Bidr. Upps.* **30**, 354–97.
PETERSEN, B., TORNBLOM, O., and BODIN, N. O. (1952). Verhaltensstudien am Rapsweissling und Bergeissling (*Pieris napi* L. und *Pieris bryoniae* O.). *Behaviour* **4**, 67–84.
PHILLIPS, G. (1962). Survival value of the white colouration of gulls and other sea birds. D.Phil. thesis, Oxford University.
PICTET, A. (1912). Recherches expérimentales sur les mécanismes du melanisme et de l'albinisme chez les Lépidoptères. *Mem. Soc. Phys. Hist. nat. Genève* **37**, 111–278.
PICTET, A. (1924). Sur les races geographiques de *Lasiocampa quercus. Bull. Soc. lepidopt. Geneve* **5**, 82–4.
PICTET, A. (1928). Les conditions du déterminisme des proportions numériques entre les composants d'une population polymorphe de Lépidopteres. *Rev. suisse Zool.* **35**, 473–505.
PICTET, A. (1931). Recherches de génetique dans les croisements de *Lasiocampa quercus* et de ses races alpina d'altitudes moyen et supérieure. *Mitt.Schweiz. ent-Ges.* **15**, 114–40.
PIERCE, F. N. (1894). Lancashire and Cheshire Entomologists' Society. *Entomologist* **27**, 359.
PLATE, L. (1910). Die Erbformeln der *Aglia tau* Rassen im Aussschluss an die Standfussschen Züchtungen. *Arch. Rass.-u. GesBiol.* **7**, 678–83.
PLOTNIKOV, V. I. (1924). Some observations on the variability of *Locusta migratoria* L. in breeding experiments. *Bull. ent. Res.* **14**, 241–3.
PLUNKETT, C. R. (1927). The experimental production of melanism in Lepidoptera. *Am. Nat.* **61**, 82–8.
PORRITT, G. T. (1906). Melanism in Yorkshire Lepidoptera. *Rep. Br. Ass. Advmt Sci.* 1906, 316–25.
PORRITT, G. T. (1916). Black pupae of *Abraxas grossulariata. Entomologist's mon. Mag.* **52**, 206.
PORRITT, G. T. (1917). Black pupae of *Abraxas grossulariata. Entomologist's mon. Mag.* **53**, 235–6.
PORRITT, G. T. (1926). The induction of melanism in the Lepidoptera and its subsequent inheritance. *Entomologist's mon. Mag.* **62**, 107–11.
POULTON, E. B. (1884). Notes upon . . . the colours, markings, and protective attitudes of certain lepidopterous larvae and pupae . . . *Trans. ent. Soc. Lond.* 1884, 27–60.
POULTON, E. B. (1884–5). The essential nature of the colouring of phytophagous larvae (and their pupae) with an account of some experiments . . . *Proc. R. Soc.* **38**, 269–314.
POULTON, E. B. (1887a). An enquiry . . . special colour relation between certain exposed lepidopterous pupae and the surfaces which immediately surround them. *Phil. Trans. R. Soc.* B **178**, 311–441.
POULTON, E. B. (1887b). On cocoons of three species of Lepidoptera. *Proc. ent. Soc.* 1887, 51–2.

POULTON, E. B. (1890). *The colours of animals.* London.
POULTON, E. B. (1892). Further experiments upon the colour-relationship between certain lepidopterous larvae, pupae and imagines and their surroundings. *Trans. ent. Soc. Lond.* 1892, 293-487.
POULTON, E. B. (1903). Experiments in 1893, 1894, and 1896 on the colour relation between lepidopterous larvae and their surroundings. *Trans. ent. Soc. Lond.* 1903, 311-74.
PREISS, J. (1929). Das *Aglia tau.* Problem nach neueren erbiologischen Gesichtspunkten. *Ent. Z., Frankf. a M.* **43**, 45-184.
PROUT, R. (1967). Selective forces in *Papilio glaucus. Science, N.Y.* **156**, 534.
PUNNETT, R. L. (1915). *Mimicry in butterflies.* Cambridge.
RAY, J. (1710). *History of insects* (p. 177), London.
REID, W. (1893). *List of the Lepidoptera of Aberdeenshire and Kincardineshire,* p.6. Pearson-Bell Hartlepool.
REID, W. (1950). Melanic *Apatele alni. Entomologist* **83**, 148.
REID, W. (1951). Lepidoptera collecting notes 1950. *Entomologist s Rec. J. Var.* **63**, 33-6.
REMINGTON, C. L. (1958). Genetics of populations of Lepidoptera. *Proc. 10th Int. Congr. Ent.* **2**, 787-805.
RICHARDS, J. E. (1965). Successes with smoke control, Manchester. *Proc. Conf. Natn. Soc. clean Air,* Eastbourne, 1965.
RILEY, N. D. (1948). The Rothschild-Cockayne-Kettlewell collection of British Lepidoptera. *Entomologist* **81**, 145-6.
RILEY, N. D. (1950). The Rothschild-Cockayne-Kettlewell collection of British Lepidoptera. *Entomologist* **83**, 19-20.
RIPPON, C. (1942). Two unusual aberrations of *Panaxia dominula. Entomologist* **75**, 75-6.
ROBER, J. (1924). *The Macrolepidoptera of the World* (ed. A. Seitz), **5**, 53-111. Stuttgart.
ROBINSON, R. (1971). *Lepidoptera genetics.* Pergamon Press, Oxford.
ROSE, A. H. (1967). *Thermobiology.* Academic Press.
ROSIE, J. H. (1958). A list of Macrolepidoptera found in north-east Caithness. *Entomologist's Gaz.* **9**, 158-62.
ROTHSCHILD, M. (1933). *Proc. R. ent. Soc. Lond.* **38**, 159.
ROTHSCHILD, M. (1961). Defensive odours and Müllerian mimicry among Insects. *Trans. R. ent. Soc. Lond.* **113**, 101-21.
ROTHSCHILD, M. (1962). Feeding insects to captive predators. Rep. 12th Congr. Brit. Ent., *Entomologist* **95**, 142.
ROTHSCHILD, M. (1963). Is the Buff Ermine (*Spilosoma lutea* Hufn.) a mimic of the White Ermine (*Spilosoma lubricipeda* L.)? *Proc. R. ent. Soc. Lond.* A **38**, 159-64.
ROTHSCHILD, M. (1964). An extension of Dr. Lincoln Brower's theory on bird predation. *Entomologist* **97**, 73.
ROTHSCHILD, M. and HASKELL, P. T. (1966). Stridulation in the Garden Tiger moth audible to the human ear. *Proc. R. ent. Soc. Lond.* A **41**, 10-12, 167-70.
ROTHSCHILD, M. and PARSONS, J. (1962). *Proc. R. ent. Soc. Lond.* C **27**, 6.

RUSSELL, S. G. G. (1943). *Argynnis paphia* L., an experiment in breeding from a melanic female. *Entomologist's Rec. J. Var.* **55**, 74–5.
SAGE, B. L. (1962). Albinism and melanism in birds. *Br. Birds* **55**, 201–25.
SAGE, B. L. (1963). The incidence of albinism and melanism in British birds. *Br. Birds* **56**, 409–16.
SANG, J. H. (1961). Environmental control of mutant expression. *Symp. R. ent. Soc.* **1**, 91–102.
SARGENT, T. D. (1966). Background selections of Geometrid and Noctuid moths. *Science, N.Y.* **154**, 1674–5.
SARGENT, T. D. (1968). Cryptic moths: effects on background selections of painting the circumocular scales. *Science, N.Y.* **159**, 100–1.
SARGENT, T. D. (1969a). Behavioural adaptations of cryptic moths. II. Experimental studies on bark-like species. *Jl N.Y. ent. Soc.* LXXVII, 2, 77–9.
SARGENT, T. D. (1969b). Behavioural adaptations of cryptic moths. III. Resting attitudes. *Anim. Behav.* **17**, 670–2.
SARGENT, T. D. and KEIPER, R. R. (1969). Behavioural adaptations of cryptic moths. I. Preliminary studies of bark-like species. *J. Lepid. Soc.* **23**, 1–9.
SCHOPF, C. and BECKER, E. (1933). *Justus Liebigs Annln Chem.* **507**, 266–96.
SCHORGER, A. W. (1949). Squirrels in early Wisconsin. *Trans. Wis. Acad. Sci. Arts Lett.* **39**, 195–247.
SCHRÖDER, C. (1908). Experimentelle und kritische Studien über den Nigrismus und Melanismus. *Z. wiss. InsektBiol.* **4**, 57–65.
SCHRÖDER, C. (1909). Zur Konstitutionellen Prävalenz der Melanismen. *Z. wiss. Insekt-Biol.* **5**, 27–9.
SCHULTE, A. (1952). Makrolepidopterologische Sammeltage in Schwedisch Lappland. *Ent. Z.*, **61**, 169–74.
SCORER, A. G. (1913). *Frankf. a. M. The entomologists' log book.* George Routledge & Sons., Ltd., London.
SCOTT-MONCRIEFF, R. (1936). *J. Genet.* **32**, 117–70.
SEITZ, A. et al. (1906–32). *The Macrolepidoptera of the World*, vol. I, *The palearctic Butterflies* (translated by K. Jordan). Stuttgart.
SEVASTOPULO, D. G. (1956). Viability of melanic and typical forms of *Gonodontis bidentata* Clerck compared. *Entomologist* **89**, 234–6.
SEVASTOPULO, D. G. (1967). Observations on Sphingidae around Cape Town. *Ent. Rec. J. Var.* **79**, 24.
SEVASTOPULO, D. G. (1968). Curious behaviour of larva of *Acherontia atropos* L. *Ent. Rec. J. Var.* **80**, 60 and 212.
SHAW, G. B. (1903). *Man and Superman.*
SHELDON, W. G. (1920). *Boarmia repandata* etc. in the Rannoch district. *Entomologist* **53**, 20–1.
SHELFORD, R. (1916). *A naturalist in Borneo.*
SHEPPARD, P. M. (1951). A quantitative study of two populations of the moth *Panaxia dominula* L. *Heredity, Lond.* **5**, 349–78.
SHEPPARD, P. M. (1952). A note on non-random mating in the moth *Panaxia dominula* L. *Heredity, Lond.* **6**, 239–41.
SHEPPARD, P. M. (1953). Polymorphism and population studies. *Symp. Soc. exp. Biol.* **7**, 274–89.

SHEPPARD, P. M. (1956). Ecology and its bearing on population genetics. *Proc. R. Soc. B* **145**, 308–15.
SHEPPARD, P. M. (1959). *Natural selection and heredity.* Hutchinson, London.
SHEPPARD, P. M. (1961). Some contributions to population genetics resulting from the study of Lepidoptera. *Adv. Genet.* **10**, 165–216.
SHEPPARD, P. M. (1964a). Protective coloration in some British moths. *Entomologist* **97**, 209–16.
SHEPPARD, P. M. and COOK, L. M. (1962). The manifold effects of the medionigra gene of the moth *Panaxia dominula* and the maintenance of polymorphism. *Heredity, Lond.* **17**, 415–26.
SHORTEN, M. (1954). *Squirrels.* New Naturalist Series, Collins, London.
SHOWLER, A. J. (1965). Industrial Melanism—some comments. *Entomologist's Rec. J. Var.* **77**, 70–1.
SKYE, E. (1968). Lichen and air pollution. *Acta phytogeogr. suec.* **52**, 123.
SMALLWOOD, J. W. (1896). On *Biston betularia*. *Entomologist* **29**, 222.
SMART, H. D. (1944). Breeding experience with *Arctia caja*. *Entomologist's Rec. J. Var.* **56**, 45–6.
SMITH, C. N. (1966). *Insect colonization and mass production.* Academic Press, New York.
SMITH, K. M. and RIVERS, C. F. (1956). Some variations affecting insects of economic importance. *Parasitology* **46**, 235–42.
SMITH, M. (1964). *British amphibians and reptiles.* New Naturalist Series, Collins, London.
SMITH, S. G. (1955). Notes on *Arctia caja* L. with a description of a new aberration. *Entomologist* **88**, 241–2.
SMILH, S. G. (1956). Experiments with a strain of *Lasiocampa quercus* ab. *olivaceofasciata* from Cheshire—*L. quercus* race *callunae* Palmer from Yorksyire (Lep. Lasiocampidae). *Entomologist* **89**, 137–8.
SMITH, S. G., and COCKAYNE, E. A. (1954). Experiments with a strain of *Lasiocampa quercus* ab. *olivaceo-fasciata* Cockerell, with descriptions of two new aberrations. *Entomologist* **87**, 225–9.
SONNEBORN, T. M. (1930). Genetic studies on *Stenostomum incaudatum*. II. The effects of lead acetate on the hereditary constitution. *J. exp. Zool.* **57**, 409–39.
SOUTH, R. (1893). Lepidoptera of the Shetland Islands. *Entomologist* **26**, 98–102.
SOUTH, R. (1904). *Entomologist* **37**, 263.
SOUTH, R. (1941). *The butterflies of the British Isles.* Warne, London.
SOUTH, R. (1961). *The moths of the British Isles*, vols. I and II. Warne, London.
SOUTHERN, H. N. (1943). The two phases of *Stercorarius parasiticus* L. *Ibis* **85**, 443–85.
SOUTHERN, H. N. (1944). Dimorphism in *Stercorarius pomarinus* Temminck. *Ibis.* **86**, 1.
SOUTHERN, H. N. and SEVENTY, D. L. (1947). The two phases of *Astur novae hollandiae* Gm. in Australia. *Emu* **45**, 331.
SOUTHWOOD, T. R. E. (1966). *Ecological methods.* Methuen, London.
SPÄRCK, R. (1950). Food of the north European gulls. *Proc. 10th Orn. Congr.* 588–91.

SPENCE, D. H. N. (1957). Studies on the vegetation of Shetland. I. The serpentine debris vegetation on Unst. *J. Ecol.* **45**, 917–45.
STANDFUSS, M. (1896). *Handbuch der paläarktischen Gross—Schmetterlinge* **107**, 340–8, Jena.
STANDFUSS, M. (1900). Synopsis of experiments in hybridization and temperature made with Lepidoptera up to the end of 1898. *Entomologist* **33**, 161–7, 283–92.
STANDFUSS, M. (1901). Synopsis of experiments in hybridization and temperature made with Lepidoptera up to the end of 1898. *Entomologist* **34**, 11–13, 75–84.
STANDFUSS, M. (1910*a*). Die alternative oder discontinuierliche Vererbung und ihre Veranschaulichung an den Ergebrussen von Zucht-experimenten mit *Aglia tau* und deren Mutationen. *Dt. ent. Natn.-Biblthk* **1**, 5, 14, 21, 28.
STANDFUSS, M. (1910*b*). *Chaerocampa (Pergesa) elpenor* L. ab. *daubi* Niep. und einige Mitterlungen über Wesen und Bedeutung der Mutationen illustriert an *Aglia tau* L. *Dt. ent. Z. Iris.* **24**, 155–81.
STANDFUSS, M. (1914). Mitteilunger zur Vererbungsfrage unter Heranziehung der Ergebrisse von Zuchtexperimenten mit *Aglia tau* L. *Mitt. schweiz. ent. Ges.* **12**, 238–308.
STANDFUSS, M. (1917). Ein eigenartiges gynandromorphes Individuum von *Aglia tau* L. *Mitt. Ent. Zürich* **3**, 154–70.
STAPLES, L. P. (1948). Further as to colour change without moult. *Bull. Br. Orn. Club* **68**, 80–8.
STAUDINGER, O. (1871). *Staudinger's Catalogue* **1**, 24, Dresden.
STAUDINGER, O. (1891). Neue arten und varietäten von lepidopteren des paläarktischen Faunengebiets. *Iris* **4**, 266.
STAUDINGER, O. (1894). *Dt. ent. Z. Iris*, **7**, 255–6.
STEINART, H. (1892). Über das Auftreten von *Amphidasys betularia* L. ab. *doubledayaria* B. in Sachsen. *Dt. ent. Z. Iris* **5**, 424–7.
STERTZ, O. (1915). Mitteilungen über eine Zucht von *A. caja*. *Dt. ent. Z. Iris* **29**, 142–3.
STEVEN, H. M. and CARLISLE, A. (1959). *The native pinewoods of Scotland.* Oliver & Boyd, Edinburgh.
STEVENSON, H. (1866). *The birds of Norfolk,* vol. I. London.
STICHEL, H. (1911). Über Melanismus und Nigrismus bei Lepidopteren. *Z. wiss. InsektBiol.* **7**, 297–302, 341–3, 369–72.
STICHEL, H. (1912*a*). Über Melanismus und Nigrismus bei Lepidopteren. *Z. wiss. InsektBiol.* **8**, 6–9, 1–3.
STICHEL, H. (1912*b*). Nachtrag und Berichtigung zu meinem Artikel Über Melanismus und Nigrismus bei Lepidopteren. *Z. wiss. InsektBiol.* **8**, 110.
STIPAN, F. (1952). *Pieris bryoniae* O. und *Pieris napi* L. *Ent. Nachr Bl., Wien* **4**, 33–8.
STIPAN, F. (1954). *Pieris bryoniae* O. und *Pieris napi* L. *Ent. NachrBl., Wien* **6**, 36–43.
SÜFFERT, F. (1924). Bestimmungsfaktoren des Zeichtrungsmuster beim Saison-Dimorphismus von *Araschnia levana-prorsa*. *Biol. Zbl.* **44**, 173–88.
SUMNER, F. B. (1929). The analysis of a concrete case of intergradation between two subspecies. *Proc. nat. Acad. Sci. U.S.A.* **15**, 110–20.

SUMNER, F. B. (1932). Genetic distributional and evolutionary studies of the subspecies of Deermice (*Peromyscus*). *Bibl-phia genet.* **9**, 1–106.
SUMNER, F. B. (1934). *Proc. nat. Acad. Sci. U.S.A.* **20**, 559–64.
TANAKA, Y. (1953). Genetics of the silkworm, *Bombyx mori*. *Adv. Genet.* **5**, 239–317.
TANSLEY, A. G. (1939). *The British Islands, and their vegetation.* Cambridge University Press.
TAZIMA, Y. (1964). *The genetics of the silkworm.* Logos Press, London.
THODAY, J. M. and BOAM, T. B. (1959). Effects of disruptive selection. II. Polymorphism and divergence without isolation. *Heredity, Lond.* **13**, 205–18.
THOMPSON, B. B. (1908). Notes on the deviations in the life histories of *Bombyx quercus* and *B. callunae*. *Entomologist's Rec. J. Var.* **8**, 125–8, 158–61.
THOMPSON, T. H. (1960). Melanins. *Comp. Biochem. Physiol.* III. Ed. Mason and Flonkin. A, 727–53.
THOMSEN, M. and LEMCHE, H. (1933). Experimente zur Erzielung eines erblichen melanismus bei dem Spanner *Selenia bilunaria* Esp. *Biol. Zbl.* **53**, 541–60.
TIMOFÉEFF-RESSOVSKY, N. W. (1933). *Arch. Naturgesch.* N.F. **2**, 285–90.
TIMOFÉEFF-RESSOVSKY, N. W. (1935). *Arch. Naturgesch.* N.F. **4**, 245–57.
TIMOFÉEFF-RESSOVSKY, N. W. (1940). Zur Andyse des Polymorphismus bei *Adalia bipunctata* L. *Biol. Zbl.* **60**, 130–7.
TINBERGEN, L. (1949). Bosvogels en insecten. *Ned. Bosch. Tijdschr.* **4**, 91–105.
TINBERGEN, L. (1960). Natural control of insects in pine-woods. (1) Factors influencing intensity of predation by song birds. *Archs néerl. Zool.* **13**, 264–343.
TINBERGEN, N. (1952). On the significance of territory in the Herring Gull; A note on the origin and evolution of threat display. *Ibis* **94**, 158–62.
TINBERGEN, N., MECUSE, B. J. D., BOERAMA, L. K. u., and VAROSSIEAU, W. W. (1942). Die Balz des Samfalters, *Eumenis (Satyrus) semele* L. *Z. Tierpsychol.* **5**, 182–226.
TONGE, A. E. (1915–16). On *Boarmia repandata*. *Proc. S. Lond. ent. nat. Hist. Soc.* 1915–16, 119.
TURNER, J. R. G. (1968). The ecological genetics of *Acleris cimariana* Zed. *J. Anim. Ecol.* **37**, 489–520.
TURNER, J. R. G. (1971). Studies of Müllerian mimicry and its evolution in burnet moths and Heliconid butterflies, in *Ecological genetics and evolution*, ed. Creed, Blackwell Scientific Publications, Oxford.
TURNER, J. R. G. and CRANE, J. (1962). *Zoologica, N.Y.* **47**, 141.
TUTT, J. W. (1890). Melanism and melanochroism in British Lepidoptera. *Entomologist's Rec. J. Var.* **1**, 5–7, 49–56, 84–90, 121–5, 169–72, 228–34, 293–300, 317–25.
TUTT, J. W. (1891). Melanism and melanochroism in British Lepidoptera. *Entomologist's Rec. J. Var.* **2**, 3–7, 31–5, 49–53, 77–80, 97–8, 145–9.
TUTT, J. W. (1902). *The natural history of the British Lepidoptera* **3**, Swann Sonnenschein, London.
URBAHN, VON E. (1971). Zunahme von Melanismus—Beobachtungen bei Makrolepidopteren Europas in neurerer Zeit. *Mitt. münch. ent. Ges.* **61**, 1–14.
VALETTA, A. (1948). *Entomologist* **81**, 102.

VARLEY, G. C. (1962). A plea for a new look at Lepidoptera with special reference to scent distribution organs of male moths. *Trans. Soc. Br. Ent.* **15**, 29–40.
VENABLES, L. S. V. and VENABLES W. M. (1955). *Birds and mammals of Shetland* (pp. 307–8). Edinb.
WAGNER, E. (1940). Über eine die Gonadien beeinflussende Mutation von *Ptychopoda seriata* Schrk. *Biol. Zbl.*, **60**, 567–89.
WALTHER, H. (1927). Über Melanismus. *Dt. ent. Z. Iris* **41**, 32–49.
WALTHER, H. (1932). Der Melanismus der Schmetterlinge. *Int. ent. Z.* **25**, 409–15.
WARBURG, J. C. (1901). On some races of *Lasiocampa quercus*. *Entomologist's Rec. J. Var.* **13**, 237–342.
WARNECKE, G. (1913). Einige Bemerkungen über die melanistischen Formen von *Cymatophora or* F. aus dem Niederelbgebiet. *Ent. Mitt.* **2** (9), 282–4.
Warren Springs Laboratory (1963–67). Investigation of air pollution. National Survey. Annual Summary Table. Ministry of Technology.
WATKINS, A. E. (1935). *Heredity and evolution.* London.
WATT, W. B. (1968). Adaptive significance of pigment polymorphisms in *Colias* butterflies. I. Variation of melanin pigment in relation to thermo-regulation. *Evolution, Lancaster Pa.* **22**, 437–58.
WEIR, J. J. (1880). The Macrolepidoptera of the Shetland Isles. *Entomologist* **13**, 249–91.
WEIR, J. J. (1881). Further Macrolepidoptera of the Shetland Isles. *Entomologist* **14**, 278–81.
WEIR, J. J. (1884). The Macrolepidoptera of Unst. *Entomologist* **17**, 1–4.
WEISMANN, A. (1882). *Studies in the theory of descent*, Part I. Sampson, Low, Marston, Searle & Rivington, London .
WENT, F. W. (1955). Air pollution. *Scient. Am.* **192**, 5, 62.
WHAYMAN, A. (1953). The Black Wood of Rannoch. *Scott. For.* **7**, 112–7.
WHITE, D. F. B. (1876). On melanochroism and leucochroism. *Entomologist's mon. Mag.* **13**, 145–9.
WICKLER, W. (1968). *Mimicry in plants and animals.* World University Library, Weidenfeld & Nicolson.
WIELAND, H., METZGER, H., SCHÖPF, C., and BULOW, M. (1933). *JustusLiebigs Annln Chem.* **507**, 226–65.
WIGHTMAN, A. J. (1931). *Entomologist's Rec. J. Var.* **43**, 105, 142, 178.
WIGGLESWORTH, V. B. (1928). Impressionist colouring among Lepidoptera. *Proc. R. ent. Soc. Lond.* **3**, 4.
WIGGLESWORTH, V. B. (1964). *The Life of insects.* Weidenfeld & Nicolson, London.
WILLIAMS, H. B. (1933). Notes on *Boarmia repandata* and *B. rhomboidaria*. *Proc. S. Lond. ent. nat. Hist. Soc.* 1932–3, 1–10.
WILLIAMS, H. B. (1949). Melanic *Boarmia repandata* at Rannoch, Perthshire. *Entomologist's Rec. J. Var.* **61**, 5.
WILLIAMSON, K. (1951). The arctic skua study, 1951. *Fair Isle Bird Obs. Bull.* **4**, 3–10.
WILLIAMSON, K. (1956). Birds and Lepidoptera in anticyclonic airstreams. *Entomologist's Rec. J. Var.* **68**, 95–7.

WILLIAMSON, M. H. (1957). Gene and chromosome frequencies in *Peronea comariana* Zell. (Lep. Tortricidae). *Entomologist's mon. Mag.* **93**, 52–3.
WILLIAMSON, M. H. (1958). Selection, controlling factors, and polymorphism. *Am. Nat.* **92**, 329–35.
WILLIAMSON, M. H. (1959). Colour and genetics of the Black Slug. *Proc. R. Phys. Soc. Edinb.* **27**, 87–93.
WILLIAMSON, M. H. (1960). On the polymorphism of the moth *Panaxia dominula* L. *Heredity, Lond.* **15**, 139–51.
WITT, M. (1933). Untersuchungen über das Zeichnungsmuster der melanistischen mutation des Schwammspinners. *Z. Morph. Ökol. Tiere* **27**, 262–93.
WOLFF, N. L. (1929). Lepidoptera, in *Zoology of the Faroes*, **39**, 1–38. Copenhagen.
WRIGHT, D. (1952). Female *Spilosoma lutea* Hufn. attracting *Arctia caja*. *Entomologist's Rec. J. Var.* **64**, 24.
WRIGHT, D. and SMITH, S. G. (1956). Experiments with a remarkable melanic strain of *Arctia caja* (L). *Entomologist's Gaz.* **7**, 119–125.
WYNN, W. G. (1933). *Entomologist's Rec. J. Var.* **45**, 6.
WYNNE-EDWARDS, V. C. (1962). *Animal dispersion in relation to social behaviour.* Oliver & Boyd, Edinburgh.
YOKOYAMA, T. (1959). *Silkworm genetics illustrated.* Japanese Society for the Promotion of Science, Tokyo.
ZEUNER, F. E. (1958). *Dating the past.* Methuen, London.
ZIEGLER, I. (1961). Genetic aspects of ommochrome and pterin pigments. *Adv. Genet.* **10**, 349–403.

Authors and Contributors Index

Adams, P. A., 16
Adkin, R., 52, 146, 293
Aitkenhead, P., 297
Alcock, A., 266
Allen, P. M. B., 121
Arkle, J., 144
Arnold, G. A., 20
Ashwell, D. A., 75
Askew, R. R., 150

Bacot, A. W., 252, 271
Bailey, N. T. J., xxii
Baker, C. R. B., 297
Barber, H. N., 27, 28
Barrett, C. G., 136, 184, 220, 294, 304
Beirne, B., xxiv
Bell, W., 228
Berry, R. J., xviii, 23, 186, 187, 193, 195, 196, 272
Bishop, J. A., 62, 87, 150
Blackman, G. E., xxi
Blair, W. F., 26
Bleasdale, J. K. A., 55
Boisduval, J. B. A., 278
Bowater, W. W., 86, 87, 88
Bowden, S. R., 270
Bretherton, R. F., 48, 49, 222, 274
Bretschneider, R., 75, 87, 221
Brett, G. A., 77
Briggs, J., 157
Brodo, I. M., 57
Brooks, G., xxii, 319
Brower, L. P., xxiii, 265, 266
Brower, J. V. Z., xxiii, 265, 266
Burns, J. M., 266, 269
Buxton, P. A., 32

Cadbury, C. J., xviii, xxii, 142, 167, 172, 223, 224, 225, 238, 243, 245, 246, 247, 252, 253, 255, 256, 295, 296, 297
Cain, A. J., xxi
Carr, F. M. B., 160
Carter, D., xxi
Caspari, E., 66
Chapman, J. W., 52
Chipperfield, R., 25
Clarke, B., xxi, 160
Clarke, C. A., xxi, 80, 87, 107, 131, 136, 141, 142, 144, 149, 186, 265, 266, 267, 317
Cockayne, E. A., xix, xx, 64, 74, 75, 77, 87, 218, 219, 221, 267, 293, 299, 304

Collinson, W. E., 246, 247, 248, 250, 251, 255, 298
Coney, G. B., 299
Conn, D. L, T., 88
Cook, L. M., 44, 83, 150, 282, 287, 320
Cooke, H., 52
Cott, H. B., 32, 33
Court, C., xxiii
Crauford, C., 151
Creed, E. R., 31, 32, 61, 137, 275
Crewdson, R. C. R., 172
Crocker, J., 20
Cunningham, D., 298
Cuno, W., 74
Curio, E., 50

Darwin, C. R., 67, 68, 155, 156
Davies, K., xxiii
Davis, P. E., 23
de Beer, G., xxi
Decker, A., 21
Demuth, R. P., 170, 182
Denham, 218
de Ruiter, L., 45, 160
de Worms, C., xxi, 94, 148
Dice, L. R., 26, 27, 30
Digby, P. S. B., 15
Dobzhansky, Th., xxiii, 35, 195
Doncaster J. P., xxi
Doncaster, L., 52, 140, 141, 144
Douwes, H., xxiv, 92, 93
Dowdeswell, W. H., 3
Drozda, A., 89
Duckett, G., 61

Edleston, R. S., 156, 228

Farrer Brown, L., xix
Faure, J. C., 50
Federley, H., 47, 93, 274
Fenton, A. F., 57
ffennell, D. W. H., 182
Fischer, E., 268
Fisher, J., 23
Fisher, R. A., xxii, 3, 7, 24, 53, 114, 140, 186, 279, 281, 302
Fitzpatrick, T. B., 25
Fletcher, S., xxi, 249
Ford, E. B., v, xix, xxi, 3, 7, 8, 24, 51, 53, 78, 87, 101, 113, 164, 173, 193, 268, 269, 271, 273, 279, 280, 281, 290, 294, 301, 302

E.M.—28

Ford, K., xxiii
Foster, M., 25
Fox, H. M., 25, 48
Fox, S., xxiii
Fraser, F. C., 47
Frings, C., 230
Fryer, J. C. F., 87, 88

Garms, B., xxiii
Garrett, F. C., 221
Garrod, A., 304
Geiger, R., 16
Gent, P. J., 157
Gershenson, S., 28
Gerschler, M. W., 89
Gibson, C., 272
Goater, B., 218
Goldschmidt, R. B., 86, 88, 268, 274, 278, 301
Goodson, L., xx, 151
Guiler, E. R., 28

Haldane, J. B. S., xxii, 24, 34, 53, 106, 137, 142, 150, 166, 186, 193, 224, 302
Hamilton, C., 319
Hamling, T. H., 77
Hare, E. J., 168
Hardy, E., 24
Harper, G. W., 168
Harper, P. S., xviii, 46, 62, 202, 257
Harris, E., 77
Harrison, J. W. H., 51–4, 76, 86, 88, 92, 121, 219, 220, 302
Hasebroek, K., 53, 78, 89
Haskel, P. T., 214
Hawksworth, D. L., 57
Hayward, J. S., xxiii
Heard, M. J., 303
Heath, J. E., 16
Henke, K., 66
Hershkovitz, P., 26
Hessel, S., xxiv
Hewson, R., 247
Heydemann, F., 89
Hoffmeyer, S., xxiv, 90–92, 229, 230
Hofmann, A., 136
Hovanitz, W., 320
Howard, G., xviii, xxiv, 91, 94, 159, 218, 264
Howarth, T. G., xxi
Hudson, G. V., 177, 179
Huggins, H. C., 169
Hughes, A. W., 53
Huxley, J. S., xxiii, 22

Isley, F. B., 20

Jackson, C. H. N., 160
Johnston, R. F., 24

Jones, E. W., 57
Julst, G., 101

Kane, W. de V., 169
Kennicott, R., 28
Kettlewell, D., xviii
Kettlewell, H. B. D., 44, 47, 51, 59, 61, 69, 70, 72, 79, 80, 83, 87, 99, 107, 109, 114, 115, 134, 142, 156, 184, 186, 187, 190, 221, 246, 247, 272, 278, 285, 287, 303, 304, 307, 308, 315, 319, 320
Kettlewell, H. M., xix, 16, 116, 251
Kikkawa, H., 66, 250
Klots, A., xxiv, 175
Kühn, A., 66

Landsborough Thomson, A., 21, 22, 23
Lane, C., 214
Lees, D. R., xviii, xxii, 57, 59, 61, 71, 76, 87, 107, 142, 246, 247, 253, 297, 317
Lemche, H., 53
Le Measurier, 228
Lempke, B. J., xxiv, 48, 89, 90, 96, 98, 230, 252
Long, D. B., 20, 49, 50
Lorimer, R. I., 64, 149, 151, 182, 183, 184, 272
Lorkovic, Z., 270
Lowe, P. R., 24
Lusis, J. J., 32, 275
Lyster, M., xxiii

McArthur, 168, 182
McDiarmid, A., 21
McGinnis, J., 21
Mackie, D. W., 20
McPhail, J. D., 30
Maddison, T., 220
Magnus, D. B. E., 178, 266, 268
Mather, K., 7
Maynard Smith, xxii
Mayr, E., 22
Mendel, G., 52
Mera, A. W., 52
Merrifield, F., 52
Michael, P., 296
Michaelis, H. N., 150
Miller, E., 80
Milsome, J., xxiii
Minnion, W. E., 75, 221
Minot, C. S., 100, 101
Monroe, E., xxiv
Moore, J., xxiv
Moreau, R. E., 24
Morley, A. M., 144
Morley, B., 247
Mosebach-Pukowski, E., 49

Müller, J., 100
Murray, K. F., 212

Newman, E., 21, 228
Newman, L. H., 75, 267, 269
Niepelt, W., 230
Nordstrom, F., 91

O'Donald, P., 23
Oertel, H., 47
Onslow, H., 17, 51, 54, 77, 78, 86, 88, 293, 294
Orel, V., xxiv
Owen, D. F., xxiv, 99–102

Parry, D. A., 13–15
Pasteur, L., 319
Pearson, J., 28
Peet, T., xviii
Perrins, C., xviii
Perrins, M., xviii
Petersen, B., 92
Phillips, G., xviii, 25
Pictet, A., 228
Plotnikov, V. I., 20
Porritt, G. T., 244
Poulton, E. B., 46
Pringle, J. W. S., xxiii
Prout, T., 266

Ray, J., 108
Reid, W., 296
Remington, C. L., xxiv, 100
Riley, N. D., xx
Rivers, C. F., 319

Robinson, R., xx, 17, 66, 73, 80, 93
Rose, A. H., 34
Rosie, J. H., 250
Rothschild, M., xxiii, 21, 31, 214, 275, 284
Rothschild, W., 24

Sabine, L. A. E., 168
Sage, B. L., 21, 22, 24
Salmon, N., 177
Salvage, W., 168
Sang, J., 220
Shapiro, A., xviii
Schörgen, A. W., 28
Schropp, K., 13
Schummer, R., 90
Scorer, A. G., 223
Seitz, A., 274
Selander, R. M., 24, 193
Sevastopulo, D. G., 50
Shaw, G. B., 36
Shelford, R., 33

Sheppard, C. A., xxiv
Sheppard, P. M., xxi, 69, 80, 83, 87, 107, 131, 136, 137, 141, 142, 144, 149, 150, 186, 265, 266, 267, 279, 280, 282, 283, 317
Shorten, M., 28, 29
Siggs, L. W., 148
Skye, E., 57
Smith, K. M., 319
Smith, M., 29
Smith, S. G., 144, 245, 246, 247, 276
Southern, H. N., 22, 23
South, R., 168
Spärck, R., 28, 238
Spence, D. H. N., 190
Standfuss, M., 86, 277, 278, 289, 301
Staples, L. P., 22
Statham, D. H., xxiii
Steinert, H., 80
Stertz, O., 74
Stevenson, H., 21
Suffert, F., 301
Sumner, F. B., 30, 114, 192, 193
Soumalainen, E., xxiv, 95, 96
Sutton, L. S., 65

Tams, T., xxi
Tanaka, Y., 66
Tanner, R., xxiii
Tazima, Y., 66, 250
Thomsen, M., 53
Thompson, B. B., 228
Tilby, F., xxiii
Timofeef-Ressovsky, N. W., 31, 195
Tinbergen, L., 45, 160
Tinbergen, N., xxi, 16, 121, 123, 125, 126, 128
Tomlinson, R., 217
Tutt, J. W., 52, 227, 228, 252

Urbahn, von E., 90

Varley, G. C., xxi
Verrall, G. H., xix
Vevers, G., 25, 48

Walsingham, Lord, 180
Walther, H., 78, 80, 89
Warburg, J. C., 252
Warnecke, G., 89, 90
Watt, W. B., 178, 180
Weir, J. J., 168, 182
Weismann, A., 52
Went, F., 55
White, D. F. B., 59
Wickett, R. M., xxiii
Wickler, W., 31

Wigglesworth, V. B., 20, 31, 45
Wightman, A. J., 304
Williams, H. B., 78, 86, 88, 172
Williamson, M. H., xxi, 34, 205

Wright, D., 148, 276
Wynn, W. G., 304

Yokoyama, T., 66, 250

Subject Index

Species of Birds and Mammels are listed under these two headings respectively.
* Identification of birds mentioned in quotation from the Bible established by A. Landsborough Thomson (*A new dictionary of birds*, Nelson, London and Edinburgh, 1964).

Abraxas, 39
Acrididae, 20
Adalia bipunctata (Coleoptera), 31, 32, 195, 275
Addison's disease, 21
Africa, 25, 176, 287, 301, 309
Agrotidae, 215
Air pollution, 5-6, 54-8, 143, 147, 149, 151, 244, 247, 256, 259
Alleles, 41, 59, 66, 76, 86, 144, 157, 186, 290-91, 317-18
Allelomorphism, 86
America
  North, 24, 28-9, 54, 94-5, 99-102, 174-6, 313-18
  South, 67
Amphidasis (= *Biston*)
Annual Review of Entomology, xviii
Ancient Melanism, see Melanism
Aposematic Melanism, see Melanism
Amphibia, 34
  frog (*Rana temporaria*), 34
Arachnidae, 6, 20, 21
  Arctosa perita, 20
  Drapetiscea socialis, 21
  Ostearius melanopygius, 21
  Red Spider, 320
  Salticus scenicus, 20, 21
Army Worms (*Laphygma exigua* and *L. exempta*), 50
Arctiidae, 39, 49, 73, 213, 214
Argynnis, 269
Artificial colonies in
  Ectropis consonaria, 288
  Lasiocampa quercus, 288
  Lycia hirtaria, 223-5
  Panaxia dominula, 277-89
Assembling in
  Biston betularia, 108
  Hepialus humuli, 295-7
  Lasiocampa quercus, 228, 238, 243
  Panaxia dominula, 278

Background choice, 65, 68-73, 88 (footnote), 157-64
Baltic Sea, 94, 181
Barley, 308
Batesian mimicry, 265-8

Beak-marks, 285
Behaviour of
  Amathes glareosa, 205-206
  Biston betularia, 107-11, 83-4
  Lasiocampa quercus, 228-9
Behavioural differences in, 65-73
  Amathes glareosa, 196, 200-206
  Biston betularia, 83-4, 88 (footnote), 107-11, 119-20
  Cycnia mendica, 293-4
  Hepialus humuli, 294-8
  larvae, 50 (footnote)
  Lasiocampa quercus, 228, 229, 241, 243
  Panaxia dominula, 284-5
  pupae, 109-11
Birds, 21, 29
  Bananaquit, West Indian (*Coereba flaveola*), 23
  Blackbird (*Turdus merula*), 286
  Bullfinch (*Pyrrhula pyrrhula*), 21
  Buzzard (*Buteo buteo*) and Rough-legged (*Buteo lagopus*), 22
  Chaffinch (*Fringilla coelebs*), 174
  Chats (*Oenanthe* sp.), 33
  Chicken
    Bantu, 20
    Buff Orpington, 21
  Corvidae, 32
  Cormorant (*Ketupa zeylonensis*), 33
  Crow
    Carrion (*Corvus corone corone*), 32
    Hooded (*Curvus cornix*), 190, 201
  Cuckoo (*Cuculus canorus*), 33, 254, 255, 281, 285
  Cuckow (*Larus* sp.)*, 33
  Curlew (*Numenius arquata*), 249
  Dunnock (*Prunella modularis*), 117, 224
  Eagle (*Neophron percnopterus*)*, 33
  Egret, Little (*Egretta grazetta*), 22
  Falcon, Eleanora's (*Falco eleanorae*), 22
  Flycatcher, Spotted (*Muscicapa striata*), 61, 126, 129, 285
  Goldfinch (*Carduelis carduelis*), 21
  Goshawk, Australian (*Accipiter novaehollandiae*) and Gabar (*Micronisus garbar*), 22
  Grouse, Red (*Lagopus lagopus scoticus*), 24, 249

Subject Index

Gull, Black-headed (*Larus ridibundus*) and Common (*Larus canus*), 25, 94, 158, 170, 190, 199, 201, 236–41, 247–9, 256, 258–60, 274, 296, 297, 314
Harrier, Montagu's (*Circus pygargus*), 22
Hawfinch (*Coccothraustes coccothraustes*), 21
Hawk, Night (*Falco* sp.)*, 33
Hen, Water (*Gallinula chloropus*), 160
Heron
   (*Ardeidae* sp.)*, 33
   Great Blue (*Ardea herodias*), 22
   Night (*Nycticorax nycticorax nycticorax*), 30
Jackdaw (*Corvus monedula spermologus*), 32
Jay
   (*Garrulus glandarius rufitergum*), 32
   Florida Scrub (*Cyanositta coerulescens*), 265, 266
Kite, Black (*Milvus migrans*)*, 33
Lark
   Desert (*Ammomanes deserti*), 24
   Shore (*Otocorys alpestris*), 170
Magpie (*Pica pica*), 249, 254
Mallard (*Anas platyrhynca*), 35
Meadow Pipit (*Anthus pratensis*), 238, 248
Merlin (*Falco columbarius*), 238, 248
Mocking bird (*Mimidae* sp.), 20
Nuthatch (*Sitta europaea*), 126, 224
Osprey (*Aegypius monachus*)*, 33
Ossifrage (*Gypaetus barbatus*)*, 33
Owls, 33, 114
   Little (*Athene noctua*)*, 33
   Short-eared (*Asio accipitrinus*), 238
Partridge (*Perdix perdix*), 24
Peewit (*Vanellus vanellus*), 190
Penguin, Galapagos (*Spheniscus mendiculus*), 30
Petrel, 23
   Fulmar (*Fulmarus glacialis*), 23
Pheasant (*Phasianus colchicus*), 24, 254, 285
Pigeon, Wood (*Columba palumbus*), 21
Plover Golden (*Charadrius apricarius*), 190, 249
Raven (*Corvus* sp.)*, 33
Redstart (*Phoenicurus phoenicurus*), 61, 122, 123, 174
Robin (*Erithacus rubecula*), 116, 117, 126, 224, 264
Skua
   Long-tailed (*Starcorarius longicaudus*), 22
   Arctic (*Starcorarius parasiticus*) (*Starcorarius pomarinus*), 22, 23

Snipe (*Gallinago gallinago*), 24
Sparrow
   Hedge (see Dunnock),
   House (*Passer domesticus*), 20, 24, 25, 116, 224, 225
Starling (*Sturnus vulgaris*), 190, 265, 286
Stork, White (*Ciconia ciconia*), 20
Swallow, (*Hirundo rustica rustica*), 32, 285
Swift (*Apus apus*), 32, 243, 248
Tern (*Sterna hirundo*), 179
Tit, Great (*Parus major*), 113, 114, 121 224, 286
Thrush, Song (*Turdus musicus*), 126, 286
Turkey (*Meleagrididae* sp.), 20
Vulture (*Milvus* sp.), 33
Wagtail, Pied (*Motacilla alba yarrellii*), 32
Wheatear (*Oenanthe oenanthe*), 190, 248, 238
White-eye, Cape (*Zosterops c. capensis*), 176
Wren (*Troglodytes troglodytes*), 128
Yellowhammer (*Emberiza citrinella*), 126

Biston betularia, 105–51, 314–22, 361–71
   (and see Species Index)
   distribution of, 131–51
   table of distribution, Appendix C
   camouflage efficiency, 113–19, 125
Black Eggars, 227, 260
Black colouration, physical effect of, 11–17
Blasket Isles, 169
Brazil, 36, 67
Breeding techniques, 319–22
British Medical Journal, 57
British Museum of Natural History, xx, xxi
Bronchitis, 3, 57
'buffering', 74, 269, 307–10
Buffonism, 52

Caithness, 231–44, 257–8
Calmytes, 54
Caledonian pine forest, 171
Camouflage efficiency (in *Biston betularia*), 113–19, 125
Canada, 165
Carcinoma of
   lung, 3
   skin, 35
Catocala, 34, 73, 157, 215–16
Cheshire (*Lasiocampa quercus*), 255–6, 259
Chromatography, 212
Cistron, 317
Cline, 89, 136, 187–90, 195–207, 231, 234–6, 272

Subject Index 415

Coleoptera, 6, 14, 20, 30, 34
   Adalia bipunctata, 31, 32, 195, 275
   Blaps mucronata, 31
   Brachinus ballistarius, 31
   Eleodes sp., 31
   Megasida, 31
   Necrophorus sp., 30
Colias, 178, 320
Colour blindness, 11
Conifers, melanism associates with, 170–76
Coniferous forests, 174–6, 313
Contrast/conflict, 72–3
Crambus, 179
Crocallis, 222
Crossing over, 9
Crowding effect (in larvae), 49–50
Crustacea (*Gammarus chevreuxi*), 308
Crypsis, 5, 67, 68, 236, 237
   adjustment of, 139
   aerial, 173–4
   nocturnal, 11
   scoring of, 125
Cryptic melanism, 21–5, 214–23

Definitions, 7–10
Denmark, 91–2, 159, 182, 230
Diptera, 32, 34
   Drosophila, 5, 13, 15, 65, 195, 236, 275
   Tachinidae, 250, 320
      Peltocarus dentatus, 250
'Disappearance distance' (in *L. quercus*), 236–7, 248
Disruptive coloration, 158, 159, 162
Distribution of
   Amathes glareosa, 187–93
   Biston betularia, 131–51, Appeneix C,
   Lasiocampa quercus ssp. *callunae*, 227–30
Dominance
   build-up and break-down of, 316–18
   incomplete, 40, 271–89, 298 313–18,
   modification, 10, 163–4, 185–6, 313–18

Earias, 221
Eetropis, 99
Electrophoresis, 212
Ennomos, 221–3
Environment,
   changes in, 54–8
   effect of, 301–309
Environmental melanism, 301–10
Erebia, 178, 264
Euchloris, 221
Europe, 89, 99
Eye-marks, 289–91

Faeroes, 166, 294
Finland, 94–6
Fish,
   melanism in, 30, 33
   Mosquito fish (*Gambusia affinis*), 30
   Sunfish (Centrarchidae sp.) 30
Fire-resistant trees, 174–6, 313
Flight, melanism associated with, 214–16
Flash colouration, 214–16
Foliage-resters, 220–23
Forestry Commission, 171

Gammarus chevreuxi, 308
Gene
   -complex, 11, 164, 166, 205, 214, 217, 224, 225, 260, 280
   -flow, 198, 222, 224
Genetics of
   Aglia tau, 289
   Amathes glareosa, 184–5
   Arctia caja, 275–7
   Argynnis paphia, 266–9
   Biston betularia, 106–107
   Cycnia mendica, 293
   Gonodontis bidentata, 87
   Hepialus humuli (race thulensis), 297–8
   Industrial melanic species, 86–8
   Lasiocampa quercus (ssp. *callunae*), 227–56
   Lycia hirtaria, 223–5
   Panaxia dominula, 278–82
   Phigalia pilosaria (syn. *pedaria*), 59–62, 87
   Polia nebulosa, 271–3
   Spilosoma lutea, 273–5
   Xylomyges conspicillaris, 298–301
Geographic
   Melanism, see Melanism
   races, 162–4
Geometra, 221
Germany, 163, 228, 230, 274
Giles Collection, 144
Gloger's Rule, 170
Gonodontis 222
Graptolitha, 99
Ground
   colour, 162
   resters, 218–20

Habit differences, see Behavioural differences
Hamsters, melanism in, 28
Hardy–Weinberg
   Law, 73, 211, 234, 271
   Tables, 375–7
Hebrides, 164
Heligoland, 274
Hemithea, 221
Heredity, Journal of, xviii, 113, 121, 165
Heterocera, 39, 263–4, 293, 320, 322
Heterosis, 8–9, 76, 77, 225
Heterozygous variability, 10, 86–9, 94–5, 100, 106–107, 182–3, 185–6, 275–7, 279–82

Holland, 96–9, 230
Homo sapiens, 34–6
Humidity, 304–305
Hybrid, 111
Hymenoptera, 20, 32
    Bombus sp., 31
    Formica rufa, 173
    Hornet (*Vespa crabro*), 32
    Ichneumonidae, 250, 320

Ice ages in
    North America, 175
    Shetland, 166
Iceland, 181
Imagines, treatment of, 322
Induced melanism, 52–4
Industrial Melanism, see Melanism
Insecta, melanism in, 19–21
Ireland, 169, 179
Isotopes, 44, 295, 287, 303–304
Italy, 277–9

Lamarckism, 52
Lapland, 90–96, 270
Larvae,
    colouration of, 44–50, 109, 247, 249, 251–5
    treatment of, 320–22
Leviticus, 33
Lichens, 55–8, 61, 148, 151
Life-history of
    Amathes glareosa, 183–4
    Biston betularia, 84, 107–11, 319–22
    Lasiocampa quercus and *L. quercus* ssp. *callunae*, 228–31, 321–2
    Panaxia dominula, 277f.
Limenidae, 34–5
Lizards, melanism in, 29
Lung carcinoma, 3

Mammals,
    Bat (*Chiroptera* sp.), 33
    Cat, domestic and Siamese (*Felis cattus*), 25, 32, 33
    Deer, red (*Cervus elaphus*), 175
    Deermouse (*Peromyscus*), 26, 27, 33, 192, 193
    Fox (*Vulpes vulpes*), 33
    Hedgehog (*Erinaceus europaeus*), 35
    Hog (*Sus scrofa*), 158
    Leopard, common (*Felis pardus*), 25
    Marmoset (*Callithricidae* sp.) 26
    Man, 35, 36
    Mouse (*Mus* sp.), 33
    Opossum (*Trichosurus vulpecula*), 28
    Rabbit (*Oryctolagus cuninculus*), 27, 28
    Rat, black (*Rattus rattus*), 28
    Shrew (*Sorex* sp.), 33
    Squirrel, 33
        Sciurus carolinensis, 28, 29
        Sciurus vulgaris, 28, 29
    Tamarin (*Callithricidae* sp.), 26
    Tiger (*Panthera tigris*), 158
    Weasel (*Mustela nivalis*), 238
    Zebra (*Equus burchelli*), 158
Mark–release–recapture experiments, 3, 321
    in *Amathes glareosa*, 181, 196–207
    in *Biston betularia*, 113–30
    in *Lasiocampa quercus*, in
        Caithness, 238–41
        Yorkshire, 245–8
    in *Panaxia dominula*, 279, 283, 287
Mating preferences, 83, 84, 268–9, 278, 280
Melanin, xvii, 16–17, 21, 34, 43
Melanism,
    Ancient, see Geographic
    Aposematic, 30, 33, 49, 213–14, 263–70
    associated with fire-resistant trees, 42, 174–6
    Barrier, 34–6
    Conifer, 42, 170–76, 313
    Cryptic, 19–30, 214–23, 248–9
    early spread of, 131–4
    Environmental, 21–2, 25–6, 40, 301–312
    Geographic, 41–2, 155–80, 269, 313–18
    Industrial, 3–7, 40–41, 51–155, 313–18, 362–72
        genetics of, 86–88, 313–17
        natural history of, 318
        origin of, 313–17
        world distribution of, 89–102
    Monomorphic, 298–9
    natural history of, 318
    Non-industrial, 6, 41–2, 51–104, 155–207, 313–14
    Northern, 41, 164–9, 181–93
    Phytoscopic, 44
    Pluvial, 42, 177
    Recessive, 42, 43, 73, 75–62, 211–63, 314–15
        in aposematic species, 213–14
        in cryptic species, 214–25
    Relict, see Geographic
    Rural, 41, 157–64
    sex-limited, 293–8
    surveys of, 134–51, Appendices B and C
    Thermal, 33, 42, 49, 178–80, 264
    Western Coastline, 42, 169–70
Melanism in
    Birds,
        cryptic, 12, 21–5
        aposematic, 32–3
    fish, 30
    hamsters, 28

Subject Index   417

insects,
  cryptic, 19
  aposematic, 30–32
  larvae, 44–9, 247–57, 276–7, 287
  lizards, 29
  locusts, 19–20
  mammals,
    cryptic, 11, 25–9
    aposematic, 33
  Man, 35–6
  rabbits, 27–8
  reptiles, 12, 29, 30
  Rhopalocera, 264–70
  shrew, 33
  slugs, 34–5
  snakes, 29–30
  spiders, 20–21
  squirrels, 28–9

  Czechoslovakia, 89, 90, 98
  Denmark, 91
  Finland, 95–6
  Netherlands, 96–9
  North America, 99–102
  Scandinavia, 90–96
  Sweden, 91–5
Mendel's Law, 52
Migration,
  Birds, 157–8, 164, 190, 205, 206
  Lepidoptera, 62, 302–10
Mimicry, Batesian, 265–7
Mortality, larval and pupal, 285–7
Multifactorial inheritance, 9, 64, 85–7, 93, 162–4, 255

Necrophorus, 30
Netherlands, 96–9
New Jersey Pine Barrens, 174–6
New Zealand, 11, 177
Nocturnal crypsis, 11–12
Nonagria, 159, 160
Nomenclature, 275
Non-industrial Melanism, see Melanism
Norway, 277
Nuffield Foundation, v, xix

Odonata (Dragonflies), 285
Ommochrome, 17
Opossum, 28,
Orkney, 159, 164, 182–4, 186, 231
Orocrambus, 179, 180
Orthoptera (grasshoppers and locusts), 14, 17, 19, 20, 214, 287
Ova, treatment of, 320
Oxford University Scientific Society, xviii, 231

Pachys (see *Biston*)
Palaeogenes, 317–18

Papilio, 266, 320
Parasites, 320
  in *Lasiocampa quercus*, 250
Phenocopies, 309
Pheromone, 83–4, 278
Phigalia, 100
Physiological
  adjustment, 77, 139–41
  differences, 65–87
  in *Lasiocampa quercus*, 236, 241–4
Phytoscopy, 44–8
Pigeons, melanic, 12
Pigments, xvii, 16–17, 301, 302–307
Pigmentation, environmental, 301–310
Pine
  Barrens, New Jersey, 174–5
  forests, 170–76
Plankton, aerial, 109
Pleiotropism, 61, 66–7, 76, 79, 85, 289
Pleurococcus, 115, 130
Pollution, 5–6, 54–8, 143, 147, 149, 151, 244, 247 256, 259
Polygenes, 9, 87, 93, 318
Polygenic inheritance, see multifactorial inheritance
Polyhedrosis, 249, 250–51, 280, 319
Polymorphism, 7–8, 19–23, 47, 67, 68, 138, 156–76, 183, 223–5, 256, 263–70, 313
  balanced, 140, 141
  transient, 7–8
  in aposematic species, 213–14, 273–80, 288, 289
  in cryptic species, 214–23
  in Recessive Melanism, 227–60
Portugal, 277
Predation,
  Amathes glareosa, 190, 198–201
  Battus sp., 265–6
  Biston betularia, 113–30
  Gonodontis bidentata, 62
  Hepialus humuli, 296–8
  Lasiocampa quercus, 158, 236–49, 253, 255, 258–60
  Lycia hirtaria, 224–6
  Panaxia dominula, 283, 284–7
  and Rural Melanism, 157–8
Predator/prey relationship, 160–65, 201, 205
Procus, 219
Pseudoips, 221
Pupae,
  Biston betularia, 109–11, 322
  Lycia hirtaria, 110

Rabbits, melanism in, 27–8
Radio-active
  isotopes, 287, 303–304, 320
  pollution, 55, 303

418 Subject Index

Recessive Melanism, see Melanism
Reflectance, 61, 63, 71
Relict Melanism, see Melanism
Reptiles, 29, 33, 34
   adder (*Vipera berus berus*), 29
   lizard, common (*Lacerta vivipara*) and sand (*Lacerta agilis agilis*), 29
   mamba, (*Dendroaspis angusticeps* and *D. polylepis*), 29, 30
Rhopalocera, melanism, 39, 49, 264–70, 293, 320, 322
Rodent ulcer, 35
Rothschild–Cockayne–Kettlewell Collection, xix–xxi, 106, 168, 218
Royal Society (Evolution Committee 1900), 52
Rural Melanism, see Melanism
Russia, 277

Scandinavia, 90–96, 158–9, 218, 228
Selenia, 221, 223
Selection experiments, 113–30
   on *Amathes glareosa* in Shetland, 181–210
   on *Biston betularia* in
     aviary, 114–15
     polluted woodland, 115–23
     unpolluted woodland, 124–30
   on *Lasiocampa quercus*, 227–62
   on *Lycia hirtaria*, 223–6
Selective predation (see predation),
Sex-ratios, in
   Biston betularia, 81–3
   Lycia hirtaria, 109–11
   hybrids, 111
Sexual dimorphism, 229–30, 293–9
Shadow, effect of, 158–9
Shetland Isles, 4, 22–3, 86, 159, 165–8, 170, 182–207, 271–272, 294–8
Shrews, melanism in, 33
Silkworms, 66
Smoke, 54–8, 61
Smokeless Zones, 149–51
Snakes, melanism in, 29–30
South Africa, 36, 174–6
Spilosoma, 273, 293

Squirrels, melanism in, 28–9
Stridulation, 241
Sulphur dioxide, 55, 61
Super-gene, 9–10, 255, 290, 317
Surveys of
   Biston betularia, 131–51 and Appendix C
   other species, Appendix B
   Phigalia pilosaria, 59–62
Sweden, 91–5, 257, 277
Syntomiidae, 212, 264

Tapinostola, 159
Tasmania, 22, 27–8
Temperature
   effects, 301–309
   regulation, 13–16
Termites (*Isoptera* sp.), 249
Theory of melanism, 313–18
Thermo-coupling, 13–16, 178–80
Tingwall Valley (Shetland), 188–9, 191–8, 203
Trichoptera, 32, 263
Triphaena, 219

Unpalatability, 30–33, 266–7, 277

Vanessa, 49, 309
Viability differences, 73–83, 212, 241–3, 283–4
Virus (see Polyhedrosis),
Voltinism, 111

Warning colouration, see Melanism Aposematic
Warren-Springs Laboratory, 55
Windscale reactor, 55–6

Xanthia, 221
Xanthopterin, 17

Yorkshire, 244–55, 257, 259

Zale, 99, 174
Zygaenidae, 39, 213, 214

Species Index (Lepidoptera)

abbreviata, Eupithecia, 349
abietaria, Boarmia, 171
abruptaria, Hemerophila, 77, 78, 86, 356, Plate 5.6
    f. *brunneata*, 77, 78
    f. *coarctata*, 77, 78
    f. *fuscata*, 77, 78, 356
    f. *knightii*, 77, 78
abyssinia, Spodoptera, 50
aceris, Apatele, 334, Plate 5.8
adusta, Eumichitis, 333
adustata, Ligdia, 351
advena, Orthosia, 342
advenaria, Cepphis, 353
aegeria, Pararge, 84
affinis, Cosmia, 340
agathina, Amathes, 329
affinitata, Perizoma, 348
aglaia, Argynnis, 178, 269
    ssp. *scotica*, 178
agrippina, Catocala, 215
albicillata, Mesoleuca, 348
albipuncta, Leucania, 341
albipunctata, Cosymbia, 343
albipunctata, Eupithecia, 97, 349
albovenosa, Simyra, 94, 181, 335, Plate 6.2
    f. *murina*, 94, 181, 335
albulata, Perizoma, 348
alni, Apatele, 333, Plates 5.7 and 6.8
amata, Calothysanis, 344
anceps (trepida), Notodonta, 324
    ab. *fusca*, 64, 324
areola, Xylocampa, 64, 313
arthemis, Limenitis astyanax, 265
asella, Heterogenea, 360
asteris, Cucullia, 50
atomaria, Ematurga, 359
atrata, Odezia, 263
aurantiaria, Erannis, 353
aurita, Setina, 179
australis, Aporophyla, 330
autumnata, Oporinia, 92, 171, 345
autumnaria, Ennomos, 75, 221, 351
    f. *schultzi*, 75, 351
    f. *brunneata*, 75, 351
avellana, Apoda, 360
aversata, Sterrha, 344

badiata, Earophila, 348
barrettii, Hadena, 169
berbera, Amphipyra, 215

betularia, Biston, 8, 34, 45, 46, 58, 59, 61–3, 68–70, 72, 76, 78–89, 92, 95, 96, 98, 99, 102, 105–51, 156, 171, 186, 198, 222, 314–18, 319–22, 356, Appendix C, Plates 5.6, 7.1, 8.1, 8.2, 9.1, 19.1, 19.2, 19.3, 19.4
    f. *carbonaria*, 8, 34, 58, 59, 68, 69, 76, 78, 79, 80–87, 95, 98, 102, 105–51, 156, 186, 198, 314–18, 356
    f. *insularia*, 95, 98, 100, 102, 107f, 114f, 134f, 156, 171, 186, 198, 317, 356
bicolorata, Plemyria, 346
bidentata, Gonodontis, 45, 46, 62, 64, 86, 87, 95, 105, 106, 171, 222, 223, 255, 352, Plates 4.2, 4.3, 4.4, 5.6
    f. *nigra*, 62, 86, 87, 95, 96, 97, 352
bilineata, Euphyia, 169, 348, Plate 10.5
    f. *atlantica*, 169, 348
bilunaria, Selenia, 52, 76, 221, 351, Plate 13.2
biselata, Sterrha, 344
bistortata, Ectropis, 52, 95, 96, 358
brassicae, Pieris, 17, 50
brumata, Operophtera, 345
bucephala, Phalera, 93, 218
    f. *tenebrata*, 93

caesia, Hadena, 169, 330
caesiata, Entephria, 167, 347, Plate 10.2
    ab. *atrata*, 167, 347
caja, Arctia, 31, 43, 73, 74, 94, 214, 271, 273, 275, 276, 307–309, 320–321, 327, Frontispiece
    f. *brunnescens*, 48, 74, 275, 276, 327
    f. *clarki*, 74, 327
    f. *fumosa*, 48, 74, 275, 276, 327
    f. *nigropennalis*, 74, 327
    f. *melanozoster*, 74, 327
    f. *obscura*, 74, 327
cambrica, Venusia, 349
camilla, Limenitis, 269
    f. *nigrina*, 269
    f. *seminigrina*, 269
cannae, Nonagria, 341
capucina, Lophopteryx, 324
carbonaria, Isturgia, 179
carmelita, Odontosia, 93
carpinata, Trichopteryx, 345
carpophaga, Dianthaecia, 169
    f. *capsophila*, 169
castigata, Eupithecia, 349

420 Species Index (Lepidoptera)

castrensis, Malacosoma, 326
cerogama, Catocala, 64, 72, Plate 5.17
   f. *ruperti*, 72
cervinalis, Larentia, 344
chamomillae, Cucullia, 343
chi, Antitype (Polia), 219, 220, 227, 333, Plate 5.9
   f. *nigrescens*, 220, 333
   f. *nigra*, 220, 333
   f. *suffusa*, 220, 333
chrystyi, Oporinia, 345
cinctaria, Cleora, 95, 357
cinerea, Agrotis, 328
citrata, Dysstroma, 92, 347
clathrata, Chiasmia, 353
clavis, Agrotis, 328
coenulata, Cidaria, 92
cognataria, Biston (Amphidasis), 84, 99, 100, 102, 111, 316, 317, 320, 321, 322, Plate 9.1
   f. *swettaria*, 99, 100, 316
comariana, Acleris (Acalla), 87
   f. *fuscana*, 87
comes, Triphaena, 164, 271, 329
   f. *curtisii*, 164, 271, 329
confusalis, Roeselia (Nola), 326
conigera, Leucania, 341
conspersa, Hadena, 167, 170, 330, Plate 10.2
consonaria, Ectropis, 54, 70, 71, 78, 171, 288, 359, Plates 5.9, 5.13
conspicillaris, Xylomyges, 298–9, 322, 330, Plate 17.3
   f. *melaleuca*, 298–9, 330
   f. *intermedia*, 298–9
convolvuli, Herse, 323
coracina, Psodos, 179
coryli, Colocasia, 97, 342, Plate 5.6
cossus, Cossus, 47
crenata, Xylophasia, 337, Plate 10.1
crepuscularia, Ectropis, 101, 102, 358
   f. *delamerensis*, 101, 358
   f. *fumataria*, 101
cruda, Orthosia, 50, 341
crystanea, Peronea, 162
cucullatella, Nola, 326
curtula, Clostera, 324

dahlii, Diarsia, 329
daplidice, Pontia, 17, 319
dardanus, Papilio, 317
defoliaria, Erannis, 162, 354, Plate 5.9
dejecta, Catocala, 215
deplana, Eilema, 326
diana, Speyeria, 265
didymata, Calostigia, 347
diluta, Asphalia, 325
dilutata, Oporinia, 86

dissoluta, Nonagria, 160, 340, Plate 10.1
dodonaea, Drymonia, 324
dominula, Panaxia, 74, 75, 83, 114, 212–14, 271, 273, 276–89, 309, 320, 322, 327, Plates 13.1, 16.3
   f. *bimacula*, 74, 75, 273, 276–89, 327
   f. *italica*, 278
   ssp. *lusitanica*, 277, 278
   f. *medionigra*, 74, 75, 83, 276–89, 327
   ssp. *majellica*, 279
   ab. *nigra*, 212, 213, 327
   f. *nigradonna*, 277, 278
   f. *paucimacula*, 309
   ssp. *persona*, 277, 278, 279
   ssp. *pompalis*, 289
   f. *pseudojuncta*, 288
   ab. *romanovi*, 289
dromedarius, Notodonta, 111, 324
duplaris, Tethea, 63, 92, 158, 325, Plate 5.7
   f. *obscura*, 63, 325

eburnata, Sterrha, 344
elinguaria, Crocallis, 95, 96, 222, 320, 352
   f. *fusca*, 222, 352
ello, Erynnyis, 50
elpenor, Deilephila, 47, 323
euphorbiae, Apatele, 334
exclamationis, Agrotis, 328
exanthemata, Cabera, 351
exempta, Laphygma, 50
exigua, Laphygma, 50
exulis, Xylophasia, 182, 338
   f. *assimilis*, 182, 338

fagaria, Dyscia, 360
fagi, Stauropus, 323, Plate 5.7
fasciaria, Ellonia, 351
favicolor, Leucania, 341
festiva, Diarsia, 167, 328, Plate 10.2
   ssp. *thulei*, 167, 328
filigrammaria, Oporinia, 345
filipendulae, Zygaena, 213, 360
   f. *chrysanthemi*, 213, 360
fimbriata, Lampra, 329
firmata, Thera, 346
flavicincta, Antitype, 333
flavicinctata, Entephria, 347
flavicornis, Achlya, 92, 96, 158, 325, Plate 6.1
   race *finmarchica*, 92, 158, 325
   f. *pseudoalbigensis*, 92, 96, 325
   race *scotica*, 92, 158, 325
fluctuata, Xanthorhoë, 167, 347, Plate 10.2
fluctuosa, Tethea, 325
fluxa, Arenostola, 341
fraxini, Catocala, 72, 92, 218, 342
   f. *mürens*, 218, 342

Species Index (Lepidoptera) 421

fuliginosa, Phragmatobia, 93, 94, 274, 327
    f. *borealis,* 93, 94, 274, 327
fulvago, Xanthia, 221
furcula, Cerura, 323
furcata, Hydriomena, 348
fuscantaria, Deuteronomos, 351

galathea, Melanargia, 268
gamma, Plusia, 20, 49, 50, 343
geminipuncta, Nonagria, 159, 162, 340, Plate 10.1
gilvaria, Crocota, 360
glabraria, Cleora, 358
glandon, Lycaena, 264
glareosa, Amathes, 77, 86, 91, 94, 95, 159, 165, 167, 181–207, 271, 272, 329, Plates 11.1, 11.2, 11.3, 11.4
    f. *edda,* 77, 95, 159, 165, 167, 168, 172, 181–207, 271, 272, 329
glaucus, Papilio, 266, 267, Plate 15.1
gnoma, Pheosia, 93
gothica, Orthosia, 50, 159
gracilis, Orthosia, 342
griseovariegata, Panolis, 342
grossulariata, Abraxas, 75, 275, 307, 350
    f. *exquisita,* 75
    f. *dohrni (lacticolor),* 75
    f. *varleyata,* 75

hastiana, Peronea, 162
henrici, Simyra, 94, 95
    f. *fumosa,* 95
hirtaria, Lycia, 75, 110, 111, 217, 223–5, 288, 320, 321, 356, Plates 13.3, 13.4
    f. *nigra,* 75, 217, 223, 228, 356
    f. *fumaria,* 223, 356
    f. *hannoviensis,* 223
hippocastanaria, Pachycnemia, 359
hispidaria, Apocheima, 356
hortaria, Epimecis, 101, Plate 6.3
    f. *carbonaria,* 101
hucherardi, Hydraecia, 221
humuli, Hepialus, 167, 168, 194–8, 360, Plate 17.2
    race *thulensis,* 294–8, 360

icarus, Polyommatus, 3, 114
incerta, Orthosia, 342
innotata, Eupithecia, 349
immutata, Scopula, 344
impluviata, Hydriomena, 348
improba, Argynnis, 264
io, Vanessa, 49
irrorella, Setina, 179, 326, Plate 10.4
    f. *andereggii,* 179
    f. *riffalensis,* 179

jacobaeae, Hypocrita, 327
juniperata, Thera, 346

kuhlweini, Setina, 179, Plate 10.4
kühniella, Ephestia, 65, 66

lanestris, Eriogaster, 305
lariciata, Eupithecia, 350
latrunculus, Procus, 92, 95, 97, 339
lepida, Hadena, 330
leporina, Apatele, 333, Plate 5.8
leucophaearia, Erannis, 97, 162, 353
leucostigma, Celaena, 340
levana, Araschnia, 301, 308
lichenaria, Cleora, 358
ligustri, Craniophora, 97, 334
ligustri, Sphinx, 323
limbaria, Isturgia, 353
limitata, Ortholitha, 344
literosa, Miana, 63, 339, Plate 5.8
liturata, Semiothisa, 97, 353, Plate 5.6
lucernea, Ammogrotis, 163, 328
lubricipeda, Spilosoma, 43, 273, 274, 326
lunaris, Phoberia, 342
lunosa, Anchoscelis, 333
lupulinus, Hepialus, 360
luridata, Ectropis, 97, 359
lutea, Citria, 221
lutea, Spilosoma, 271, 273, 326, Plate 16.2
    f. *deschangei,* 274
    f. *zatima,* 271, 274, 326
Lutosa, Rhizedra, 341
lutulenta Aporophyla, 330

machaon, Papilio, 266, 267, Plate 15.1
macularia, Pseudopanthera, 353
mandarina, Bombyx, 66
marginata, Lomaspilis, 351
marginepunctata, Scopula, 344
meadii, Colias, 178
megacephala, Apatele, 97, 334, Plate 5.9
    f. *nigra,* 97, 334
mendica, Cycnia, 88, 230, 293, 294, 298, 327, Plate 17.1
    f. *rustica,* 293, 294, 327
    f. *standfussi,* 293
menyanthidis, Apatele, 334
mi, Euclidimera, 342
monacha, Lymantria, 86, 92, 96, 163, 171, 326, Plate 5.19
mori, Bombyx, 65, 66, 250
monoglypha, Xylophasia, 97, 156, 335, Plate 10.1
montanata, Xanthorhöe, 347
mucronata, Ortholitha, 345
multistrigaria, Calostigia, 347
muralis, Cryphia, 333

nana, Hada, 330
nanata, Eupithecia, 349
napi, Pieris, 178, 270
  ssp. *bryoniae*, 178, 270
nastes, Colias, 178
nebulosa, Polia, 40, 97, 271, 272, 273, 329, Plate 5.7, 16.1
  f. *robsoni*, 97, 272, 273, 330
  f. *thompsoni*, 272, 273, 329
neurica, Nonagria, 340
nigricans, Euxoa, 327
noctuella, Nomphila, 302, 303, 304
nubeculosa, Brachionycha, 111, 305
nupta, Catocala, 64, 72, 215, 342
  f. *nigra*, 64, 342
  f. *brunnescens*, 215, 342

obeliscata, Thera, 346
obfuscaria, Gnophos, 359
oblonga, Hydraecia, 337
obscura, Catocala, 215
obscurata, Gnophos, 162, 163, 359
occulta, Eurois, 329
octogesima, Tethea, 62, 92, 97, 324, Plate 5.7
  f. *fusca*, 62, 97, 324
  f. *frankii*, 62, 97, 325
oleracea, Diataraxia, 20, 50
olivacearia, Phigalia, 100, 101
  f. *mephistaria*, 100, 101
or, Tethea, 63, 92, 95, 96, 325, Plate 5.7
  f. *albingensis*, 63, 96, 325
ornitopus, Graptolitha, 313
oxyacanthae, Allophyes, 45, 97, 98, 105, 156, 313, 331, Plate 10.1

pallens, Leucania, 341
paphia, Argynnis, 178, 266, 268, 269, 320
  f. *valesina*, 178, 266, 268, 269
  f. *melaena*, 269
parthenias, Brephos, 343
pavonia, Saturnia, 250, 290, 326
pedaria, Phigalia, see *pilosaria*
peltigera, Heliothis, 302, 304–9
pendularia, Cosymbia, 343
pennaria, Colotois, 353
perla, Cryphia, 333
persicariae, Mamestra, 50
philenor, Battus, 265, 267, Plate 15.1
phragmitides, Arenostola, 341
pilosaria, Phigalia, 59, 60, 61, 62, 65, 71, 76, 87, 97, 98, 100, 105, 106, 162, 314, 317, 319, 321, 354–6, Plates 5.4, 5.5, 5.9
  f. *monacharia*, 61, 76, 87, 354–6
pinastri, Hyloicus, 90, 96, 98, 217, 218, 323
  f. *nigrescens*, 90, 97
  f. *brunnea*, 90, 218
  f. *unicolor*, 90, 218

piniarius, Bupalus, 359
plagiata, Anaitis, 345
plantaginis, Parasemia, 83, 327
plumbaria, Ortholitha, 345
polycommata, Trichopteryx, 345
polyxenes asterius, Papilio, 265, 266
populata, Lygris, 167, 346, Plate 10.2
populi, Laothoe, 323
populi, Orthosia, 342
porata, Cosymbia, 343
porcellus, Deilephila, 323
potatoria, Cosmotriche, 254
procellata, Melanthia, 348
progemmaria, Erannis, 353
promissa, Catocala, 72
pronuba, Triphaena, 161, 215, 329
protea, Dryobota, 333
prunaria, Angerona, 353
psi, Apatele, 97, 333
  f. *suffusa*, 97, 333
pudibunda, Dasychira, 48, 92, 96, 98, 325, Plate 5.7
  f. *concolor*, 48, 92, 98, 99, 325
  f. *obscura*, 48, 325
pulchella, Utetheisa, 307
punctinalis, Boarmia, 54, 78, 97, 358
  f. *humperti*, 78, 97, 358
punctulata, Ectropis, 212, 359
pupillaria, Cosymbia, 343
pusaria, Cabera, 351
puta, Agrotis, 328
pygmina, Arenostola, 341
pyramidea, Amphipyra, 215

quadripunctaria, Euplagia, 286, 320, 327
quercifolia, Gastropacha, 45, 221
quercinaria, Ennomos, 221, 351
  f. *perfuscata*, 221, 351
quercus, Lasiocampa, 16, 43, 48, 49, 16, 75, 91, 158, 227–60, 288, 314, 315, 320, 321, 326, Plates 14.1, 14.2, 14.3, 14.4, 14.5, 14.7, 17.4
  ssp. *quercus*, 228, 229, 255, 256, 259
  ssp. *callunae*, 48, 49, 228–55, 257–60, 288
  f. *alpina*, 228
  f. *lurida*, 227, 230, 245, 326
  f. *meridionalis*, 252
  f. *olivacea*, 43, 48, 49, 75, 227, 230–55, 257–60, 326
  f. *olivaceo-fasciata*, 227, 230–55, 257–60, 326

rapae, Pieris, 268
ravida, Spaelotis, 328
rectangulata, Chloroclystis, 92, 350
remissa, Xylophasia, 337
repandata, Cleora, 54, 78, 86, 92, 95, 96, 97, 157, 171–4, 357, Plate 5.6, 10.6

f. *nigra*, 78, 157, 172, 358
f. *nigricata*, 78, 86, 157, 172, 173, 357
retecta, Catocala, 215
revayana, Sarrothripus, 162, 342
rhomboidaria, Cleora, 78, 86, 97, 171, 318, 357, Plate 5.6
f. *perfumaria*, 318, 357
f. *rebeli*, 78, 86, 357
ribeata, Cleora, 54, 78, 357, Plates 5,8, 5.9
ridens, Polyploca, 325
roboraria, Boarmis, 63, 92, 95, 96, 97, 358, Plate 4.1, 5.9
f. *infuscata*, 63, 358
f. *melaina*, 63, 358
f. *varia*, 63, 358
roscida, Setina, 179
ruberata, Hydriomena, 348
rubricollis, Atolmis, 263
ruficornis, Drymonia, 64
rumicis, Apatele, 95, 334

sacraria, Rhodometra, 47, 302, 306, 307, 309, 344
salicata, Calostigia, 347
scabriuscula, Dypterygia, 335
secalis, Celaena, 161, 219, 338
segetum, Agrotis, 328
sesamus, Precis, 301, 308
simulans, Rhyacia, 328
socia, Lithophane, 331
sordens, Apamea, 65, 338, Plate 5.17(a)
sparganni, Nonagria, 341
sparsata, Anticollix, 350
sphinx, Brachionycha, 64, 313, 320, Plates 5.10, 5.11, 5.12
sponsa, Catocala, 72
stellatarum, Macroglossa, 323
straminea, Leucania, 341
strataria, Biston, 76, 97, 356, Plate 5.14, 5.18
f. *robiniaria*, 76, 356
f. *melanaria*, 356, Plate 5.18
strigilis, Procus, 92, 97, 338
strigillaria, Perconia, 360
strigosa, Apatele, 334
subsericeata, Sterrha, 344
suffumata, Lampropteryx, 158, 347
f. *piccata*, 158, 347
sylvata, Abraxas, 351
sylvata, Hydrelia, 349

tau, Aglia, 86, 271, 273, 289–91
f. *fere-nigra*, 290
f. *melaina*, 290
f. *weismanni*, 290

tenebrata, Panemeria, 342
tenuiata, Eupithecia, 349
testacea, Luperina, 339
testata, Lygris, 346
tetralunaria, Selenia, 221,351
titea, Phigalia, 100, 102
transversata, Philereme, 346
trapezina, Cosmia, 340
trifolii, Lasiocampa, 184
trifolii, Zygaena, 213, 360
trigrammica, Meristis, 213, 227, 257, 339, Plate 10.1
f. *obscura*, 213, 339
trilinearia, Cosymbia, 343
tripartita, Abrostola, 343, Plate 5.8
tritici, Euxoa, 328
tulbaghata, Cleora, 176, Plate 10.3
truncata, Dysstroma, 346
typhae, Nonagria, 95, 159, 160, 340
f. *fraterna*, 95, 159, 340

undularia, Zale, 174
f. *umbripennis*, 174
unipuncta, Leucania, 341
umbratica, Rusina, 335
urticae, Spilosoma, 273, 274
urticae, Vanessa, 49, 309

versicolor, Procus, 339
versicolora, Endromis, 75, 93, 257, 326
venosata, Eupithecia, 167, 349
vestigialis, Agrotis, 163, 328
vidua, Catocala, 215
villica, Arctia, 83, 327
viminalis, Bombycia, 7, 330
vinula, Cerura, 48, 93, 323
f. *minax*, 93
f. *phantoma (arctica)*, 93
f. *vinula*, 93
virgularia, Sterrha, 344
vitellina, Leucania, 341
vulgata, Eupithecia, 349

wauaria, Itama, 353

xanthographa, Amathes, 167, 181, 329, Plate 10.2
xanthomista, Antitype, 167, 170

ypsilon, Agrotis, 302, 303

zelicaon, Papilio, 267
ziczac, Notodonta, 93
zonaria, Nyssia, 111, 356